STUDIES IN ENVIRONMENT AN

Nature and the English Dias

This book offers a comparative history of ideas about nature, particularly native nature as a part of the culture, in the Anglo settler countries – the United States, Canada, Australia, and New Zealand. It begins with the rise of natural history in the late eighteenth century and examines its development in the great nineteenth-century expansion of settlement. It explores settlers' adaptations to the end of expansion and scientists' shift from natural history to ecology as a way of understanding visible nature, processes that were largely complete by the 1940s. Finally, it analyzes the diffusion of ecology through the Anglo world and to the general population, as well as the rise of environmentalism. Addressing not only scientific knowledge but also popular issues such as hunting, common names for plants and animals, landscape painting, and nature stories, this book explores the ways in which English-speaking settlers looked at nature in their new lands and the place they gave it in their societies.

Thomas Dunlap is Professor of History at Texas A&M University, where he has taught since 1991. He is the author of two previous books, *DDT: Scientists, Citizens, and Public Policy* and *Saving America's Wildlife*, and of about two dozen articles and book chapters. He has won the Theodore Blegen Award, given by the Forest History Society for the best article on forest and conservation history, three times. He has delivered papers before a variety of professional organizations, including the Agricultural History Society, American Historical Association, American Society for Environmental History, American Society for Mammology, International Theriological Congress, International Canadian Studies Conference, North American Wildlife and National Resources Conference, Organization of American Historians, Social Science History Association, and the Western History Association. He is currently Chairman of the Board of Directors of the Forest History Society.

NATURE AND THE ENGLISH DIASPORA

ENVIRONMENT AND HISTORY IN THE UNITED STATES, CANADA, AUSTRALIA, AND NEW ZEALAND

Thomas R. Dunlap

Texas A&M University

CAMBRIDGE
UNIVERSITY PRESS

PUBLISHED BY THE PRESS SYNDICATE OF THE UNIVERSITY OF CAMBRIDGE
The Pitt Building, Trumpington Street, Cambridge, United Kingdom

CAMBRIDGE UNIVERSITY PRESS
The Edinburgh Building, Cambridge CB2 2RU, UK   www.cup.cam.ac.uk
40 West 20th Street, New York, NY 10011-4211, USA   www.cup.org
10 Stamford Road, Oakleigh, Melbourne 3166, Australia
Ruiz de Alarcón, 13, 28014 Madrid, Spain

© Thomas Dunlap 1999

First published 1999

Printed in the United States of America

Typeface New Baskerville 10/12 pt.   *System* QuarkXPress [BTS]

A catalog record for this book is available from the British Library.

Library of Congress Cataloging-in-Publication Data
Nature and the English diaspora : environment and history in the
United States, Canada, Australia, and New Zealand / Thomas R.
Dunlap.
p. cm. – (Studies in environment and history)
ISBN 0-521-65173-5 (hb). – ISBN 0-521-65700-8 (pb)
1. Natural history – History. 2. Ecology – History.
3. Environmentalism – History. 4. Nature. I. Series.
QH15.N26 1999
508'.09 – dc21

                                                                98-43736
                                                                   CIP

ISBN   0 521 65173 5   hardback
ISBN   0 521 65700 8   paperback

To my daughter,
Margaret Miller Dunlap,
with love

CONTENTS

Photo Section appears following page 127

FIGURES

PREFACE AND
ACKNOWLEDGMENTS

This book continues the exploration of a subject I began with my master's thesis on the history of DDT – the impact of science on ideas about nature in modern industrial society. The part that led to this book began, insofar as it had a definite beginning, in my research at the Canadian National Archives in the early 1980s. This work, much of which went into my book *Saving America's Wildlife*, suggested that a comparative view might be useful, though it was not clear to me then just what was to be compared. I considered a study of American and Canadian wildlife policy; some thought and investigation led me to add Australia and New Zealand (in the process ruling out several other countries for reasons I discuss in the Introduction). Preliminary research in Australia and New Zealand caused me to look more broadly at the ways in which the Anglo settlers of these lands came to understand the natural worlds they found and shaped. They had a common British and European culture, which they continued to see as a model for their own societies, but they confronted very different lands, and so they developed distinctive societies and reshaped their lands in their own ways.

That story is in my text, but before we come to that I have to thank the people and institutions that helped me over the years. Among them are the staffs of various libraries and archives. These include the national collections in the Canadian National Archives in Ottawa, the Turnbull Library and the New Zealand Archives in Wellington and the National Library of Australia in Canberra, the state libraries of New South Wales, Victoria, Tasmania, and Queensland, and the Queensland State Archives in Brisbane. Several agencies opened their libraries or archives: New Zealand's Department of Scientific and Industrial Research in Lower Hutt, New Zealand; the Commonwealth Scientific and Industrial Research Organization (CSIRO), Canberra; and the Tasmanian Department of Lands, Parks, and Wildlife. People in the Australian Conserva-

xiv   *Preface and Acknowledgments*

tion Agency and Landcare New Zealand sent me documents. To all the staff members who helped me, and to those behind the scenes who made and kept these collections, I extend my thanks.

Several institutions provided financial support. A grant from the National Science Foundation (DIR-8921746) allowed me to make an extensive initial trip to New Zealand and Australia; a summer stipend from the National Endowment for the Humanities (FT-36290) financed another. Jane Marceau, Professor of Public Policy, Research School of the Social Sciences, Australian National University, invited me to speak at a conference on science and technology policy and technological change in December 1992, giving me another chance, and Texas A&M University paid my way to the Sixth International Theriological Congress in Sydney in 1993, where I presented a paper and stayed to look at some records.

A variety of people offered information and advice, saved me time and gave me insights. Some offered home-cooked meals, oases in the desert of restaurant food, and Ross Galbreath, his wife, Susan Grant, and their son, Grant Galbreath, put me up for a night in their home in Naike, New Zealand, and gave me a tour of their local bush. I owe thanks well as to John and Meg Flux, also of New Zealand. In Australia a number of historians helped me: Linden Gillbank, Tom Griffiths, Rod Home, Roy MacLeod, Joseph Powell, Libby Robin, and Boris Schedvin. Tom and Libby deserve special thanks. Geoff Law of the Tasmanian Wilderness Society provided some insight into that group's activities. Among the scientists, I owe the most intellectual stimulation to Graeme Caughley (now, sadly, deceased) of CSIRO. For guidance, introductions, tours of the countryside around Canberra, ecological observations, and hospitality, I offer special thanks to Dan Walton of the Australian Conservation Agency. There was also Joan Dixon of the Victoria Museum, Patricia Mather of the Queensland Museum, and – at CSIRO's Division of Wildlife and Ecology – Robin Barker, John Calaby, Alan Newsome, and Mike Young. Various people at bed-and-breakfast places, on buses and trains, and in parks told me the names they gave various birds or trees, and in general let me probe their knowledge of nature.

All academics rely on the schools that employ them, and I am no exception. The students in my environmental history class at Virginia Tech (1988–1991) and Texas A&M University (since 1991) listened to earlier versions of my ideas with the patience they extend to any professor with an enthusiasm. They have my thanks. Larry Hill and Julia Blackwelder, heads of the History Department at Texas A & M during my years here, have with my colleagues made the university a good place to live and work. Among them Harold Livesay and Cynthia Bouton have been not only colleagues but treasured friends.

My wife, Susan Miller, has continued to support my work, listen to my ideas, and in general make life wonderful. After I dedicated *Saving Amercia's Wildlife* to her, I promised our daughter, Margaret Miller Dunlap, that the next one would be hers. So it is, and here it is. With much love, Dad.

INTRODUCTION

INVADERS, SETTLERS, INHABITANTS

> You know, I think if people stay somewhere long enough –
> even white people – the spirits will begin to speak to them.
> It's the power of the spirits coming up from the land. The
> spirits and the old powers aren't lost, they just need people
> to be around long enough and the spirits will begin to
> influence them.
>
> A Crow elder, as reported by poet Gary Snyder[1]

We Anglos – whites, whitefellows, pakeha – do not usually think in these
terms. The land is something we possess, not something that possesses us.
We know it and we shape it; it does not know or shape us. But even in our
own tradition there is that other current. The land was ours, said Robert
Frost, before we were the land's, and a multitude of others have said it too,
in poetry, paintings, stories, and reports. This book is about the ways in
which the Anglo settlers of Australia, Canada, New Zealand, and the
United States have in the past two centuries sought to understand their
lands and find their place in them by the use of their culture's organized
nature knowledge – science. It is not as visible or colorful a tale as the epic
of conquest that has become so much a part of national identity, but in the
long run it is at least as significant. The current environmental crisis suggests that unless we learn to live with the land we might not live on it at all
and certainly will not continue to live well.

These countries are my subject because of their common history,
common demography, and interconnections. They are, in Geoffrey C.
Bolton's terms, the Anglo "colonies of settlement," in Alfred Crosby's,
the Anglo part of the "neo-Europes."[2] Unlike the "colonies of empire,"

---

[1] Gary Snyder, *The Practice of the Wild* (San Francisco: North Point Press, 1990), 39.

[2] Geoffrey C. Bolton, *Britain's Legacy Overseas* (London: Oxford University Press, 1973), 5;
Alfred Crosby, *Ecological Imperialism* (Cambridge University Press, 1986). The latter
appeared about the time I began research on this project and I am much in Crosby's debt.

where a small foreign ruling class dominated a much larger population of "natives," here the Anglos were not only masters but by far the largest group of powerful actors. Here, and only here, predominantly English-speaking Europeans dispossessed and almost exterminated the earlier inhabitants, allowing the illusion that the lands were "vacant" or "wilderness." Here they could speak of creating a "new England" – a dream as marked in New Zealand, founded on the Wakefieldian vision of a trans-planted and purified British society in the South Seas, as in Crevecouer's America. Read through their literature, newspapers, legislative debates, and speeches. They were new nations populated by new men. (Women were physically present but rhetorically almost invisible.) Everywhere there were the same appeals to the "conquest of nature," "progress," a particular kind of civilization, and until recently the virtues of an agri-cultural life and a society of independent farmers. Land laws had in common the aim of individual independence and self-sufficient small farms. (There is even a depressing similarity in the desire to evade these laws, accumulating more acres, and in the methods used to do it.)[3] Statutes made the same kinds of animals "game" and set standards for a "fair chase." "Bad" animals were everywhere marked for destruction, and by changing the names of the species and adding or deleting ref-erences to "the Queen's most excellent majesty" to suit the jurisdiction, the same mammalian pest control laws could have been used from Perth to Fredricton.

These countries also form a group because of the connections they developed to each other. They had, to be sure, other ties. Australian con-nections to southern Africa, for one example, began when the First Fleet picked up plants and animals at the Cape on its way to Botany Bay, and continued into the twentieth century, when Australians looked to the region for everything from pasture grasses and farm stock to orna-mental plants. An elite group of experts that conspicuously included continental Europeans as well as British scientists circulated among all the settler countries and the colonies of empire. German naturalists were not only travelers and explorers but directors of agencies and museums. German foresters staffed the South Indian Forest Service, which was the administrative model for the U.S. Forest Service, and they went from there to Australia. Others went to North America. Nor is the

---

[3] Joseph Powell, *Environmental Management in Australia, 1788–1914* (Melbourne: Oxford University Press, 1976); Manning Clark, *A Short History of Australia* (New York: Penguin, 1987), 140–6; Keith Sinclair, *A History of New Zealand* (Auckland: Penguin, 1980), 151–71; Fred Shannon, *Farmers' Last Frontier* (New York: Holt, Rinehart, & Winston, 1945), 51–75. A recent discussion of this topic is John C. Weaver's "Beyond the Fatal Shore: Pastoral Squatting and the Occupation of Australia, 1826–1852," *American Historical Review, 101* (October 1996), 981–1007.

demographic separation between the colonies of empire and settlement airtight. South Africa had two European populations, and Argentina an even more heterogeneous mix of Europeans.

These four countries, though, are at the far end of the demographic spectrum, and a common language, culture, and settlement experience insulated, if it did not isolate, them. Ties were closest between Canada and the United States – too strong and too one-sided for many Canadians – but enough Americans went to Victoria in the mid-nineteenth century that, a generation later, an Australian naturalist could complain that too many of the common names for animals and plants were not really Australian but American.[4] The eucalyptus that form a distinctive part of California landscapes are the most visible evidence of an extensive set of connections – botanical, zoological, and intellectual – that began between that state and southeastern Australia in the nineteenth century.[5] New Zealand legislators debated American ideas of conservation and imported American and Australian plants and animals. All the settler countries imported natural history's ideas, institutions, and practice from Britain in the nineteenth century. The American idea of vast wild country reserves and the British model of urban open space shaped the idea of a national park in the other three. In the twentieth century, ecology developed as a discipline in academic centers in Britain and the United States, and went from there to the others. American ideas and action influenced the early environmental movement elsewhere.

The ground of Anglo settlement has been the continuing process of discovery. It began with entries in ships' logs and continued through the measured prose of army and navy officers surveying coasts and interiors. Each generation of settlers added knowledge and lore, and had its maps, which mixed named and fixed features with ones observed and others conjectured or wished for, all embedded in the white space that gave scope for dreams. In Europe people lived in country they knew; these societies lived on land they were discovering. Our period is the nineteenth and twentieth centuries because this is when formal nature knowledge – bodies of knowledge that were also ways of organizing the world – guided and shaped that process. Such knowledge had profound effects on the settlers' understanding of their lands and their relation to them. Natural history provided the settlers with pictures of the land in maps and reports, and placed their local knowledge in a universal system that ordered plants and animals around the world. It also gave

---

[4] J. A. Leach, *An Australian Bird Book* (Melbourne: Whitcombe & Tombs, 1911), 1, 72–73, 74.
[5] Ian Tyrrell of the University of New South Wales has in press a major work on the connections between Australia and California in the late nineteenth century.

individual settlers the chance to participate in the advance of knowledge. Collecting specimens, forming local societies, and building museums, they gave the world evidence of their societies' growth and maturity, and established social ties with each other and the elite of Europe. Ecology, which developed in the early twentieth century, was equally universal, but it had a different social context and presented to the settlers a different understanding.

Three terms need definition before we go on: "Anglo," "nature," and "science." The first is certainly something of a misnomer. Settler populations included entire groups – Africans and Chinese – who were not European at all, and many Europeans from the Continent. In places these last formed separate colonies – Quebec and New Mexico are obvious examples – ones the Anglos overran. A large proportion of those who traced their roots to the British Isles would not thank you for calling them either British or Anglo. On the other hand, the Atlantic and Indian Oceans were not the waters of Lethe or impassable gulfs, and the settlers who formed the governments and societies came from Britain and looked to it as home or at least a model. This was true even in the United States, which, for all its political heresies, was a cultural colony well into the nineteenth century. Continuity persisted, despite immigration, because newcomers found it advantageous to assimilate the dominant ideas and attitudes. "Anglo" is no worse a cultural tag than most, and it has the merit of fixing attention on the common cultural base and the ideas and aspirations of the people with the money and the guns.

"Nature" is another sticky term. We use it for everything from the essence of human psychological identity (human nature) to the physical universe. Here we will take it as the culture's understanding of the land and the living creatures on it at the level of "unaided observation." It was what people saw without telescopes or microscopes, felt, smelled, fixed in memory, and thought of as their "direct experience" with the world around them. Certainly viruses and galaxies are as much a part of "nature" as kangaroos and oak trees, but it requires experts with specialized equipment to place the first two in our picture of the world and everyone understands the others before they encounter formal education and even if they never do. This, admittedly, involves a certain analytical and philosophical sleight of hand. The existence of a "natural world," separate from society, the ideas of "direct experience" and "unaided observation," and the mental constructs that result from them are as layered and theoretical as anything philosophers have produced, even if the assumptions are not as clearly articulated. For this analysis, though, we can take all that for granted.

Nature on this level can be roughly but usefully divided into plants, animals, landscape, and climate. The first two pose few problems. The principles people use to decide what these are and to arrange them are so common that a taxonomist from New Jersey and a hunter from New Guinea would agree down to the level of what the scientist would call a species.[6] Landscape, which Donald Meinig calls "an attractive, important, and ambiguous term," we will take as the culture's picture of the land. It may include the landscape of geographers, which John Stilgoe defines as "shaped land, land modified for permanent human occupation," but more commonly it is the picture of the land people see as having significance for the nation and their culture.[7] It is what is presented in national myths of the "new country," in the landscape paintings hung in national galleries of art, the poems and stories printed in cheap paperbacks and taught to schoolchildren and found in exiles' recollections and memoirs of childhood. It is a continuing construction, shaped by each generation from the land, the culture, and experience.

Climate has something to do with temperature, rainfall, sunshine, and other atmospheric variables, but it has several meanings. It is, to start with, an economic and social reality. Our countries are what they are because European crops flourished there, and climate in this sense still dictates settlement patterns. It is also part of individual experience, a popular idea tied up with frosty mornings or harsh and sun-baked noons, wind, fog, snowstorms, and the rhythm of seasons. For much of our period it was also a physiological and even moral matter, for the Anglos retained classical beliefs about the links between climate and temperament. In the late nineteenth century Canadians saw the country's cold as a shield against moral dangers from their southern neighbors, and fifty years later Australian authorities fretted about the difficulties of settling their tropical North with white families.[8]

"Science" involves almost as many tangles as "nature." We apply the

---

[6] On the common basis of our construction of nature see Scott Atran, *Cognitive Foundations of Natural History* (Cambridge University Press, 1993), 15–80.

[7] Donald Meinig, "Introduction," in Donald Meinig (editor), *The Interpretation of Ordinary Landscapes* (New York: Oxford University Press, 1979), 1; John Stilgoe, *Common Landscapes of America, 1580–1845* (New Haven, Conn.: Yale University Press, 1982), 3.

[8] Clarence Glacken, *Traces on the Rhodian Shore* (Berkeley: University of California Press, 1967), 80–115, gives the ancient foundations of this. For late examples see I. Clunies Ross, "Blanks on the Map," in J. C. G. Kevin (editor), *Some Australians Take Stock* (London: Longmans, 1939), 83, and A. Grenfell Price, *White Settlers in the Tropics* (New York: American Geographical Society, 1939). This is late, though, even for Australia. See Warwick Anderson, "Geography, Race and Nation: Remapping 'Tropical Australia,' 1890–1930," *Historical Records of Australian Science*, *11*, 4 (1997), 457–68. On Canada see Carl Berger, "The True North, Strong and Free," in Peter Russell (editor), *Nationalism in Canada* (Toronto: McGraw-Hill, 1966), 4–26.

term to systematic attempts to acquire and organize knowledge, from ancient Greek philosophy and Babylonian astronomy to modern research, or confine it to a particular set of disciplines using a particular method of analysis and standards of proof. There is the additional complication that while the intellectual activity of modern European science dates from the seventeenth century, "scientist" as a social role, job, and profession developed only in the nineteenth. It was not until 1833 that William Whewell coined the word "scientist," and most of the disciplines we recognize, along with professional standards, degrees, and formal training, date from the second half of the century. Strict usage, therefore, would require us to call Newton a "natural philosopher" and Darwin a "naturalist" (which is in fact what he was called and what he called himself).[9] Since the history of science and the status of scientists are not central to this argument we need only note the variety of meanings and the distinction between doing science and doing it for a profession (which will be part of this study).

Here we can take "science" as the organized, written knowledge of plants and animals and the land, supported by social institutions, that developed within European culture in the early modern period. It took two forms, which differed in their perspective, methods, and relation to society. The first was natural history, a distinct field by the eighteenth century and the organizing principle for the study of visible nature to the late nineteenth. The second was ecology. It developed as a discipline in the late nineteenth century and solidified, institutionally and intellectually, in the years between the world wars. Our primary concern, though, is not the development of these fields, but their use in the culture. They were successor, supplement, and complement to the settlers' folkbiologies, guiding and affecting but not displacing that unwritten nature knowledge that the settlers brought with them, developed in the new lands, and passed on to their children. They helped people understand. "The role of science, like that of art," said E. O. Wilson, "is to blend exact imagery with more distant meaning, the parts we already understand with those given as new into larger patterns that are coherent enough to be acceptable as truth."[10] It is this use that is our central concern – the settlers' continuing journey from knowledge of nature to an understanding of their place in the land.

All this was in the settlers' minds. The lands, though, were not, and they had their own imperatives. If settler dreams were all that mattered,

---

[9] On words associated with science see the *Oxford English Dictionary*, Second Edition (Oxford: Oxford University Press, 1989).

[10] E. O. Wilson, *Biophilia* (Cambridge, Mass.: Harvard University Press, 1994), 51.

North American wheat fields would extend north to Great Slave Lake; the Centre of Australia would be farms and pastures; and there would be any number of rural utopias scattered from the Canterbury Plains to Saskatchewan. People can think about the world in many ways and change it in many more, but it is not infinitely plastic. The settlers spoke of "new lands," but they were new only to them. In parts of Australia you can walk on the rocks of the vanished supercontinent of Gondwana, and even the Canadian Arctic, where the ground is still rising from the just-removed weight of the glaciers and all the plants and animals are pioneers, is in its ecology far older than any human records. Nor were the lands vacant or "unsettled." Except for New Zealand, where the Maori had landed less than a millennium before, humans had been shaping the lands in myriad ways for thousands of years. The Anglos saw lands before time and outside history, but the opposite was more nearly the case. They were people with little history, coming to lands that had much.

What was there? Let us review the ground – glance briefly, that is to say, at the current social construction. We will start with the area the Anglos settled first, North America. The main line of their expansion ran east to west. In what became the United States the beachheads were on an open coast, dotted with harbors, in well-watered and forested country, rich in game. The soil and climate supported familiar farming. In Canada the land was colder, rockier, and entered not along a broad coast but through the narrow passage of the St. Lawrence River. Americans also had an easier time reaching the central valley. They had only to cross the Appalachians, relatively low mountains whose passes had been used by humans and animals for thousands of years. Canadians faced the Shield, a thousand miles of Precambrian granite so forbidding that until railroads were built almost all traffic to the west detoured south around it into the United States.

Between the Appalachians (or the Shield) and the Rockies two gradients, temperature and rainfall, shape the country. They run at right angles. One, temperature, falls as we go north. South Texas is subtropical, and the bulk of the central valley is squarely in the temperate zone. At the Canadian border we are in cold temperate conditions. A few hundred miles north European agriculture dwindles out in the oat and canola fields of Alberta and Saskatchewan. Beyond is the boreal forest, then the tundra that stretches to the shores of the Arctic Ocean. Rain falls off as we move west. The eastern prairies are well watered, but as we climb the great outwash plains of the Rockies vegetables give way to corn, and corn to wheat, and only cattle graze in the mountains' rain shadow. In the United States the mountains divide to form the Great Basin, an arid region around the Great Salt Lake, and in the southwest there are deserts. West of the mountains the Pacific stabilizes the temperature of the coastal plain and in much of it produces a Mediterranean climate, whose rainy winters and dry summers are much like those of southeastern Australia. From Oregon north there are vast temperate rain forests, stretches of fir and spruce that are the last strongholds of wilderness and the lumber industry.

Descriptions of North America commonly start with the continent in place, but Australia is so much a product of geology that we must begin with the breakup of Gondwana, some 60 to 80 million years ago.[11] It has drifted since, and for millions of years neither volcanoes nor plate collisions have thrown up new mountains or made new soil. It is the lowest and flattest of continents, and its soils, leached by sun and rain, are often deficient in minerals. The drift has largely been outside the global rain belts and it is, except for Antarctica, the driest continent. The result is a unique suite of plants and animals, adapted to drought, great variation in rain, fire, and poor soil. Eucalyptus dominates the plant communities, marsupials the fauna (for species larger than rats or mice). Until the Anglos arrived, there were no hoofed mammals to compact the soil and no carnivores larger than the dingo, itself an Aboriginal introduction.

The climate is dramatically different from that of Britain or North America. Only the southeast has what Anglos and their crops would consider sufficient rainfall, and it falls off quickly as we move off the coast. Most of the continent is arid or semi-arid, and the Centre stony and sandy desert. The northern edge, reaching into the tropics, is another world, with rain forests and a dramatic two-season year – the Wet and the Dry – but that was outside the mainstream of Anglo settlement.[12] Even today 80 percent of the population lives within fifty kilometers of salt water, most in a strip in the southeast running from Adelaide to Brisbane, and enormous areas are still marked on maps as "sparse fluctuating population" or "virtually uninhabited." Perth is the only city in the

[11] Two accessible examples of this modern narrative are Stephen Pyne, *Burning Bush* (New York: Henry Holt, 1991), 1–11, and Tim Flannery, *The Future Eaters* (London: Secker & Warburg, 1996), 20–52.

[12] Bureau of Meteorology, *Climate of Australia* (Canberra: Australian Government Printing Service, 1989). In no other country was exploration so hard or explorers so lauded. Lewis and Clark and MacKenzie are minor figures in North American history; Burke and Wills, who died on the return leg of the first south-to-north crossing of Australia, became national heroes. A. L. Burt made this point in commenting on the political development of frontier societies: "If Turner Had Looked at Canada, Australia, and New Zealand When He Wrote about the West," in Walker D. Wyman and Clifton B. Kroeber (editors), *The Frontier in Perspective* (Madison: University of Wisconsin Press, 1965), 59–77.

western half, Darwin (population 68,000 in 1988) the largest in the north. Canberra, established as a national capital, is the only inland city.[13] Even in imagination the Anglos have not ventured far inland. Movement, said a modern Australian novelist, Thomas Keneally, "is not westward to the center but eastward to the coast. Australia is periphery. It dreams of and yet abandons the core."[14]

Plants, the lack of land mammals, and the suite of unique flightless birds show that New Zealand is also part of Gondwana, but it is a very different part, not a continent but two small islands. They do not lie in Australia's latitudes but across the great trade winds of the Southern Hemisphere, which makes them temperate and well watered. Too well watered in places – parts of the west side of South Island average thirty feet of rain a year. New Zealand is geologically active; the grinding of tectonic plates shakes the land, springs and geysers dot the countryside, and the mountains are still rising. More than any of the others it has conventionally scenic landscapes, coastal plains backed by snow-capped mountains. On North Island there is the perfect cone of Mt. Egmont (Taranaki) and the peaks of Tongariro, while the great chain of the Southern Alps stretches down South Island.[15]

That is the almanac view. Let us, in imagination, get a little closer, take a mental tour of now-vanished landscapes – another current social construction. Our first stop, on a warm June day a few centuries before Columbus, is the woods of what will be central New Jersey. In the eighteenth century the Anglos will start turning it into farmland, lacing it with fences and dirt roads, and a century later the railroad will connect it to New York and Philadelphia. After World War II wheat will give way to potatoes and vegetables, and at the end of the twentieth century they will yield to ranch houses. Now we walk under enormous trees, slathering ourselves with mosquito repellent. Those are oak, says the ecologist with us, and that is a beech. The bird-watchers in the party ignore her to focus on the scarlet tanagers, Baltimore orioles, and warblers in

---

[13] "Physical Geography and Climate of Australia," 202–56, population density map after page 256; P. Laut, "Changing Patterns of Land Use in Australia," 547–56, all in Australian Bureau of Statistics, *Year Book Australia, 1988* (Canberra: Australian Bureau of Statistics, 1988). Canada is demographically comparable, its population clustered in a few strips along the border with the United States, but agriculture is possible in much more of the land. On the ecology of these inland pastures see Graeme Caughley, "Ecological Relationships," in Graeme Caughley, Neil Shepherd, and Jeff Short (editors), *Kangaroos: Their Ecology and Management in the Sheep Rangelands of Australia* (Cambridge University Press, 1987), 159–87. Caughley claimed in an interview with the author, June 1990, that equilibrium models of ecosystems were inappropriate and misleading in the chaotic Australian system. His viewpoint is not universal, but it is as least defensible.

[14] Thomas Keneally, *Woman of the Inner Sea* (New York: Doubleday, 1993), 179.

[15] *New Zealand Official Yearbook* (Wellington: Bureau of Statistics, 1992), 1–11.

the branches overhead. The ground smells like any forest, woody and damp, but this is not plowed land gone back to woods. For centuries trees have been falling and rotting in place, and it is as lumpy as an old mattress in a cheap motel. It seems the forest primeval, but our ecologist says that the dense patch we hiked around yesterday afternoon was a Lenape cornfield, abandoned fifteen or twenty years ago.

Skipping across the continent, we find everything changed at each stop. Along the Ottawa River we camp beneath spruce and giant white pine – the latter evidence of dry years and Indian fires a century and a half ago. The woods are aromatic, but with the piney odor of decaying conifer needles, that thick layer of duff we walk on and kick up under the trees. We have a gray jay ("camp robber," as it is known) and wood ducks on the river. At dusk moose come to drink, and at dawn a loon wakes us with a cry that really does sound crazy. A few hundred miles west of the Mississippi and well south of the Canadian border, we pitch our tents on a low hill amid grass that stretches to the horizon like some green ocean, waving and rippling in the wind. The children flush chunky brown birds that sail off on stiff wings – meadowlarks. Piles of dung and a scraped-out wallow show that buffalo are here, even if today they are over the horizon. In the afternoon there is a prairie thunderstorm, as near a timeless spectacle as the land affords. Great black thunderheads loom overhead in a sky that, off to the side, still shows blue. They swell over us, then comes a cool, hard wind, smelling of rain, a few scattered warning drops, and a downpour. In an hour the clouds disperse and the sun shines again. We are lucky – no hail with this one. Along the Virgin River in what will be Zion National Park we camp on a floodplain below sandstone cliffs, hear coyotes and a mountain lion at night. Early risers get to see the sun paint clouds and cliffs deep red and watch a golden eagle prospecting for rodents. The rest of us have to be content with the harsher light and the washed-out colors of the desert day. On the West Coast we camp under gigantic Douglas fir and Sitka spruce, and have around the camp a handsome bird with a crest, metallic blue on the back and brown on the head and breast. The easterners call it Steller's jay, the westerners just "jay." To the Kiwis and Aussies in the party it is another oddity. In bright sunshine on the beach we watch great swathes of fog drift off the sea and into the trees, visible evidence of the Japan current just offshore.

Crossing the Pacific we arrive in Australia at twilight and set up camp near a billabong (water hole to the Americans) in what will be western New South Wales. It is winter here, and we gratefully take to our sleeping bags after supper to study the "wondrous glory of the everlasting stars" (a line from Banjo Paterson, author of "Waltzing Matilda," contributed by one of our Aussies).[16] We can all pick out that emblem of these lands, the Southern Cross – a striking, kite-shaped formation – but the Aussies have to tell us the rest. When the moon comes up, two days past full, the North Americans eye it with suspicion. Yes, it does look different. We are in the Southern Hemisphere and seeing its face at a different angle. In the morning the Aussies and the family of Kiwis with us remind us that the sun is in the north, not the south. Walking downslope to the billabong, our westerners find the land familiar at first glance but strange on inspection. What looks like bunch grass is not; the soil is powdery and strange-smelling; and the trees seem thin and straggly, with the merest scattering of foliage. No kangaroos today, but we see a flock of galahs, huge parrots with dusky gray backs, rose breasts, and absurd topknots, circling and landing in great noisy groups.

Along a small stream on the south coast plain, in what will be Anglo farmland and then Melbourne suburb, we get a wake-up call from a kookaburra. The maniacal laugh is as bad as the loon's. It is also known, our Aussies tell us, as Bushman's Alarm Clock. Someone mutters that it looks like a kingfisher on steroids. In the woods the trees are scattered, strings of bark hang on their trunks, and the leaves crunching underfoot have a faintly medicinal smell. On a nearby stream a swan is swimming, graceful and swanlike, but coal black. The exploring children, silent for once, have located something even more interesting. A small furry animal, its webbed feet sticking out at absurd angles, is stirring up the mud at the bottom of the stream. It is a platypus straining the mud for its food. Notice, our ecologist says, that sparrow-sized kingfisher (an azure kingfisher) perching on a nearby twig. It is beautiful; deep blue on the back, orange on the breast and belly. It is watching, she goes on, for minnows the platypus disturbs. Sure enough, there is a loud "plop" as it dives, and before the rings have spread across the stream it is back on its perch. A quick wiggle, a swallow, and it resumes its watch. The platypus surfaces to lie spread-eagle, chewing and swallowing its mouthful. Then it heads down again, diving with a grace that seems out of place in so odd-looking a creature. Odd to you, say the Aussies.

In New Zealand we camp on a hillside fifty miles south of Auckland. The Anglos will make this pasture, but now it is the forest primeval; the Maori have not yet landed. At night we can hear a moa crashing around in the bush, and our (human) Kiwis confidently identify a set of whistles and hoarse calls as the real thing. In the morning we find another landscape superficially "normal" to North Americans, strange on closer

---

[16] It comes from "Clancy of the Overflow."

acquaintance. Some of the trees look "cabbagey," as someone puts it, and the giant ferns remind the middle-aged members of the party of schoolbook illustrations of the Age of Dinosaurs. The hiking is the hardest since New Jersey; the bush is thick and tangled, and brushing against one of the drooping vines or a fern means a shower. A moa track makes for easier going – as long, someone grumbles, as we don't meet the moa. A hundred meters on the slope and a few hundred on the flat bring us to a small stream, where we jump a pair of large, white-headed ducks (paradise shelducks). Their "zeek-zeek" and "zonk-zonk" alarm calls echo over the valley as they flap off. We can see a plume of steam where a small hot spring empties into the stream and smell the hot, mineral-laden water. There are no aquatic plants for some yards down-stream of the junction, and a crust of minerals lines the bank and bottom. Otherwise it looks all right. The adventurous find a spot that is warm but not hot and declare that all the place needs is a proper pool and a bathhouse. On the way back our guide finds a silver fern. Sure enough, it is just like the emblem on his All-Blacks rugby shirt, and the underside of each leaf is as silver as if it were painted.

This exercise reminds us of what was, but it also illustrates one of the basic themes of this work: that we construct our world. We can imagine hiking in a forest that does not exist by extrapolating from our experiences of hiking in those that do. We have all seen the stars in the night sky, watched and listened to birds, frogs, and insects, run our hands over rocks and trees, felt the breeze on our skin and the ground under our feet, and smelled woods, swamps, streams, and dust. We all added to that information from the culture. Those who had not seen a platypus or a moose no doubt called up pictures from books, television, or films. We automatically incorporated as well the more specialized knowledge of science. Who thought of geology when the word "floodplain" was mentioned? Even those who did immediately translated that concept into a mental picture. We do this not only with landscapes we imagine. We invest landscapes we see with significance because of what we know. The thrill of wilderness is not just the trees or prairie in front of us but the knowledge that the land stretches for miles without human habitation – or that there are wolves out there. The beauty of the Grand Canyon comes from color and form but also from knowing that this is the river's work over ages and that the rocks reaching into the depths show ages before that. The culture does the same thing, though on a different scale and through a different, social process. People agree on what the land "really" looks like and what it "really" is. They put themselves into this picture, not as tourists but as active forces and as people shaped by the land. They ask what this land and their life on it means. They listen, in the Crow elder's phrase, to the spirits' voices.

So much for lands and settlers. What about the topography of the book? The main line is not quite the historians' beaten track. We have customarily focused on the nation or a group within it and have been suspicious of work on larger units, unless they had some obvious political or social unity. There are good reasons for that, and they raise questions about this enterprise. New Zealand has a land area 3 percent of Australia's (270,000 square kilometers against Australia's 7,682,000). Canada (9,970,000) and the United States (9,363,000) are even larger. Population has a similar range: there are now some 3 million people in New Zealand, 16 million in Australia, 26 million in Canada, and 260 million in the United States. People are fairly evenly distributed in the United States and New Zealand, but most Canadians live near the U.S. border and most Australians by the sea in the southeast. New Zealand comprises two islands, Australia has its own continent, and the other two divide most of North America. New Zealand is temperate, and so is most of the United States, but almost all of Australia is arid or semi-arid, and Canada stretches from the edge of the Temperate Zone to the Arctic. The Anglo history of North America goes back two hundred years before Anglos arrived in Australia, and the United States was a nation sixty years before the Treaty of Waitangi established a British colony in New Zealand. What comparisons can we make among countries so disparate in size, population, and history? They have, as well, internal divisions. Can we speak of "an" Australian response to the landscape or "the" American attitude toward wilderness, even at a given time, without identifying culture with nationality, falling into geographic determinism, or producing conclusions that are only truisms?

We should be able to and there are good reasons to try. We can avoid the pitfalls of essentialism – the fallacy that there is some "real" core of the nation or group – by recognizing that national attitudes are a matter of statistics. Each Anglo society had the full range of ideas and attitudes, but in different proportions. There may, for example, have been some late-nineteenth-century Australian seeking transcendence in the bush with the same fervor with which John Muir sought it in the Sierra, but Muir had many followers and our hypothetical Australian none. As for trying, it is abundantly clear that there are many discussions about nature in these countries that are variations on common themes. Everywhere people spoke of parks, wilderness, wildlife, and the environment. Even without their references to events and ideas from elsewhere, it is clear they were talking about the same things, but only the culturally tone-deaf could confuse an Australian discussion of wilderness with an American one, find the New Zealand environmental movement just like the Canadian, or think the term "national park" meant the same thing in any two of these nations. Taking these societies as a group allows us to make comparisons. Seeing how the same idea met different fates in different lands can help us separate the influences of land and society, what is unique to each one, what is part of a common cultural inheritance. It also allows us to deal with topics that not only cross borders but exist because of them. National studies miss or slight the networks that ran among these countries and are part of their histories.

Science is a more familiar trail, but we will not go down the familiar turnings. The subject here is not science as such, but the interaction of popular knowledge and expert knowledge. This is a tangled topic, for people have always named plants and animals, and Linnaean taxonomy built on the concepts of folktaxonomy, which, in fact, remain at the base of popular and scientific ideas today. Besides, the division between expert and popular knowledge has never been complete. Natural history was an accessible science. Everyone could understand its central task – the arrangement of life's forms – and join in collecting specimens. Ecology built on that and, despite its use of all the apparatus of modern academic specialization, has closer ties to popular understanding, and is more part of common culture, than sciences that examine nature on other levels. It almost has to, for nature on this scale is the world we live in and learn about as small children. We have no common human experience at smaller or greater scales, and so physicists, chemists, and astronomers can develop theories and concepts with no apparent ties to common sense. There is also the common human tendency to use nature at this level as a model and moral guide. There is little use saying people should not do this; the practice goes back to the Preacher's exhortation to "go to the ant, thou sluggard" and forward to modern environmentalist appeals to nature's processes as a model for responsible action.

The development of knowledge, though, has created some barriers between bodies of knowledge. Folkbiology was local and instrumental knowledge, passed on in bits and pieces to the young in the process of acculturation. Natural history, a more extensive and formal learning, was forced to resort to institutions – private societies – and its own means of communication, their journals. Its theories, although quite sophisticated (Darwinian evolution), remained close enough to the public's that educated people could read and understand the field's major works (which helps account for the uproar over Darwin's ideas). Being a natural historian, however, was not a career, and there was no professional training. Ecology was, intellectually and institutionally, a step beyond, a specialized discipline, housed in university departments, its research the profession of people with advanced training, its theories increasingly couched in technical terms. It was less open than natural history even on the national level. All the settler societies established societies and museums to pursue natural history, but only the United States and Britain had the money to support ecology's infrastructure of education and research.

This line, like any other, requires neglecting some things or treating them in part or in passing. The most obvious are the full histories of natural resources and nature policy in any of these countries. The same is true of the sciences. We see formal knowledge as it bore on the settlers' efforts to understand and live with their lands and as they interacted with popular ideas. Concern with formal nature knowledge means there is little on the period before 1800. Putting the dominant group at the center leaves out the influence the earlier inhabitants had on the Anglos or the contributions Africans made to popular North American ideas. The histories of other European enclaves, notably the French in Quebec and the Spanish in the American Southwest, are omitted. Geographic coverage is necessarily uneven. Canada gets short shrift, not because it is not important but because it is so entangled with the United States (U.S.–Canadian interaction deserves, incidentally, far more serious attention from environmental historians than it has received). So does New Zealand, which was too small a society to support an independent dialogue on many of these matters. It appears as developments in the islands show themes in settler development. Writing one book, though, requires not writing six or seven others, and there are good reasons to start with this one. Understanding what the Anglos thought and did is a prerequisite to a full environmental history of each country and of the group they formed, and seeing them as a group points to an important (and underappreciated) part of their history.

Let us leave topography and what topics I do not consider and look at the main lines of the argument. That is the layering of knowledge and the shifts from one system to another against the backdrop of settlement. The first two chapters deal with natural history in the great expansion of the nineteenth century. One takes up the field as a science and the uses to which the settlers put it – organizing their knowledge, placing it in the growing body of European nature knowledge, revealing the land and its resources, offering individuals the opportunity to contribute to their nation's greatness and form connections to the metropolis. The next discusses the settlers' less organized attempts to come to terms with the country, programs not closely tied to formal knowledge but based in natural history's perspective and ideas. The focus is on two important enthusiasms of the second half of the century, the fad for importing birds and mammals and the fashionable recreation of sport hunting. These had biological and social consequences, but they also show deep-seated attitudes toward nature and ways of relating to it.

The next three chapters deal with the end of expansion and the generation that grappled with the aftermath during, roughly, the years from 1880 to 1930, a period in which the settlers had to face limits on their action and knowledge and in which they began to ask more consciously what value the land had for them. The limits of their power appeared in debates that began in the 1860s and culminated in the conservation movement of the early twentieth century. A second current, the use of native nature for national identity, began to flow in earnest toward the end of the century. It was more diffuse, appearing in everything from the use of native species as national symbols, to nature literature and landscape painting, formal and informal nature education, the boom in outdoor recreation, and the development of national parks. A third development was the change in science, which was intellectual, institutional, and social. With the exhaustion of natural history there was a shift from the observation of organisms to laboratory study of the processes of life, and even in the field attention shifted from nature's parts to their relation on the land. Academic departments replaced museums and societies as the locus of research, and professionals replaced amateurs. Nature knowledge became more and more the province of experts.

Three more chapters, overlapping with the last set, develop changes in popular and scientific ideas in the first half of the twentieth century. There was a renewed debate about human power over nature, which ran the gamut from vast optimism (in the 1920s) to depression and despair (in the 1930s). It was in some respects like the battles of a half-century before, but underneath were new currents. Another continuing theme is the exploration of nature's value and place. Some elements of this, like the boom in Australian bushwalking, harked back to ideas from the boom of the late nineteenth century. Others, particularly the debate in North America over the place of predatory mammals, showed new ideas. The final chapter of the section treats the emergence, in the interwar years, of a new perspective on nature among a small group of ecologists, game managers, and enthusiasts and tells how a few pushed beyond policy to apply this view to humans' use and treatment of the land.

The final two chapters treat the impact of their ideas on the public's, the rise of environmental consciousness and an environmental movement. After World War II the institutional apparatus of ecology and applied ecology – a network of academic departments and government agencies – spread the field's perspective to scientists throughout the Anglo world. School courses, books, movies, and television shows did the same for an increasingly interested public. In the late 1960s the union of knowledge and concern put the defense of visible nature and its value to the settlers on a new foundation. We are now picking over the experience and legislation of the first generation of environmental action, in the early stages of what promises to be a revolutionary transformation of our attitudes toward the land.

This narrative necessarily dwells on the destruction the settlers brought to their lands, but it seeks to point to a deeper, less visible story: the ways in which the lands shaped the settlers. The Anglos came as conquerors. Seeing the land in European terms, they tried to make it like their old homes. They remained to become settlers and to value the land for what it was, or had been. Now we debate whether we must, should, or can become native, learn to live with as well as from the land. If we do not and the ecological systems of the lands collapse, the settlers' search for a place will be but a minor note in their history – assuming anyone has the leisure to write history. If, on the other hand, we do learn wisdom, the tale of the land speaking to the Anglos and their listening to it may become the centerpiece of the settlers' stories.

SECTION ONE

MAKING THE LAND FAMILIAR

In the nineteenth century the settlers made themselves at home on the land by making it like home. They used European plants and animals and the tools of industrial civilization – axes, rifles, plows, poisons, and railroads – to transform the countryside with a speed and thoroughness never seen before and on a scale that will never be repeated. Learning about the land meant thinking about it in familiar ways. Learning its wisdom was a hidden current in the heady days of conquest.

Natural history as science is the subject of the first chapter, for it was the settlers' primary intellectual tool and their guide to conquest. It was a powerful tool, organizing all the earth's "natural productions" and the earth itself into a formal system of learning that included institutions intended to hand on this knowledge. It was, however, open to anyone with a smattering of education and interest, and its concepts were close to popular ones. As a result its perspective, that of nature as pieces, was part of the larger culture. It was hobby as well as study, social activity as well as science, and connected as well to rural nostalgia and people's view of the landscape.

The second chapter deals with the settlers' attempts to make the land into something familiar. The destruction of native ecosystems is central. This was so vast and obvious that contemporaries and the first generation of environmental historians have seen destruction as the only story to be told, but there were others, efforts to remake the land or to find familiar recreations within it. One was the enthusiasm for importing mammals and birds for sentiment and sport, the other the activity of hunting. Both had biological and social consequences. Introductions of animal species reshaped the landscape of Australia and New Zealand and, to a lesser degree, affected North America. Sport hunting helped change ideas about wildlife, and it laid the institutional groundwork for wildlife preservation. These trends, though, also show the settlers' early struggles to create connections to the land, the differences among the lands, and the emerging differences in their societies.

1

NATURAL HISTORY AND THE CONSTRUCTION OF NATURE

> ... Nature, that universall and publik Manuscript, that lies
> expans'ed unto the eyes of all. ...
>
> Thomas Browne, 1642[1]

During the great expansion of the nineteenth century, the settlers
learned by trial and error and from the peoples they dispossessed, but
natural history was their common and most important guide to "that
universall and publik Manuscript" which lay before them. Now it seems
outdated, a matter of odd facts, Latin names, and amateur observation.
Then it was the leading edge of European understanding, an immensely
powerful intellectual tool that was as expansive as the visible world it
studied. It shaped the settlers' understanding of nature in several ways.
It, obviously, organized their local knowledge and placed the "natural
productions" and curiosities of their countries in a comprehensive
system that was part of European high culture. It also laid out the land
before them. Scientific expeditions mapped coasts and mountain
ranges, traced rivers, and put names on the land. Measurements of rainfall, temperature, frost, and sun – translated into lines on maps – showed
weather and climate. Geological studies (this field was the earliest part
of natural history to become an independent speciality) showed what
was beneath their feet. Natural history also informed books like Thomas
Jefferson's *Notes on the State of Virginia*, which we hardly think of as
science, and it affected even the search for beauty and significance in
the landscape.[2] It was part of society. Collecting specimens and taking
measurements, individual settlers could form connections to the

---

[1] Thomas Browne, *Religio Medici* (1642; Oxford: Oxford University Press, 1964), 15.
[2] On names see especially Paul Carter, *The Road to Botany Bay* (Chicago: University of
Chicago Press, 1987); on natural history as construction of nature, Pamela Regis, *Describing Early America* (Dekalb: Northern Illinois University Press, 1992).

metropolis and contribute to the progress of their society, and the societies, museums, and expeditions they supported were visible evidence of the maturity of their new societies. (This was particularly important in the early days of the United States, where every national accomplishment justified the republican experiment.)

It was not that nature study was new, or even that taxonomy, the arranging of plants and animals in some order, was an intellectual leap. Every culture had an accumulation of knowledge about the world and a folkbiology – names and an arrangement of local plants and animals. The settlers soon had their own in each new land, and an Anglo popular understanding developed alongside natural history. They also had natural history, a much more formal system with a set of social institutions – societies, journals, and collections – to compile what was known, add what was being found, and pass on the whole to the next generation. It was one element in what Keith Thomas called "a whole cluster of changes in the way in which men and women, at all social levels, perceived and classified the natural world around them." Between 1500 and 1800 there was "generated both an intense interest in the natural world and those doubts and anxieties about man's relationship to it which we have inherited in magnified form."[3] These include a shift from seeing humans as the center of the world to viewing them as one species in a complex whole; the growing knowledge that humans could affect the world in ways that would, in the long or possibly the short run, destroy them; the shift in nature knowledge from a body of information accessible to all to one that was the province of experts; and the creation of social structures to carry and develop that knowledge.

Natural history was based on the Enlightenment faith in humans' ability to understand the universe. Everything would be subjected to the tests of reason, and all would become clear. This confidence reached its peak in the nineteenth century, when men like the German naturalist Alexander von Humboldt could see natural history as encompassing everything. After the process of division by classification would come comprehension of all as a great whole. We would know the mountains and the butterflies on their slopes.[4] It appealed, though, to other elements. Particularly in Britain one of its great supports was the rural nos-

[3] Keith Thomas, *Man and the Natural World* (New York: Pantheon, 1983), 15. On the background of European ideas about nature see Clarence Glacken, *Traces on the Rhodian Shore* (Berkeley: University of California Press, 1967).

[4] On Humboldtian science and its influence see Susan Faye Cannon, *Science in Culture* (New York: Dawson and Science History Publications, 1978); Margarita Bowen, *Empiricism and Geographical Thought* (Cambridge University Press, 1981); and Michael Dettelbach, "Humboldtian Science," in N. Jardine, J. A. Secord, and E. C. Spary (editors), *Cultures of Natural History* (Cambridge University Press, 1996), 287–304.

talgia that developed with early industrialization. The study of nature was to lead us back to a better past and a simpler world. We find this mixing of formal knowledge, technique, philosophical perspective, and social critique odd, but that is only because we have a different perspective. We distinguish between expert, objective knowledge and subjective reactions to the world in a way the Victorians did not. They simply drew different lines. Our own, as we shall see, have their own drawbacks.

In discussing the development of natural history as a way to know and understand the world we begin with folkbiology, the popular system of knowledge that preceded natural history and is still everyone's first nature knowledge. Its interactions with natural history are the first examples of a continuing theme, the negotiation between popular and formal knowledge. Then we come to natural history as a system of thought, an arrangement of the world, and part of the high culture. It was, though, one thing in the metropolis, quite another at the periphery of European civilization. As they did with every other part of European culture, the settlers shaped natural history to their society and their situation. Here is another recurring theme.

Folkbiology and the Visible World

> And out of the ground the Lord God formed every beast of the field, and every fowl of the air; and brought them unto Adam to see what he would call them: and whatsoever Adam called every living creature, that was the name thereof.
>
> Genesis 2:19

Everyone is Adam. The urge to name and arrange what we see around us is universal, and its expression is common, if not uniform, across cultures. Anthropologists have found that societies usually recognize between six hundred and a thousand distinct things, most of them of no economic or cultural use, and for similar environments have about the same level of specific knowledge, whether such knowledge is prized or not. More than that, people use the same rules to name and arrange what they find.[5] All cultures recognize a distinction between living crea-

---

[5] This discussion draws heavily on Scott Atran, *Cognitive Foundations of Natural History* (Cambridge University Press, 1990). See also Brent Berlin, *Ethnobiological Classification* (Princeton, N. J.: Princeton University Press, 1992). I do not assert here that folkbiology is biologically based, though that can be argued, nor do I take any particular philosophical position on the existence of the world. Neither is required for this argument: that there are strong common trends, apparently quite deeply rooted.

tures and nonliving things, a division even very young children make. There is, usually, no common word for the group. In English, for instance, we can say "plants and animals" or resort to the technical term "biota," but there is no popular word for "all living things as distinct from nonliving ones." Every culture divides living things into a few large classes by their form, cuts these into smaller groups by finer distinctions, and so on down to the level a modern taxonomist would call a species. In more technical language, people construct nested hierarchies. A sugar maple, for instance, is part of a larger group of all maples, then successively larger ones of deciduous trees (roughly ones that shed their leaves in the fall), trees, and plants. Strategies for naming are similar. At higher levels, for instance, the type name for the large class also serves for a part of it. In English, again, "animal" designates living things that are not plants but also the smaller class of things within that which have fur and mammary glands. This produces some apparent contradictions that in common usage we all pass over. We all know frogs are animals rather than plants, but not animals when the conversation turns to cats and dogs.

What marked natural history off from folktaxonomies was its aim of going beyond the local biota to arrange all life on earth in a single system, one that would show the natural relationships among all its forms. This was a great intellectual project, but it grew out of practical circumstances. The age of exploration produced a flood of oddities from Asia, Oceania, and the Americas that overwhelmed the Europeans' systems, both folktaxonomies and Aristotle's categories (the basis of such formal taxonomy as there was). But while each thing was strange the flood was not chaotic; each thing had affinities to known forms. The Europeans set out to arrange everything. It took them almost two centuries to make a system that seemed adequate for the task, and it required, beyond intellect, institutions that would collect, arrange, and organize this knowledge, provide a forum for deciding hard cases, and transmit the information to the next generation.

Settlement was going on even as the scientists worked, and the settlers did not wait for the savants to tell them what to call what they found, nor did they always respect expert judgments. Names came from their history and experience. Names from Europe were common, applied to something in the new land that looked familiar. The most obvious result was the "robins" scattered from New Brunswick to Australia (sixteen species there). They do not look much like the British bird or each other, and they have no close genetic relationship. What they do have is a bright or at least pale breast and a dull back (though one New Zealand species with brightly colored relatives is itself dark all over – the black robin).[6] In the early years of settlement the colonists added the adjective "native" or the name of the new country to distinguish their species from ones back "home." Thus what North Americans now call the "robin" was earlier known as the "American robin" or (after another European bird) the "American fieldfare."[7] In Australia the koala was the native bear or native sloth (borrowing a South American name attached for the animal's supposed characteristics) and the wombat the native badger (for its burrowing). A group of marsupial carnivores (the dasyurids) were called native cats, some very small marsupials native or pouched mice.[8] Australians called one eucalyptus "mountain ash," and New Zealanders tagged a smooth-skinned tree the "southern beech." An early-twentieth-century Australian naturalist spoke of English names being applied to native species "in the most curious and often fanciful way."[9] An old name for the platypus is "duck-mole," a blending from its "bill" and fine fur. There are cuckoo-shrikes, common birds that are built like a cuckoo and patterned like a shrike, and magpie-larks, which have a lark's form and pied feathers.

North American settlers found it easy to use British names, for their species were often closely related to European ones. Oaks and maples were different species, but clearly part of the same families. In some cases there was no difference. Wolves were wolves from Asia to North America. On the other hand, the settlers called their population of red deer (*Cervus elephas*) "elk" and, like Australians, used familiar names in nonapproved ways. In Britain, day-flying birds of prey were divided into "hawks," "buzzards," "falcons," and "harriers." The North American settlers lumped them together as various kinds of "hawk," applying "buzzard" to the large, soaring carrion-eating birds the British called vultures. "Turkey buzzard" is confusion compounded. The modifier was a name first applied to the African guinea fowl, then to the ground-dwelling bird (whose domestic form graces American tables at Thanksgiving) before being tacked on to a large soaring species no one ate.[10]

---

[6] The *Oxford English Dictionary*, Second Edition (Oxford: Oxford University Press, 1989), is the most accessible etymological source. See also any collection of field guides to birds.

[7] *Wilson's American Ornithology* notes Redbird, Blackbird, and American fieldfare for this bird. Edited by T. M. Brewer, with notes by Jardine and other material by Brewer (1840; reprinted New York: Arno, 1970).

[8] Edward E. Morris, *Austral English: A Dictionary of Australasian Words, Phrases, and Usages* (London: Macmillan, 1898), lists almost three pages of names that start with "native" and have an English noun attached. On the practice see also Frederic Wood Jones, *The Mammals of South Australia* (Adelaide: Government Printer, 1923), 1: 26–27.

[9] Baldwin Spencer, *Wanderings in Wild Australia* (London: Macmillan, 1928), 1: 151.

[10] A useful reference, showing some of the complexities of the subject, is Ernest A. Choate, *American Bird Names* (Boston: Gambit, 1973).

When something was outside their experience, the settlers often asked the original peoples. For example, they sometimes called a very compact and bad-smelling member of the weasel family "polecat" (from a European weasel that smelled) and employed euphemisms like "woods pussy" and "essence peddler," but the name that stuck was the Indian one: "skunk." When something was very unusual, the native name might be the only one. No one seems to have called kangaroos or kiwis anything else. Descriptive names were also popular; the whitehead (New Zealand) and the red-headed woodpecker (North America) are examples. Birds were often named after words fitted to their calls. There is the bobwhite. In New England the species known formally as the white-throated sparrow is called the "Peabody bird," for it says, "Old Sam Peabody." North of the border it sings, "Oh, sweet Canada." Australia and New Zealand have the variously named "mo-pork," "more pork," and "boo-book" owl. Some species got their names from their actions, as the fantails or wagtails. Others received familiar names. A common and fearless Australian flycatcher is called "Willie Wagtail," and the kookaburra (an Aboriginal name that is now the popular Anglo one) was also called Jack, Jacky, John, and Johnny. Like other common and conspicuous species, it also has a number of other names – alarmbird, breakfastbird, and bushman's, settler's, or shepherd's clock.[11]

Scientific names sometimes passed into popular use. This was not common in North America, where several generations had been named things before anyone arrived with Latin and specimen tags, but the downy and hairy woodpeckers have English names translated from Linnaeus's Latin, magnolia is the scientific name for this lovely tree, and so is sequoia (from a genus name honoring an Indian leader). The practice was more marked in Australia and New Zealand, where plants and animals were unlike those at home, naturalists came before the settlers, and natural history was a popular pursuit in the early years. The major plant groups – eucalyptus, acacia, and banksia (from the genus name that honored Joseph Banks) – are scientific names, and among the animals there is the platypus. Scientific names were so common that in the early twentieth century an advocate of popular nature education (and Australian nationalism) called for "[g]ood descriptive names for our varied and beautiful birds – more children's and poets' names, and less of the deadly formal" ones the experts had provided.[12]

Scientific practice, it might be noted, followed some of the same lines. Linnaeus used the Latin words for European species as their scientific

---

[11] Graham Pizzey, *A Field Guide to the Birds of Australia* (Sydney: Collins, 1980).

[12] J. A. Leach, *An Australian Bird Book* (Melbourne: Whitcombe & Tombs, 1911), 72–3, 74.

names, and the official nomenclature prefers descriptive names. The two systems, however, had different goals and uses. Scientists, seeking to communicate across oceans, languages, and generations, insist on a unique name for each form. The settlers, though they might recognize science, sought only understanding with their neighbors, and any name locally known would do. The result was that common and widespread species had many different names. A modern guide lists twenty-two for the laughing kookaburra, and a nineteenth-century American ornithologist found a hundred for the flicker (a brownish woodpecker with yellow or red wing linings).[13] Popular names also came and went, while science had well-defined rules for changing names, and there have been large shifts over the generations. British names with adjectives attached were most popular in the first generations of settlement and fell out of favor as ties to "home" weakened or when people sought to enlist native nature as part of national identity. Around 1900, for example, Australian naturalists and educators attacked English names as a vestige of a colonial past unsuited to a new nation.[14] In New Zealand, Maori names, even for forms the settlers knew, have become more common in the twentieth century and (though this is hard to measure) the pace of replacement seems to have risen in the past fifty years.[15] The many different names and their changes over time (only hinted at here) illustrate something important for the popular construction of visible nature. It was not an event but a process or a set of processes that reflected and shaped the settlers' relation to the land. Scientific names or official popular ones affirm a universal system. Local ones officially adopted speak to the heritage of settlement; ones from the earlier people seek identification with the land.[16]

Science as a Way to Nature

But what occupied me chiefly and became more and more the solace and delight of my lonely rambles among the moors and mountains, was my first introduction to the variety, the beauty, and the mystery of

[13] Pizzey, *A Field Guide to the Birds of Australia*; Frank M. Chapman, *Color Key to North American Birds* (New York: Appleton, 1903; 2d ed., 1912, same text), 5.

[14] Leach, *An Australian Bird Book*, 1, 72–4.

[15] On Maori names see H. W. Williams, "Maori Bird Names," *Journal of the Polynesian Society*, *15* (December 1906), 193–208. Among modern lists are those of R. A. Falla, R. B. Sibson, and E. G. Turbott, *Collins Guide to the Birds of New Zealand* (Auckland: HarperCollins, 1981), and Malcolm Foord, *The New Zealand Descriptive Animal Dictionary* (Dunedin: Malcolm Foord, 1990), which includes mammals, birds, insects, and other forms.

[16] On the persistence of old names and the shift in "standard" ones see various editions of Roger Tory Peterson's *Field Guide to the Birds* (Boston: Houghton Mifflin, 1934). On persistence the author also relies on personal observation.

nature as manifested in the vegetable kingdom. . . . I first named the species as nearly as I could do so, and then laid them out to be pressed and dried. At such times I experienced the joy which every discovery of a new form of life gives to the lover of nature.

Alfred Russell Wallace, 1905[17]

Here was the great attraction of natural history for the Victorian public – the pleasure of personal discovery that contributed to a social body of knowledge. If it was not available to all, it was at least widespread and accessible. Natural history had no abstruse theories and could be practiced without special equipment or training. Inexpensive handbooks like the one Wallace used poured off the presses of Victorian Britain, and there were local societies and enthusiasts in every district. So little had been done, particularly in the early nineteenth century, that new species or new facts could be found almost anywhere. The amateur was hardly limited. He or she (natural history was popular with women as well as men) could collect many things or one, and the activity could be confined to vacations or become a lifelong avocation pursued with vigor. For a few it could lead to adventure and honors. Wallace collected in South America and Southeast Asia, made observations on the distribution of plants and animals that helped divide the world into biogeographic regions (one division, in Indonesia, is marked by Wallace's Line), and wrote a paper on competition in nature so close to Darwin's that it forced the latter to publish, lest he lose credit for the theory now associated with his name. Darwin, of course, had an even more illustrious career.

At the heart of natural history was the foundation of modern scientific taxonomy, Linnaeus's system. Biology has built so much and so well on it and education has made it so familiar (if not well loved) that it is hard now to see its importance. It was important, the greatest achievement in the study of visible nature before Darwinian evolution. It not only allowed unambiguous identification of forms but encouraged and guided thought on fundamental questions about the relationships among them. The triumph of the system developed by the Swedish naturalist Carolus Linnaeus was a matter of intellect, but also of social conditions and luck. Linnaeus's technical accomplishment was to construct a comprehensive system that could be easily learned and used and that could incorporate novelties. It was not to arrange everything. Even at his death there were important gaps. The smaller forms were still confused, some of the higher taxa vague, and the sexual system for classi-

---

[17] Alfred Russell Wallace, *My Life* (London: Chapman & Hall, 1905), 1: 192, 195.

fying plants was completely artificial (and so at odds with the goal of classification by natural relationships). Its social advantage was that it was well known. The successively larger editions of the *Systema Naturae* did much to spread the system, but Linneaus also had a core of disciples and students who devoted their careers (and a few lost their lives) to the great project. He was also an active correspondent and in the course of a long career made many friends. His British admirers provided, after his death, a crucial boost. Sir James Edward Smith bought the great man's private collection, and he and others formed the Linnaean Society in London in 1788. It was the start of natural history in Britain, and the society's journal became one of the foundations of the scientific literature in biology.[18]

Linnaeus is conventionally seen as a pioneer of science, breaking with the past and blazing a new trail. He was and he did, but the story is more complex than that. His success was in part due to his using familiar techniques and concepts. The system did not require special apparatus. To classify it was necessary only to observe, count, and measure. Conceptually, his system is not so much a break with popular ideas as a refinement of them. It organized creatures in the same way that folkbiologies did, by exhaustive division into a nested hierarchy. The largest division was the kingdom, dividing plants from animals. (The modern system has refinements, including super-kingdoms based on the presence or absence of cell nuclei, but we can ignore them.) Within the animal kingdom, to pursue a familiar part of the taxonomic tree, things were divided into phyla by their basic body plans. Humans, to take an example at hand, are in the phylum of chordates (roughly things with backbones, spinal cords, and central nervous systems). Here we split off from starfish, sponges, and clams, which are "animals" but belong to other phyla. Phyla, in turn are divided into classes. Within chordates, for example, birds constitute one class, fish another, mammals a third.[19] These in turn are arranged in orders. Among mammals there is one for meat eaters (Carnivora), another for marsupials (initially at least; taxonomists now argue about splitting this into two), and others for rodents, hoofed animals, and so on. We are among the primates. Orders are divided into families (ours is Hominidae), then genuses, then species. A Linnaean binomial, in what is by now a specialized version of

---

[18] David Allen, *The Naturalist in Britain* (London: Allen Lane, 1976), 46.
[19] This class (Mammalia) is named for the common method of feeding the young. Linda Schiebinger discusses the social roots of this scientific name in "Why Mammals Are called Mammals: Gender Politics in Eighteenth-Century Natural History," *American Historical Review, 98* (April 1993), 382–411.

the Latin language, gives each species a unique, two-part italicized name. The first word, capitalized, is the genus, the second, not capitalized, the species. Humans are *Homo sapiens,* Latin for "wise man." (This should give that classic detached observer, the Martian anthropologist, some clue as to which organism devised the list.)[20]

The system, though, is more than a catalog. It is a way to think. Indeed, by its formal division and explication of categories, it forces thought about basic issues. Placing bats and whales in the class Mammalia, for example, requires deciding what characteristics are basic, which peripheral, to the idea of "mammal." Here taxonomists decided that physiology meant more than form. The platypus is an even better example. European naturalists condemned the first specimen as a fake – the work of a sailor with some body parts, needle and thread, and too much time on his hands – but dissection showed it was real. Where did it belong? It had fur, like all mammals and only mammals, but it lacked obvious mammary glands. It had poison spurs, like some reptiles but no known mammals. There was a single body opening for elimination and reproduction, an arrangement common to birds and reptiles but not mammals. It was finally labeled a mammal, but one so different that a new order, Monotremata (one-hole, referring to its combined reproductive and excretory functions), had to be added to the class Mammalia to accommodate it.[21] That placed it, but it also more closely defined what a mammal was.

In other ways Linnaeus did break with popular ideas and earlier systems. This is clearest in his treatment of humans. They had always been in some way outside the system. Some earlier taxonomies, including Aristotle's, had used the human form as the standard by which all others were judged. Others had classified creatures in relation to humans, arranging them by usefulness or uses, or dividing them between wild and tame. The *scala naturae* went beyond that to make a single hierarchy from worms to angels. Here humans were the link between the spiritual and material realms, the "great and true *amphibium.*"[22] Linnaeus took account of all biological life, including humans, on the basis of form. His only concession to our pride (failure of nerve, if you will) was to substitute for the usual physical description of the specimen the Socratic injunction – Know thyself.

[20] On the modern system see, e.g., Ernst Mayr, *Principles of Systematic Zoology* (2nd ed., New York: McGraw-Hill, 1991); on modern divisions among mammals, Ronald Nowak, *Walker's Mammals of the World* (5th ed., Baltimore: Johns Hopkins University Press, 1991).

[21] On the platypus's place and the arguments over it see Harry Burrell, *The Platypus* (Sydney: Angus & Robertson, 1927).

[22] Browne, *Religio Medici,* 33 (Sec. I.34).

The Linnaean system affected science but it was also a key to popular natural history. Now anyone who could read could participate in the great adventure. There were other reasons, but taxonomy was one reason why natural history societies sprang up across Britain in the early nineteenth century like mushrooms after a rain. Botany was particularly popular, for plants did not move and Linnaeaus's system for classifying them by the number of stamens and pistils in the flowers allowed anyone with a hand lens and an inexpensive handbook to pin the "proper" name on a specimen. There were also enthusiasms for collecting beetles, butterflies, mineral specimens, ferns, fossils, birds, and birds' eggs. People made fern gardens, then turned to terraria and aquaria.[23] Technology helped. Factories provided geologists' hammers, butterfly nets, telescopes and binoculars, chemicals to kill insects, pins to mount them, and arsenical soap to preserve bird and mammal specimens. Faster printing presses, coupled with increasing literacy, brought a flood of books, ranging from inexpensive handbooks with woodcuts to multi-volume sets of the natural wonders of the Empire. These last, sold by subscription, were the nineteenth-century equivalent of Ansel Adams coffee-table productions. Railroads provided access to the country even from the heart of London, and the improving postal system let people trade specimens and communicate with one another across the country.[24]

Natural History as Culture

Natural history's concepts were popular ideas refined, but its social structure was not. Folkbiology was local knowledge passed from one generation to the next in the context of daily life. Natural history was organized knowledge of the world passed on by institutions devoted to that task. They developed within the social structures of science, which by the early nineteenth century was becoming a recognized field and profession. Scientific bodies were numerous enough that in 1831 a group formed the British Association for the Advancement of Science (BASS) as a national coordinating body. Two years later, at its 1833 meeting, William Whewell coined the term "scientist."[25] This was not, however, modern, professional science, the domain of the expert. Natural history in particular was largely an amateur pursuit, and people had a variety of roles. Some simply wanted to join societies in order to learn about the latest novelty. Others contributed specimens and gave papers, acquiring

---

[23] Allen, *Naturalist in Britain*; Lynn Barber, *The Heyday of Natural History*, 1820–1870 (London: Jonathon Cape, 1980).
[24] Ibid.   [25] The *Oxford English Dictionary* traces this term.

modest local reputations. The more adventurous or committed collected abroad, and they might cement social ties by giving new specimens to wealthy or noble patrons or by naming a new species after them.
At the scientific pinnacle were those who, by virtue of experience and
access to large reference collections, could recognize new forms and
describe them properly for the scientific literature. They might be at the
center of an extensive network of collectors and protégés.[26] On a slightly
different plane were the rich, powerful, and noble, who gained and conferred prestige by their patronage of science. The BAAS, two historians
of this period note, was "central to early Victorian culture. Its members
included earls, marquises, and viscounts, while politicians of the calibre
of Sir Robert Peel and Lord Palmerston were pleased to accept office
within it; the Prince Consort himself was President in 1859."[27]

Just as each settler country had its own local collectors and societies,
each had its own version of this social web. In the United States, where
the community was large and political independence cut Americans off
from British honors (such as membership in the Royal Society), ties were
more within the country than they were in the political colonies. Still,
ties led to Europe. Asa Gray, for instance, was the American authority in
botany. He had a network of protégés and collectors and was the authority for people across the continent. He was also one of Darwin's correspondents (furnishing information on North American flora) and
became Darwin's principal defender in the United States.[28] In Australia,
Canada, and New Zealand political loyalties and small populations kept
communications more directly oriented to the cultural center. Royal governors organized the early natural history societies, and local collectors
sent specimens to experts at "home." The reward of effort and luck was

---

[26] On this see, e.g., Sally Gregory Kohlstedt, "The Nineteenth-Century Amateur Tradition," in G. Holton and W. A. Blanpied (editors), *Science and Its Public* (Dordrecht:
Reidel, 1976), 173–90. The power relationships here are central to ideas of colonial
science, discussed later.

[27] Jack Morrell and Arnold Thackray, *Gentlemen of Science* (Oxford: Clarendon Press,
1981), xxi; Richard Yeo, *Defining Science* (Cambridge University Press, 1993), 110–11,
32–5. On the union of science and society in a central figure of the period see John
Gascoigne, *Joseph Banks and the English Enlightenment* (Cambridge University Press,
1994).

[28] A. Hunter Dupree, *Asa Gray* (Cambridge, Mass.: Harvard University Press, 1959),
233–63. On the general process see Nathan Reingold and Marc Rothenberg (editors),
Scientific Colonialism (Washington, D.C.: Smithsonian Institution Press, 1987). Examples
of this process in Australia are described in Robert A. Stafford, "The Long Arm of
London: Sir Roderick Murchison and Imperial Science in Australia," in R. W. Home
(editor), *Australian Science in the Making* (Cambridge University Press, 1988), 69–101,
and others in this collection. On the amateur level see William Lines, *An All-Consuming
Passion* (Berkeley: University of California Press, 1996).

admission to scientific societies in the metropolis. A select few became fellows of the Royal Society.

The type example of a colonial using science for social position was Walter Buller. Born a missionary's child in New Zealand in 1838, he combined his interest in natural history and talent for social climbing into a stellar career. At the periphery of the British sphere, without access to experts or patrons, he relied on boldness. Ignoring the deficiencies in his education and the lack of collections in New Zealand with which to compare his specimens, he nevertheless described them, including some new ones.[29] This drew criticism, and European natural-ists eliminated many of his finds, but he had made his mark. He sought membership in the Linnaean Society by sending in his money and asking for a waiver of the rule requiring that candidates be nominated by three members, explaining that he knew only one. He neglected to note that he was two years under the minimum age. The society found sponsors who lacked personal knowledge but were willing to sign. (One was Joseph Hooker, of the family that dominated English botany and ran Kew Gardens.) In 1859 Buller became secretary and curator of the revived New Zealand Society, the national scientific group, positions that involved some drudgery but brought him into contact with high society. He "found another powerful patron" in Sir George Gray, the governor, and elbowed his way to greater prominence by writing an essay and exhibiting a collection of birds at the 1864 Dunedin Exhibition. He con-tinued to seek membership in European societies and cemented the relationships he established through them by sending specimens of New Zealand birds to prominent collectors. In 1874 he was nominated to the Royal Society, and his sponsors included John Gould, expert on Aus-tralian birds and a major publisher of natural history works, and Charles Darwin. He was then in London, working on his major publication, *A History of the Birds of New Zealand*, with leave and money from the New Zealand government. He was not elected that year, but in 1879 he was, and New Zealand newspapers hailed "the first man born and educated in the colonies" to have "received the highest compliment and distinc-tion which the scientific world of Great Britain has in its power to bestow."[30] He died in 1906 on his estate in Britain.

Most people sought other satisfactions, finding in natural history an irresistible combination of industry, self-improvement, piety, and recrea-tion. For the pious it was an ideal hobby – the adoration of God by exam-ining the wonders of His creation. The less religiously inclined simply

---

[29] Ross Galbreath, *Walter Buller* (Wellington: GP Books, 1989), 90–2. Information on Buller, unless otherwise cited, comes from this useful biography.

[30] Ibid., 33–4, 78–80, 129.

found beauty in the structures and ordering of plants and animals. For everyone it was a way to learn. This mixture of emotional appreciation and rigorous study may strike us as odd. We tend to see a contradiction between what Donald Worster, in *Nature's Economy*, described as the "arcadian" holistic tradition and the "imperial," reductionist strain, the one leading to appreciation of nature, the other to scientific study of it.[31] This is a useful analytical distinction, but it does not hold in life. Wallace's "joy [at] every discovery of a new form of life" lay not, or not just, in "conquering" nature or increasing his prestige. It was also pleasure in finding a new aspect of nature's beauty. The line between cold, rational practice, dissecting and analyzing, and empathic identification with nature does not run between people. It runs through each one. For many, from John Ray in the seventeenth century to the present, intellect and emotion fed on and reinforced each other. Consider the reaction of Frank Chapman, a curator at the American Museum of Natural History at the turn of the century, on sighting a rare specimen. In his *Autobiography of a Bird-Lover* (1935) he spoke of "two forces struggling within me. One, of the hunter and naturalist, bade me shoot this animal, excellent to eat and rare in collections; the other, of the nature lover, prompted me to spare it. It seemed murder outright to fire at this defenseless creature." Still, he had to do it, and not just for the sake of his employer. "[O]nly the bird in the hand will satisfy the desire for intimate, exact knowledge of [the bird's] characters which obsesses the naturalist."[32]

This union of opposites is at the heart of nature writing, another aspect of natural history that was part of the general culture. Its central aim since the days of natural theology has been to induce awe and wonder by presenting in accurate detail the marvels of nature, but, since people have always used nature as a contrast and example for society, it has also been a vehicle for social criticism. This is a strong, if implicit, element in the genre's first best-seller, Gilbert White's *The Natural History of Selbourne* (published in 1787 and still in print). White, a country clergyman, called his book "a parochial history," and that was a literal description. It was an account of the life of his parish, with notes on birds, animals, plants, and the weather, comments on local lore, personalities, superstition, history, and legend. It spoke as well of a settled society in which class and place were fixed and people knew each other, of work and life tied to the round of the seasons. Longing for a simpler and more harmonious society, in tune with nature, was a strong current

---

[31] Donald Worster, *Nature's Economy* (1977; reprinted Cambridge University Press, 1985).
[32] Frank Chapman, *Autobiography of a Bird-Lover* (New York: D. Appleton-Century, 1935), 119, 49.

in late-eighteenth-century English literature, a reaction to the urban, industrial society that was visible in the cities. Natural history, unifying place, nature, and the social order, became one vehicle for that sentiment, and by the 1820s Selbourne had become a place of pilgrimage. It served others as well. "The early cheap editions were carried all over the world by expatriate Englishmen, sick for the dells and hillocks of home."[33] The homesick settlers carried its conventions as well; it was one prominent model for settler nature writing.

Natural History in the Settler Lands

The settlers brought natural history with them or imported it as part of the high culture they wished to emulate or reproduce, but like everything else it changed in its passage. Each settler society stressed some of its elements, downplayed others. Recreation was less important in all the settler lands than in Britain, if only because it was harder to get equipment and books or to find enough sympathizers to form a society. On the other hand, natural history as a way to understand the land was more important. The British and other Europeans had mapped and occupied their lands. They knew the land, even what was under their feet.[34] In the early years, maps of the settler countries often traced little more than some coastline and a few uncertain features inland, and through the nineteenth century each generation added to its knowledge, reducing the blank spaces of ignorance and romance. There were as well more specialized surveys. Topographic mapping, geological studies, the reports of botanical and zoological studies, and the compilation of statistics on temperature, rainfall, and wind revealed new aspects of the country. From the late eighteenth century, natural history was the intellectual framework of exploration and settlement, the way to knowledge about unknown lands. Individuals also took part. In backcountry districts everywhere there were some for whom natural history was "an all-consuming passion" (the title of a recent study of an early Western Australian naturalist).[35] They commonly, however, acted alone, sustained by a few ties to the metropolis rather than a local network.

[33] Richard Mabey, Introduction to the Penguin edition of Gilbert White, *The Natural History of Selbourne* (1787; New York: Penguin, 1977), viii. The genre has been remarkably long-lived and widespread. A late example is H. Guthrie-Smith's *Tutira: A New Zealand Sheep Station* (London: Blackwood, 1921).

[34] David R. Oldroyd, *The Highlands Controversy* (Chicago: University of Chicago Press, 1990), provides an example of this process of construction of nature in British geology.

[35] Lines, *An All-Consuming Passion.* Isolation was a common fate. For a late example see Loren Eiseley's account of his childhood in *All the Strange Hours* (New York: Scribner's, 1975), 162–71.

This gave natural history purposes it lacked in Europe. It was "a fundamental part of the quest for a national identity in societies where the cultural differentiation from Britain was insecure and the sense of the land correspondingly important for self-awareness." It also gave individuals a niche in science and a "psychological identification with pioneers" that rooted them in their new culture.[36] In Australia and New Zealand, it was at the base of settlement. Cook's reports were part of the record used in planning the first Anglo settlements there. Surveys gave them the first impressions of the land, and in Australia were a source of information and local pride. In the United States, government topographic surveyors helped fix the idea of the Great American Desert, then the riches of the West. In Canada, natural history was particularly important. Australia and New Zealand had ocean borders and distinctive landscapes, plants, and animals. The United States had its republican experiment in government. Canada was a scattered collection of colonies divided by geography and united only by its determination not be part of the United States. It acquired a political identity – and that within the Empire – only in 1867. Natural history fostered a vision of a continental, northern nation, rich in resources. The "tasks of identification, inventory, and mapmaking gave form to the idea of a transcontinental national existence; they imparted to Canadians a sense of direction, stability, and certainty for the future."[37]

Natural history also gave the settlers a chance to prove their cultural worth. The arts required a settled society and patronage, which, as a wealth of colonial complaints made clear, were in short supply. The same problems told against contributions in the physical sciences. Not only could natural history be done on the spot, it had to start there. In addition, the settler lands were treasure houses of novelty. The specimens collected by the guards and convicts at Sydney cove had created a sensation in London. New Zealand's birds were in demand throughout the century, and at its end there was keen competition to get the last huia and other vanishing species. Even in North America, more extensively explored and less strange, there was much to be found.

People and resources were still important, as the rapid development of the field in the United States showed. Its citizens were no wiser or more interested in natural history, but there were many of them and

---

[36] Donald Fleming, "Science in Australia, Canada, and the United States: Some Comparative Remarks," *Actes du Dixième Congress International d'Histoire des Sciences* (Paris: Hermann, 1962), 183.

[37] Suzanne Zeller, *Inventing Canada* (Toronto: University of Toronto Press, 1987), 9.

their society was, by virtue of long settlement, the best established.[38] Peter Kalm, collecting in North America for Linnaeus between 1748 and 1751, found people to aid him, and by the time the United States was independent there was in Philadelphia the nucleus of organization: a society (the Junto), Peale's Museum, and the farm, nursery, and collections of John Bartram and his son William.[39] By the 1830s American printers had progressed from pirating and adapting less expensive British books (though this practice continued) to producing local ones like John Godman's *American Natural History* (1826) and Thomas Nuttall's *A Manual of the Ornithology of the United States and Canada* (1832).[40] The others continued to look abroad. Canadians relied on Americans, who were close at hand and studying the same plants and animals. Australians and New Zealanders depended on the British. Most of their major works were published in London, often as part of the literature of empire. So strong were the connections that the Australian equivalent of the American Audubon Society – the Gould League of Bird Lovers – was named after John Gould, a London-based artist, naturalist, and entrepreneur; New Zealand's first compendium of plant life was arranged along lines suggested by the English botanist Joseph Hooker "for a uniform series for the British colonies"; and Walter Buller's monumental volumes on New Zealand's birds, which became national icons, were written and printed in London.[41]

Americans also led (the way) in supporting scientific expeditions and institutions. Lewis and Clark's journey to the Pacific began less than twenty years after the establishment of the constitutional government, well before any of the other countries could support this sort of work. In the 1840s the Wilkes expedition undertook a global survey on the model of Cook and Vancouver. By then the United States also had a national museum – the Smithsonian Institution – which was becoming the center for a national scientific community, and the states were funding geological surveys or setting up offices to carry on natural history work. These were usually modest, but they were beginnings on which later generations would build. The typical geological survey was conducted by a single expert with a few assistants, who prepared

---

[38] The U.S. population in 1790 (the first census) was higher than the Canadian population in 1870, the Australian population in 1880, and the New Zealand population today.

[39] Peter Kalm, *Travels in North America* (1753; reprint of English translation, New York: Dover, 1966), 60–75.

[40] John Godman, *American Natural History* (Philadelphia: Carey & Lea, 1826; reprinted New York: Arno, 1974); Thomas Nuttall, *A Manual of Ornithology of the United States and Canada* (Cambridge, Mass.: Hilliard & Brown, 1932; reprinted New York: Arno, 1974).

[41] Leonard Cockayne, *The Vegetation of New Zealand* (Leipzig: Engelmann, 1921), 5–6, reviews this development. On Buller see Galbreath, *Walter Buller.*

a report and a map, and natural history studies were much the same. When states established an office, it was commonly the responsibility of a professor at the state university, who got a title but not much else.[42]

Elsewhere institutions developed more slowly, hindered by lack of population and political dependence that caused talent to look abroad. The societies that Anglo-Canadians formed in Quebec and Montreal in the 1820s were gone a generation later.[43] There were surveys, but they were usually the work of a British officer stationed in the region, and the settlers could do little to exploit what knowledge was gathered. As late as the 1860s American collectors, often connected to the Smithsonian, did more work in the Arctic and among the Hudson's Bay Company factors than did Canadians, and the Smithsonian organized the exchange of scientific periodicals between Canadian museums and those abroad.[44] Studies of the high Arctic were the province of the Royal Navy.[45]

Australia's first natural history group, the Philosophical Society of Australasia, was established in Sydney in June 1821. Despite the patronage of the governor, Sir Thomas Brisbane, a fellow of the Royal Society (London), it collapsed after fifteen months of irregular meetings. The Van Diemen's Land Scientific Society, organized in 1829 by a former military surgeon, lasted two years. Its successor, the Tasmanian Society for Natural History, lost momentum when Lieutenant Governor Sir John Franklin returned to Britain (eventually to lead the most famous lost expedition of the century in the Canadian Arctic). His successor, Sir John Eardley-Wilmot, founded a competitor, the Royal Society of Van Diemen's Land for Horticulture, Botany, and the Advancement of Science. Instability was built into the system. With "society" divided between those in power and those who wished to be, the support of royal officials was as much hindrance as help, for it drew science into the

---

[42] On geology see Anne Marie Millbrooke, "State Geological Surveys of the Nineteenth Century" (unpublished Ph.D. dissertation, University of Pennsylvania, 1981). On natural history, Robert Allyn Lovely, "Mastering Nature's Harmony: Stephen Forbes and the Roots of American Ecology" (unpublished Ph.D. dissertation, University of Wisconsin, 1995), provides a good discussion of an early effort, the Illinois Natural History Survey, which still exists and supports ecological research.

[43] Carl Berger, *Science, God and Nature in Victorian Canada* (Toronto: University of Toronto Press, 1983), 4–5. Zeller, *Inventing Canada*, deals with the early activities, though mainly after this period.

[44] Debra Lindsay, *Science in the Subarctic: Trappers, Traders, and the Smithsonian Institution* (Washington, D.C.: Smithsonian Institution Press, 1993); Berger, *Science, God, and Nature*, 5, 23–7.

[45] Trevor H. Levere, *Science and the Canadian Arctic: A Century of Exploration, 1818–1918* (Cambridge University Press, 1993).

political intrigues that divided each colony.[46] In New Zealand colonists arrived in substantial numbers only in 1840, but interest in natural history was at its peak, and they were quick off the mark. Before landing, they organized the Literary and Scientific Institute of Nelson and, soon after arrival, the Wellington Horticultural and Botanical Society. In 1851, probably following the example of Franklin's society in Tasmania (several New Zealanders had been members), they formed the New Zealand Society and asked the governor, Sir George Grey, to preside. It lapsed two years later, was revived in 1858, became moribund again in the early 1860s, and was revived – permanently as it turned out – a decade later.[47]

This seems the second phase of George Basalla's model of scientific development outside Europe, the period of "colonial science." Where locals, rather than transient Europeans, did the collecting but could not identify specimens or place them in the scientific literature.[48] Historians have criticized this scenario extensively, and with good reason, but it has a core of truth. The settlers as a group occupied, and felt they occupied, an inferior place.[49] It assumes, though, that the settler countries sought to build their own scientific establishments. The United States did, but elsewhere science, like nationalism, developed within the context of the British Empire. Australia, Canada, and New Zealand built their scientific communities within and as part of that larger entity. This pattern is worth noting, for it is a feature of Anglo history. In the twentieth century

---

[46] Colin Finney, *Paradise Revealed: Natural History in Nineteenth Century Australia* (Melbourne: Museum of Victoria, 1993), 21 et seq. See also Ann Moyal, *'A Bright and Savage Land': Scientists in Colonial Australia* (Sydney: Collins, 1986).

[47] C. A. Fleming, *Science, Settlers and Scholars* (Wellington: Royal Society of New Zealand, 1987), 6–10.

[48] George Basalla, "The Spread of Western Science," *Science, 156* (1967), 611–22. In the first phase science was a matter of Europeans coming out, collecting, and going home. In the second residents collected specimens for experts in the metropolis to identify. In the third a country produced its own scientific establishment, trained people, supported their research and careers, and provided professional rewards. For its application to Australia see R. W. Home's introduction to Home, *Australian Science in the Making*, vi–xxvii, and the editors' introduction to R. W. Home and Sally Gregory Kohlstedt (editors), *International Science and National Scientific Identity: Australia between Britain and America* (Dordrecht: Kluwer, 1991), 1–17. There is a longer critique in Reingold and Rothenberg, *Scientific Colonialism*.

[49] In 1884, for example, an English scientist created a scientific sensation by reporting that monotremes (platypuses and echidnas) laid eggs. An Australian's report, that same week, was ignored. Kathleen G. Dugan, "The Zoological Exploration of the Australian Region and Its Impact on Biological Theory," in Reingold and Rothenbeng, *Scientific Colonialism*, 93–7. See also Elizabeth Dalton Newland, "Dr. George Bennett and Sir Richard Owen: A Case Study in the Colonization of Early Australian Science," in Home and Kohlstedt, *International Science*, 55–74.

the smaller countries came to rely on the larger for training and ideas in ecology, adapting their institutions to take advantage of the training and experts available in the United States and Britain. Our models of scientific development need to take this into account.

Natural History and Sentiment

Science and discovery were much the same across the Anglo world and indeed around the world. Here natural history was a matter of applying the same method to different lands to fit them into accepted categories. The opposite is true of natural history as sentiment. The aim was not to fit the land into an organized body of knowledge but to discover what the land meant. Here differences in the lands and settler histories were decisive. The settlers' problems began with finding beauty in nature. For the Americans it was relatively easy. William Byrd, writing of the survey of the Virginia–North Carolina border in 1728, invoked the sublime to describe his reactions (though he found the emotion easier to summon in the leisure of rewriting than in his original journal). A generation later William Bartram recorded his rapture at the scenery of the southern colonies, so strong that at times it almost distracted him from his specimens.[50] Romanticism was quickly accepted. North American scenery had helped define the concept, and educated Americans quickly assimilated it. This is evident in their incorporation of it into their view of the primitive life and wilderness. Filson's biography of Daniel Boone (1784) shows the full outlines of the American natural man, the wilderness hunter who is America's pathfinder and trailblazer in the conquest of nature. So are his contradictions. His virtues are the product of his life in the wilderness, and he uses them to change the wilderness. He is honest, forthright, brave, and (though he has no creed) religious. In woodcraft he can match the Indians. His delight is to be alone in the wilderness; he flees contact with his civilized fellows. And in all the canonical biographies he declares that he was ordained by God to open the wilderness. He destroys what he loves in the service of what he cannot tolerate. The first popular, fully fictional counterpart is James Fenimore Cooper's Natty Bumpo, the hero of the *Leatherstocking Tales*, which began appearing in the 1820s.[51]

---

[50] Roderick Nash, *Wilderness and the American Mind* (3d ed., New Haven, Conn.: Yale University Press, 1982), 52–6.

[51] The literature on this phenomenon is grounded in Henry Nash Smith, *Virgin Land* (Cambridge, Mass.: Harvard University Press, 1950), 54–63. The most recent study of Boone and his legend is John Mack Farager's, *Daniel Boone* (New York: Henry Holt, 1992).

These sentiments were also at the core of the first popular American paintings, those of the Hudson River school. In the 1830s these painters, led by Thomas Cole, began creating a national landscape of the United States that incorporated the American pioneer myths. Cole's *The Oxbow* (1836) is a striking example. A diagonal runs across the painting, dividing wilderness and storm, the chaos and fury of nature from sun and smiling fields. The artist's umbrella in the mid-ground is the only bridge.[52] Nature appears here as a sublime but passing spectacle. Where the subject is not obviously the march of progress there is in the corner the settler's cabin, the clearing, and the stumps of trees. The genre of conquest runs through American art of this period in everything from high culture to cheap illustrations. It is apparent even in the viewpoint. Nature is seen from a commanding height, with what one art historian termed the "magisterial gaze." The apotheosis of this genre is the twenty by thirty foot mural in the U.S. Capitol Building, Emmanuel Leutze's *Across the Continent, Westward the Course of Empire Takes Its Way.* Here Anglo settlers gaze from a height into the trans-Appalachian west like the Chosen People looking into Canaan.[53]

There was also a countercurrent, drawing on the natural virtues formed by the wilderness. This, apparent by mid-century and by 1900 a dominant theme in American nature appreciation, was that nature was a necessary part of life, the touchstone of reality, and a refuge from civilization. Thoreau's is the canonical statement: "I went to the woods because I wished to live deliberately, to front only the essential facts of life."[54] He stayed only two years, had few followers, and *Walden* was a commercial failure, but as wilderness retreated, his reputation and influence grew. By the end of the century Walden Pond had become a place of pilgrimage, *Walden* a bible for American nature lovers.[55] In articles that appeared in general interest magazines with large circulations, John Muir hymned the glories of the Sierra Nevada, John Burroughs the prosaic beauties of nature in the Catskills. The frontier

---

[52] See the analysis in William Cronon, "Telling Tales on Canvas," in Jules Prown et al. (editors), *Discovered Lands, Invented Pasts* (New Haven, Conn.: Yale University Press, 1992), 40–1.

[53] Prown et al., *Discovered Lands, Invented Pasts,* particularly Cronon, "Telling Tales on Canvas," 37–86; Stephen Daniels, *Fields of Vision* (Princeton, N.J.: Princeton University Press, 1993), chaps 5 and 6. On the paintings' view see Albert Boime, *The Magisterial Gaze* (Washington, D.C.: Smithsonian Institution Press, 1991). A longer discussion of landscape paintings and nature can be found in Barbara Novak, *Nature and Culture* (1980; rev. ed., New York: Oxford University Press, 1995).

[54] Thoreau, *Walden,* 81.

[55] On Thoreau and this vision see Lawrence Buell, *The Environmental Imagination* (Cambridge, Mass.: Harvard University Press, 1995).

vanished, the wilderness dwindled, but the appeal of Romantic nature only grew.

The antipodean settlers faced a more daunting task. Their landscapes had not helped shape European concepts and were much further from the European ideal. Australia was the extreme case. Here the settlers had to struggle to see what was in front of them. Early drawings of kangaroos show things with mouse's muzzles, horses' ears, and forepaws like strange hands. Echidnas looked like modified hedgehogs; platypuses were oddly shaped furry creatures with ducks' bills.[56] Early landscape paintings show the bush as a distorted version of conventional English scenery or the home of freakish vegetation. (This is true of all countries but was very marked in Australia.)[57] There was also Australia's reputation. It had entered European consciousness as a "land of contrarieties," and further acquaintance only made it seem stranger. Here, as a Mr. J. Martin famously complained in the 1830s,

> trees retained their leaves and shed their bark instead, the swans were black, the eagles white, the bees were stingless, some mammals had pockets, others laid eggs, it was warmest on the hills and coolest in the valleys, even the blackberries were red.[58]

The landscape was no help. The convicts and guards at Botany Bay knew what constituted beauty in a landscape, but they could not find it. There were no settled districts, so no harmonious rural scenes. None of the settlers had childhood associations with eucalyptus and acacia. Natural beauty was inevitably associated with flowing streams, deep green plants, and rugged, snow-capped mountains. Australia was dry, its plants twisted and sparse, its animals freakish, its mountains rounded and worn. In the early 1790s Thomas Watling, transported for forgery,

---

[56] Bernard William Smith, *European Vision and the South Pacific, 1768–1850* (London: Oxford University Press, 1960), deals with the general process, as does Carter, *The Road to Botany Bay*. On the kangaroo see R. M. Younger, *Kangaroo Images Through the Ages* (Melbourne: Hutchinson, 1988). The Mitchell Library, Sydney, New South Wales, has an invaluable collection of Anglo-Australian views of the land. Distorted images are a feature of early exploration. For a North American parallel see Larry Barsness, *The Bison in Art* (Fort Worth, Tex.: Amon Carter Museum, 1977), 40–7.

[57] On art see Bernard Smith, *Australian Painting, 1788–1970* (Melbourne: Oxford University Press, 1971), 82–5. The Australian collections in the national galleries of Victoria, Melbourne, and Australia, in Canberra, are excellent. More accessible to non-Australians are William Splatt and Barbara Burton, *Masterpieces of Australian Painting* (Melbourne: Viking O'Neil, 1987), and William Splatt and Susan Bruce, *Australian Landscape Painting* (Melbourne: Viking O'Neil, 1989).

[58] Quoted in Alfred Crosby, *Ecological Imperialism* (Cambridge University Press, 1986), 7. See also F. G. Clarke, *The Land of Contrarieties* (Clayton: Melbourne University Press, 1977), chap. 9.

wrote home to Scotland of the "Arcadian shades" and "romantic banks" around Sydney harbor, but in the next sentence was forced to admit that this was but "the specious external."[59] Progress inland led to such a series of disappointments that one scholar felt compelled to describe nineteenth-century Australia as a "diminishing paradise." The first generation dreamed of a great inland sea or at least verdant pastures within the great blank outlined by the coastal surveys. They found that the land grew harsher and drier as they approached the Centre. By the early twentieth century, disillusion was complete. When in 1906 J. W. Gregory spoke of the deep interior as "the dead heart of Australia," there were few to protest.[60]

Different but no less daunting was the Canadians' problem – the existence of a northern frontier even more hostile to settlement. But while the Australians turned their backs on the Centre, Canadians made the North into a national myth. They did not embrace it, though. It was a dominant, threatening presence. Canadian literature, Margaret Atwood wrote, was not about conquest but survival. Death in nature, death by nature, was its great theme. Another scholar saw Canada as a fearful society, huddling in its fort, hiding from the menace of the forest.[61] In neither country did people look to the wild for enlightenment or refuge.[62]

Some of this seems an odd mixture, but the Anglos of those years drew different boundaries than we do. Neither Byrd surveying nor Bartram botanizing felt a contradiction between exactness and sublimity, between measuring and classifying on the one hand and being enrap-

---

[59] Quoted in Ross Gibson, *The Diminishing Paradise* (Sydney: Angus & Robertson, 1984), 50. Watling is a favorite, appearing as well in Bernard Smith, "The Artist's Vision of Australia," in *The Antipodean Manifesto* (Melbourne: Oxford University Press, 1976), and Alec H. Chisholm (editor), *Land of Wonder* (Sydney: Angus & Robertson, 1964).

[60] J. W. Gregory, *The Dead Heart of Australia* (London: John Murray, 1906). See also Gibson, *Diminishing Paradise*. For comparison see A. L. Burt, "If Turner Had Looked at Canada, Australia, and New Zealand When He Wrote about the West," in Walker D. Wyman and Clifton B. Kroeber (editors), *The Frontier in Perspective* (Madison: University of Wisconsin Press, 1965), 59–77.

[61] Margaret Atwood, *Survival* (Toronto: Anansi, 1972). See also Gaile McGregor, *The Wacousta Syndrome* (Toronto: University of Toronto Press, 1985). For an introduction to the North see Louis-Edmond Hamelin, *About Canada: The North and Its Conceptual Referents* (Ottawa: Minister of Supply and Services, 1986).

[62] Charles Barrett, an early-twentieth-century Australian nature writer, admired Thoreau, but Thoreau's thought seems to have had no influence on him. Barrett's intellectual roots were in the English tradition; he relied on Richard Jeffries and W. H. Hudson. On Barrett and his school see Tom Griffiths, "'The Natural History of Melbourne': The Culture of Nature Writing in Victoria, 1880–1945," *Australian Historical Studies*, 23 (October 1989), 339–65. See also Richard White, *Inventing Australia* (Sydney: Allen & Unwin, 1981), 117–19.

tured by nature on the other. This mixture was, in fact, characteristic of natural history. Consider Alexander von Humboldt. His account of his travels in South America (1799–1804) inspired a generation of naturalists, including both Darwin and Wallace, and helped create the image of the scientist-naturalist, the heroic explorer of the mental and physical unknown. On the one hand he was quite "scientific," using instruments, measurement, and analysis to serve the demands of a reductionist, rational account of nature. He was a pioneer in mapping information and developed now-common devices like isotherms (lines of equal temperature). His work on the distribution of plants and animals in relation to climate and latitude foreshadowed ecologists' work on bioregions. He put this, though, in the service of what now seems a mystical ideal. All these details were to show the harmony and order in all. His great work, *Cosmos*, was an attempt to unify the world and reconcile art and science, reason and sentiment. His appeal was quite broad. American army officers attached his name to rivers, mountains, bays, and a swamp in the American West. The natural history of the last decade of Thoreau's life was quite Humboldtian. Frederic Church, the mid-century American landscape painter, was an ardent disciple. Traveling in South America, he stayed in the house in Quito that had sheltered Humboldt sixty years before. He painted plants in exquisite and accurate detail, and arranged the scenery to illustrate Humboldt's life-zones. His enormous painting, *Heart of the Andes*, was "an emblem of an entire zone of the earth, Humboldtian both in its epic scope and its firm grasp of physical reality," and he was planning to ship it to Germany for the master's approval when Humboldt, at the age of ninety, died.[63]

Conclusion

The settlers had to develop a folkbiology in the new lands, if only to put names on the new things they found, and they had as well to develop that larger body of folk-knowledge that told them what they could use and eat, what plants and animals were best avoided, what land could be farmed, and how it could best be done. They added to this a new kind of knowledge, natural history, whose reach and power ran well beyond earlier systems. It cataloged not a local biota but all the life of the world,

---

[63] Quote from Franklin Kelly, *Frederic Edwin Church and the National Landscape* (Washington, D.C.: Smithsonian Institution Press, 1988), 99. See also Stephen Jay Gould, "Church, Humboldt, and Darwin: The Tension and Harmony of Art and Science," in Franklin Kelly (editor), *Frederic Edwin Church* (Washington, D.C.: Smithsonian Institution Press, 1988), 94–107; Novak, *Nature and Culture*, 66–77.

placing the settlers' knowledge in a different context. It also had greater scope, considering besides plants and animals soil, minerals, rocks, and the fossils embedded in them – things that folk-knowledge treated lightly and in an instrumental fashion. (There is no folkgeology comparable to folktaxonomy.) Its reach gave it different social functions as well. It was a much more extensive guide to settlement and knowledge, and it connected the settler periphery to the European metropolis on every level from individual to government.

The rise of natural history marks the start of a gradual and never-complete drift of the culture's authoritative knowledge about the world from the general population to a small group of experts. What marks nature knowledge on this level off from other scientific studies is the persistence and power of popular understanding. People, of necessity, learn about nature on this level as part of their passage from infancy to childhood, and continue to have contact with it in this way even if they acquire expert knowledge. The settlers, though, were not primarily interested in learning in these years. The "conquest" of nature was their principal concern. The turning of ancient ecosystems into fields and pastures was its most obvious effect, but there were others. One was the popular enthusiasm for introducing birds and large mammals that flourished just after mid-century, the other sport hunting, a fashionable British recreation the settler societies adopted at about the same time. Both had significant biological or social consequences, and their histories provide another window on the settlers' attempts to understand and relate to the worlds they found and made. They also show, sometimes quite dramatically, the ways in which the land was shaping the settlers.

2

REMAKING WORLDS
EUROPEAN MODELS IN NEW LANDS

> [T]here is no fascination in life like that of the amelioration
> of the surface of the earth . . . to fence, to split, to build, to
> sow seed, to watch your flock increase – to note a countryside
> change under your hands from a wilderness.
>
> H. Guthrie-Smith, 1921[1]

This was the leitmotif of nineteenth-century settlement, the dream of remaking the land. (That the epigraph to this chapter comes from a minor classic of nature writing, *Tutira: The Story of a New Zealand Sheep Station*, suggests the complexity of settler visions.) The settlers destroyed and re-created, appreciated the beauties of the land, and sought to bring it closer to their own ideal, and they did it on a grand scale. In less than a century Americans and Canadians built towns and farms across two-thirds of a continent, Australians and New Zealanders pastoral empires across their lands. Plat maps replaced sketches, tax assessors' records the descriptions from explorers' journals. The settlers saw all this as the march of civilization, but that considerably overstates the orderliness of the process. It was a frantic rush. Farmers and ranchers pushed into new country with high hopes and little information. Some were ruined by drought, frost, or heat, others failed as their farming or grazing techniques exhausted the land. In marginal regions like South Australia and the western edge of North America's Great Plains, population ebbed and flowed with wet and dry years.

More than farms were involved, however. The settlers also wanted to make the land beyond the fencerows familiar or use it in familiar ways. They brought in birds and large mammals, particularly in the second

---

[1] H. Guthrie-Smith, *Tutira: The Story of a New Zealand Sheep Station* (London: Blackwood, 1921).

half of the century, and they adopted the fashionable and popular re-creation of sport hunting. The fad for introductions was strongest in the southern countries, where many remembered Britain and found the bush poorly stocked with creatures they could either admire or shoot. North Americans with the money to indulge in this hobby were more likely to be native-born and had fewer sentimental ties to British nature. The biological impact of these actions was also much larger in the South, for these lands were more open to invasion. Their climates were temperate and the fauna sparse. (In the scientific phrase they were depauperate, meaning there were fewer species and kinds of species than would be expected for the size of the area.) New Zealand was the extreme; its only native land mammals were two species of bats. In both North and South, the large and varied suite of introductions fundamentally changed the face of the land. Hunting was most popular in the North, and there it had its greatest social effects. The rise of sport hunting among the elite led to a movement for wildlife preservation and on to the agencies that became the basis for wildlife research and management in the twentieth century. Hunting also found a small niche in New Zealand, where Parliament gave the acclimatization societies, established to import game, the duties of wildlife law enforcement. Australia was the odd country out; hunting never became well established.

The limits on agricultural expansion that the settlers reached toward the end of the century we will put off to the next chapter. This one deals with expansion, the obvious effects of industrial expansion on the land, the attempt to remake the country into a more familiar form, and European ideas of sport in non-European lands. These were all part of a single enterprise, to find a place in the new lands by making them like the old or acting as if they were the old. More even than natural history, these activities brought the settlers into contact with the land, and in the process of changing it they were themselves changed. The days of conquest were a prelude to something that would become more conscious and important as the century went on, the search for what it meant to be part of the new lands.

Transforming the Land

The great expansion of the nineteenth century destroyed existing ecosystems and made new ones on a scale and with a thoroughness unprecedented in human history. These were not paper empires or the work of a few adventurers dispossessing a ruling class. The earlier peoples were nearly exterminated. So were many of the plants and animals, and new ones were introduced to establish a familiar agricultural economy over vast areas. In 1800 some 4 million Europeans (and their African slaves) lived in the newly independent country of the United States, whose settled land was a strip along the Atlantic coast barely 150 miles wide. Beyond was the unknown. "No European foot," Hector St. John Crevecoeur said in 1782, "has as yet travelled half the extent of this mighty continent."[2] At the end of the century 76 million settlers lived in a country that stretched from sea to sea. Canada was similarly transformed – from a few settlements and garrisons on the Atlantic coast and the Saint Lawrence River to a continental nation of 5 million whose authority (and, Canadians were coming to believe, destiny) stretched to the Arctic. Australia, at the start of the century a penal colony on a poorly charted coast, was at its end a set of colonies covering the whole of the continent, and in 1901 they would form a nation of 4 million people. New Zealand, in 1840 the land of the Maori, supported in 1900 almost a million Anglos, governing themselves within the British Empire and living, like the Australians, from the produce of the land.

Americans think of "conquest" as their national epic, and it was, but theirs was a variant on a common story that stretched over the neo-Europes. Even in so different a place as the Atlantic forest of Brazil, the same process is visible.[3] The settlers were Portuguese, the land tropical rather than temperate, and the pace of settlement, investment, and destruction slower, but much is familiar. The Portuguese began by exploiting local resources, often gathered with the help of the indigenous people.[4] Extractive industries followed – sugarcane (using slave labor and destroying the forests), gold mining, then diamonds. In the early nineteenth century coffee became king, and the forests fell more rapidly with cattle raising and railroad construction. Late in the nineteenth century the devastation became apparent enough that the urban middle class began to demand action. The technology and the world market that allowed people to bring nature to the metropolis were not unique to the United States or even the Anglo world.

In the Anglo settler countries assumptions, ideas, and goals were far more common than the myth of American exceptionalism allows. Legislatures from Perth to Fredrickton echoed with impassioned rhetoric about sturdy yeoman and the virtues of farm life. Land laws aimed (whatever the reality) to establish a country of small farms owned by those who worked them. There was the common conviction that what was on

---

[2] J. Hector St. John Crevecoeur, *Letters from an American Farmer* (1782; reprinted New York: Doubleday, n.d.), 47.

[3] This account is drawn from Warren Dean, *With Broad Ax and Firebrand* (Berkeley: University of California Press, 1995).

[4] For Anglo comparisons see Timothy Silver, *A New Face on the Countryside* (Cambridge University Press, 1990), and Peter Wood, *Black Majority* (New York: Random House, 1974).

the land would be, must be, supplanted. In North America it was an article of faith that wild animals, the forests, and the Indians were destined to vanish before white civilization. Speeches, paintings, and cheap lithographs extolled the process of conquest, the glorious national destiny, and the Indians' sad but inevitable decline (without dwelling on the roles of whiskey, disease, and firearms). Australians and New Zealanders had the same ideas and made the same connections between conquest, dispossession, and civilization, though with fewer artworks and their own biological twist. Independent of specifically Darwinian ideas they held that the native species were doomed because they were "primitive" creatures, less "fit" than European plants and animals, and so handicapped in the "struggle for life." They would fall before the invaders, just as the white man would inevitably displace the Aborigine and the Maori.[5]

Settlers everywhere spoke of individual effort and of themselves as rugged pioneers, but their progress was the expansion of an industrial society serving a world market. They used guns, traps, and poison to kill the wildlife, steel axes and plows to clear the land and turn the soil. They relied on ships, canals, and railroads to send to the metropolis what they found and raised. They were not isolated, only at the far ends of the social and physical structures that put steaks from Texas cattle on New York dinner tables and turned wool from New Zealand sheep into clothes for the British gentry. Wheat farmers in South Australia and Saskatchewan, sheepmen on New Zealand's Canterbury Plains and in Nevada's mountains read their fate in futures prices on the Liverpool markets, brought to them by submarine telegraph, and so specialized was this economy that, to transfer wool from the sheep's back to their own, Australians and New Zealanders sent it through British mills. Settlement was less an escape from civilization into the wilderness than a way to take nature to the market.

The settlers usually began by killing the animals and cutting down whatever trees were around. An environmental history of Australia titled a chapter on pioneer land clearing "They Hated Trees," and this might serve for all the settler countries.[6] Cutting was so thorough that by the 1870s only Canada was not discussing forest conservation.[7] The improve-

---

[5] A comparison of Anglo attitudes toward native peoples is outside the scope of this book, but it ought to be done. Much of what passes for "American" attitudes toward Indians is not limited to that country or the continent.

[6] Geoffrey Bolton, *Spoils and Spoilers: Australians Make Their Environment, 1788–1980* (Sydney: Allen & Unwin, 1981), chap. 4.

[7] Andrew Hill Clark, *The Invasion of New Zealand by People, Plants and Animals: The South Island* (New Brunswick, N.J.: Rutgers University Press, 1949). There is a good description of New Zealand land practice in Guthrie-Smith, *Tutira*. On Australia see John

ment of weapons and transportation in the second half of the century made it possible, and profitable, to kill everything, and one result was the slaughter of wildlife on a fantastic scale. The clumsy flintlock of 1800 took about a minute to load, was accurate to no more than fifty yards, of no use in bad weather, and barely capable of killing large animals. The only advantage it had over bows and arrows was that it required a relatively low level of strength and skill. By 1880 the average settler could afford a rifle that would fire several rounds a minute in any weather and kill the largest game at several hundred yards. Shotguns underwent similar improvements. When railroads connected the backcountry to the market, they set off a slaughter that became fixed in popular memory. The fate of the buffalo and the passenger pigeon became cautionary tales recalled a century later and as far away as Australia.[8] Other species vanished with less publicity. The steel leg-hold trap increased the speed with which the beaver, marten, fisher, and other fur-bearers vanished in North America, and the settlers used rifles, traps, and poison to get rid of wolves, bears, mountain lions, and other predators to make their pastures safe. Better shotguns in the hands of market hunters and an army of sportsmen brought a dramatic decline in ducks, geese, and shorebirds. In the 1890s a fad for birds' feathers as decoration on women's hats and dresses all but destroyed egrets and herons in North America and affected bird species around the globe. In Australia the fur trade reduced koala and platypus to remnant populations. In New Zealand the native birds fell to hunting, land clearing, and imported predators, but also to the demands of science and amateur collectors. The huia went extinct amid an undignified scramble for the last specimens.[9]

The law allowed and even encouraged this, for the settlers saw wildlife as, at best, a transient resource. Until well into the twentieth century, regulations on killing things in nature were few, poorly enforced, and

Dargavel, *Fashioning Australia's Forests* (Melbourne: Oxford University Press, 1995); on the United States, Samuel P. Hays, *Conservation and the Gospel of Efficiency* (Cambridge, Mass.: Harvard University Press, 1959).

[8] References to the buffalo, for example, occur in the speeches of E. C. M. Fox, a member of Parliament, when he urged new measures to preserve Australian wildlife. See the Parliamentary record, the *Hansard*, 58, 2 May 1968. In the United States the literature on the buffalo and its slaughter is almost as large as pre-conquest herds. A recent discussion is Dan Flores, "Bison Ecology and Bison Diplomacy," *Journal of American History*, 78 (September 1991), 465–85.

[9] On the plume trade see Robin Doughty, *Feather Fashions and Bird Preservation* (Berkeley: University of California Press, 1975); on New Zealand, Ross Galbreath, *Walter Buller: The Reluctant Conservationist* (Wellington: GP Books, 1989); on Australia, J. M. Thomson, J. L. Long, and D. R. Horton, "Human Exploitation of and Introductions to the Australian Fauna," in Bureau of Flora and Fauna, *Fauna of Australia* (Canberra: Australian Government Publishing Service, 1987), 1A: 227–49.

applied to only a few species.[10] The law was active only where animals threatened humans, crops, or livestock, and there it encouraged destruction. The bounty system began in North America when Massachusetts Bay, in 1630, and Virginia, in 1632, started paying for wolf scalps, and it became universal among American states and Canadian provinces and persists today.[11] As sheep raising became important the Australians followed suit. In 1830 the Van Diemen Land Company began paying for scalps of the thylacine, the so-called marsupial wolf, and in 1852 New South Wales passed an Act to Facilitate and Encourage the Destruction of Native Dogs (No. 44). Legislators put bounties on a wide variety of other species and, as in North America, kept them there.[12]

There were bounties on almost anything that walked, flew, swam, or crawled – Western Australia, for instance, paid bounties on 284,724 emus between 1945 and 1960 – but the main targets were mammals that killed livestock.[13] The method of choice was poison. Strychnine, an alkaloid extracted from an East Indian plant, had been known in Europe from the middle of the seventeenth century, but in the 1830s cheap shipping and factory processing made it common. Two biologists with experience in predator control said that the destruction of wolves "by this . . . poisoning campaign . . . hardly has been exceeded in North America unless by the slaughter of the passenger pigeon, the buffalo, and the antelope."[14] Americans were not alone; Australians

---

[10] This conclusion is based on the statutes of all four countries. Accessible sources include T. S. Palmer, "Chronology and Index of the More Important Events in American Game Protection, 1776–1911," Biological Survey Bulletin 41 (Washington, D.C.: U.S. Government Printing Office, 1911); Michael Bean, *The Evolution of American Wildlife Law* (New York: Praeger, 1983). On Australia see Thomson, Long, and Horton, "Human Exploitation of and Introductions to the Australian Fauna," 227–49, and George R. Wilson, "Cultural Values, Conservation and Management Legislation," 250–60, both in Bureau of Flora and Fauna, *Fauna of Australia.*

[11] T. S. Palmer, "Extermination of Noxious Animals by Bounties," in U.S. Department of Agriculture, *Yearbook of the United States Department of Agriculture, 1896* (Washington, D.C.: U.S. Government Printing Office, 1897), 57. On wolves see Ronald Nowak, "The Gray Wolf in North America," a preliminary report submitted to the New York Zoological Society and the U.S. Bureau for Sport Fisheries and Wildlife, 1 March 1974, 221. Copy courtesy of Nowak. The contemporary situation is covered in Ruth S. Musgrave and Mary Anne Stein, *State Wildlife Laws Handbook* (Rockville, Md.: Government Institutes, 1993).

[12] Eric R. Guiler, *Thylacine: The Tragedy of the Tasmanian Tiger* (Melbourne: Oxford University Press, 1985), 16. See Australian state statutes; I have relied mainly on the collection in the National Library, Canberra. On the dingo see Roland Breckwoldt, *A Very Elegant Animal: The Dingo* (Sydney: Angus & Robertson, 1988).

[13] On emu bounties see Vincent Serventy, *A Continent in Danger* (Worcester: Bayliss, 1966), 109.

[14] Stanley Paul Young and Edward A. Goldman, *The Wolves of North America* (Stackpole, 1944; reprinted New York: Dover, 1964), 335.

poisoned "[e]verything that seemed at all likely to be troublesome." Victoria's 1876 Act for Regulating the Sale and Use of Poisons provides a glimpse into this world. It declared that "large quantities of arsenic, strychnine and other poisons are in use in the colony for pastoral, agricultural, and other purposes." The effects seem not to have been confined to wildlife, for the act went on to say that wide and careless use of these materials "often leads to fatal accidents and the commission of crime."[15]

Everywhere the settlers had the same goals and used the same technologies, but the lands and the animals were different. (New Zealand we ignore here. It had no predatory wildlife and early and strong pressure not to bring in anything likely to eat a sheep.) In North America settlement was continuous enough that by the early twentieth century the large, slow-breeding predators – wolves, grizzlies, and mountain lions – survived in substantial numbers only in the rugged lands along the spine of the Rockies, and most of them were in Canada rather than the United States. In Australia the thylacine, confined to Tasmania, was almost gone by then. Smaller, quicker-breeding animals survived. The coyote (fifteen to twenty pounds against the wolf's sixty or more) began at this time to occupy the wolf's former range, spreading out of the Southwest. Moving across the Canadian prairies, it arrived in Ontario in the interwar period and followed the Appalachian chain into the eastern United States. It is now as far south as North Carolina.[16] In Australia the dingo's population dropped, but it survived in hill country and the interior deserts. Like the coyote, it continues to plague graziers. There is now a fence, several thousand kilometers long, barring the inland pastures from the Centre.[17]

Old Worlds in New

Changing the land was not an event or a process but a set of actions that created a suite of landscapes. Around settler homes the transformation was most complete. European grasses spread to picket fences. Lilacs and roses bloomed in North American dooryards, primroses and other

---

[15] Eric Rolls, *They All Ran Wild* (Sydney: Angus & Robertson, 1969), 18. The law is No. 559, 1876, Victorian Statutes.

[16] On coyote control see Stanley Paul Young and Hartley H. T. Jackson, *The Clever Coyote* (Washington, D.C.: Wildlife Management Institute, 1951), 171–228. I gleaned information on recent events from conversations with biologists at the 1985, 1989, and 1993 International Theriological Congresses; see also Robert Chambers (editor), *Transactions of the Eastern Coyote Workshop* (n.p., internal evidence suggests 1975).

[17] Breckwoldt, *Very Elegant Animal*, 195–227.

English flowers by Australian and New Zealand homes.[18] In parks from New York to Sydney people walked on European grass growing in imitation English meadows, and the commonest birds they saw were starlings, pigeons, and English sparrows (*Passer domesticus*, outside the United States commonly called "sparrow"). In some areas – southern California being a preeminent example – the settlers built exotic landscapes, but these, as much as the "ordinary" ones, were the product of their dreams. In rural areas European crops filled the fields and European weeds the roadside ditches, but native species persisted and some flourished. Stone walls in (American) New England created habitats for chipmunks. The snake fences, small fields, and dirt roads of the South and Midwest did the same for the bobwhite. Telegraph and then telephone poles, coupled with roadside grasses, created new space for the red-tailed hawk across North America.[19] Farther out, the wild was changed as plants and feral animals spread before the settlers. In some cases this dramatically changed the land, in others it had little visible effect.[20]

Beyond the fencerows the most organized and popular attempt to shape nature was the enthusiastic mid-century importation of game animals and songbirds that swept through Europe and the neo-Europes. It was a distinctive movement but also part of a long-running trend in settler history. The first explorers had ransacked the world, bringing to Europe whatever struck their fancy – flowers, foods, minerals, odd animals, spectacular birds, the tools of the people they found, the people themselves. By the late eighteenth century governments began to convert this casual looting into a systematic effort in the service of empire. They established experimental gardens in the tropics and at home as centers for botanical studies and a source of novelties for the gardens and hothouses of the rich.[21] They tested new species and shifted

---

[18] On Australia see David Irwin, "An English Home in the Antipodes," in John Hardy and Alan Frost (editors), *Studies from Terra Australia to Australia* (Occasional Paper No. 6 Canberra: Australian Academy of the Humanities, 1989), 195–210.

[19] The best exposition of the way in which shifting cultivation patterns affected a wildlife population is Herbert Stoddard's, *The Bobwhite Quail* (New York: Scribner's, 1931).

[20] Guthrie-Smith, *Tutira*, noted the tendency of exotics to cluster near farm buildings and in intensively worked areas (244), as did G. M. Thomson, *The Naturalization of Plants and Animals in New Zealand* (Cambridge University Press, 1922), 509, 514. On Australia see Tom McKnight, *Friendly Vermin: A Survey of Feral Livestock in Australia* (Berkeley: University of California Press, 1976); on New Zealand, Carolyn King (editor), *Handbook of New Zealand Mammals* (Auckland: Oxford University Press, 1990), and Clark, *The Invasion of New Zealand*, 359–76.

[21] Richard Grove, *Green Imperialism* (Cambridge University Press, 1995); Lucille Brockway, *Science and Colonial Expansion: The Role of the British Royal Botanic Gardens* (New York: Academic Press, 1979). Alfred Crosby, *The Columbian Exchange* (Cambridge University Press, 1990), discusses the early phases.

promising ones to new lands. (The *Bounty*, recall, was transporting breadfruit plants from Tahiti to the West Indies when a breakdown in labor relations cut short the voyage.) In the nineteenth century the increasing speed and volume of shipping made voyages shorter, and people found by experiment how to ship plants and animals. What had been difficult and chancy gradually became routine. By mid-century commercial gardeners were stocking British greenhouses with tropical orchids and sending British flowers to homesick colonials in the antipodes.[22]

In the 1860s Europeans and settlers, but particularly the settlers, formed private societies to improve their countries by stocking them with the "useful and beautiful productions of the world." Parisians formed the first group, the Société Zoologique d'Acclimatation, in 1854. Five years later, after a dinner of African eland, a group of British naturalists and enthusiasts organized an English counterpart. Edward Wilson, editor of the *Melbourne Argus*, was in London at the time and knew Frank Buckland, one of the organizers. He took the idea back to Australia and helped form the Victoria Acclimatisation Society in 1861. Others followed. The societies concentrated on species from "home," but the Australians also imported alpaca, llamas, ostriches, camels, peafowl, guineafowl, mynahs, and red-whiskered bulbuls, the New Zealanders more than a hundred species of birds and a wide variety of mammals, including zebras and gnus.[23] Enthusiasm was strongest in the

---

[22] Geoffrey Blainey, *The Tyranny of Distance* (rev. ed., South Melbourne: Macmillan, 1982). On plant transport and technology see Lynn Barber, *The Heyday of Natural History, 1820–1870* (London: Jonathon Cape, 1980), 111–14; Robert Dixon, "Nostalgia and Patriotism in Colonial Australia," in Hardy and Frost, *Studies from Terra Australis to Australia*, 211–17.

[23] On France see Michael A. Osborne, "The Société Zoologique d'Acclimatation and the New French Empire" (unpublished Ph.D. dissertation, University of Wisconsin, 1987). The revised, published version is *Nature, the Exotic and the Science of French Colonialism* (Bloomington: Indiana University Press, 1994). On Great Britain see Christopher Lever, *The Naturalized Animals of the British Isles* (London: Hutchinson, 1977), 30–3, and *They Dined on Eland* (London: Quiller, 1992). I thank Sir Christopher for lending me a copy of his manuscript in 1990. On Australia see Linden Gillbank, "The Acclimatisation Society of Victoria," *Victorian Historical Journal, 51* (1980), 255–70, and "The Origins of the Acclimatisation Society of Victoria: Practical Science in the Wake of the Gold Rush," *Historical Records of Australian Science, 6* (December 1986), 359–74. On the successful acclimatization of birds and mammals see Christopher Lever, *Introduced Birds of the World* (London: Longman, 1987) and *Introduced Mammals of the World* (London: Longman, 1985); on French–Australian cooperation, Michael A. Osborne, "A Collaborative Dimension of European Empire," in R. W. Home and Sally Gregory Kohlstedt (editors), *International Science and National Scientific Identity* (Boston: Kluwer Academic, 1991), 97–119; on Australian species, Lever, *They Dined on Eland*, 99–129; on New Zealand, Thomson, *Naturalisation of Animals and Plants*, 25.

islands. A network of local societies eventually covered the country; Parliament made them the legal owners of game in their districts; and they evolved into local game law enforcement and wildlife management authorities. They were renamed fish and game councils in 1990 but retain their powers.[24] Interest was much weaker in the United States and Canada. Within a generation the fad had passed, but seldom have so few done so much over so large an area with so little effort (or understanding). Now there are English skylarks in Tasmanian fields, European rabbits across Australia and in New Zealand, and red deer in New Zealand's forests. Across North America English sparrows fight in the gutters and starlings squabble in the trees. Everywhere the shock waves from these silent biological explosions continue to reverberate.[25]

We now see the acclimatizers as simply reckless, but that is because we have suffered the consequences of their actions, and we see nature in different terms. They focused on species, which they saw as things living on the neutral backdrop of the land. Within a wide range of climate, plants and animals would adapt – in the word of the day, acclimate – and take their place as part of the country. This conception had its roots in popular views of the world, but it was more directly the product of European science and experience. The center of authoritative understanding, natural history, was taxonomy, and it treated nature's pieces.[26] Besides, for two centuries the settlers had lived by bringing livestock, grasses, crops, flowers, and fruit from Europe and elsewhere. Their agriculture and the economy of empire rested on transplanted plants and animals. If so much could be done, what could not? And why should it not? Changing the land was evidence of progress and civilization.

The acclimatizers' dream of a familiar landscape had its greatest appeal in Australia and New Zealand, where many were British-born and the literate were familiar with rural traditions and Gilbert White. Later generations would find the songs of Australian birds delightful, but this one yearned for English songbirds, "those delightful reminders of our English home." One enthusiast hoped that these birds, if introduced, would spread through the plain, bush, and forest, which would "have

---

[24] Their status and evolution can be traced through the New Zealand statutes and records. On recent developments see Ross Galbreath, *Working for Wildlife* (Wellington: Bridget Williams, 1993), 224.

[25] On imported birds and mammals see Lever, *Introduced Birds of the World* and *Introduced Mammals of the World*. There are no comparable directories for plants. Local and national flora give some idea, but generally do not cover all species. One example is Woodbridge Metcalf, *Introduced Trees of Central California* (Berkeley: University of California Press, 1968). Success is impossible to measure with any accuracy, for records are fragmentary.

[26] Scott Atran, *Cognitive Foundations of Natural History* (Cambridge University Press, 1990).

their present savage silence, or worse, enlivened by those varied touching joyous strains of Heaven-taught melody."[27] The pull of home is evident in the lists of animals. Of twenty-four exotic mammals established in Australia fifteen are from Europe, as are thirteen of the twenty-six birds.[28] In New Zealand climate and the creation of fields and pastures weighted the scales in favor of European immigrants, but even so the figures are striking. A survey of New Zealand mammals, done in 1950, found just over half from England, 11 percent from North America, 14 from Asia, 16 from Australia, and 3 from Polynesia (the dog and rat brought by the Maori). A more recent study found that "all but one of the most widespread and successful" exotics were from Britain and Europe.[29]

The English countryside had little appeal in North America, where those with the money to indulge in the hobby were mostly native-born, and often of native-born parents. Where they did seek a familiar landscape, it was often a remembered homestead "back East." Anglophiles and homesick immigrants were active, but on a small scale: the Natural History Society of America and the Brooklyn Institute imported English sparrows, and a group of German-Americans in Cincinnati spent $9,000 to bring in some 4,000 specimens of twenty European species. In the 1890s one enthusiast tried to establish every bird mentioned in Shakespeare's works (adding the starling to North America's bird list), and the success of the ring-necked pheasant about this time brought on a low-grade enthusiasm for exotic game birds.[30] That was about it. North of the border Anglophilia was stronger but the climate more demanding. Where it did not discourage the introducers, it usually killed off their introductions. Skylarks took hold on the south end of Vancouver

[27] Quoted in H. J. Frith, *Wildlife Conservation* (rev. ed., Sydney: Angus & Robertson, 1979), 138.

[28] The accidents of trade and commerce led in the same direction. A modern Australian geographer noted that the "*Illustrated British Flora* is still the best availabl book for the identification of most temperate Australian weeds." See also Jamie Kirkpatrick, *A Continent Transformed: Human Impact on the Natural Vegetation of Australia* (Melbourne: Oxford University Press, 1994), 84. Statistics on exotics are from Lever, *Introduced Mammals of the World*, 434–36, and *Introduced Birds of the World*, 591.

[29] Kazimierz Wodzicki, *Introduced Mammals of New Zealand* (Wellington: Department of Scientific and Industrial Research, 1950), 11. See also King, *New Zealand Mammals*, 15, who lists forty-six species, including Maori introductions.

[30] The ring-neck, chukar, and Hungarian partridges are the only survivors. Lever, *Introduced Mammals*. See also Palmer, "Introduction," and John C. Phillips, "Wild Birds Introduced or Transplanted in North America," U.S. Department of Agriculture, Technical Bulletin 61 (Washington, D.C.: U.S. Government Printing Office, 1928).

Island, an unusually temperate area, and ring-necked pheasants strut through Prairie Province cornfields, but there is little else to see.[31]

Sport hunting was the other great motive for imports, and it too was most popular in the antipodes. North America was well stocked with deer, bear, ducks, and upland birds, but Australia had few suitable species and New Zealand none. Australians were already trying rabbits and foxes by the time acclimatization became a fad, and they brought in seventeen species or subspecies of deer. (About a dozen were tried in the wild, and half of these survived, though in restricted areas.)[32] New Zealanders rejected foxes, but did bring in rabbits and almost a dozen species or subspecies of deer from Asia, Europe, and North America. They also continued well after the others had given up. Just before World War I the New Zealand Tourist Board, pursuing the dream of "New Zealand, Sportsmen's Paradise," brought in thirteen more species, stocking the Southern Alps with chamois from Europe and thar from the Himalayas, the forests and meadows with white-tailed deer and moose.[33]

By the 1880s, though, the seductive visions of a land stocked with "useful and beautiful productions of nature" or transformed into the image of "home" had largely faded. Societies disbanded, took over zoos and botanical gardens, or pursued less ambitious plans, such as stocking streams with game fish or bringing in new plants.[34] One reason was the unexpected difficulty of the task. Many species failed. Everywhere the settlers tried nightingales and skylarks, the archetypal visible species of the English countryside. Almost everywhere they were disappointed. In 1922 G. M. Thomson found some two dozen exotic birds established in New Zealand, most of them in limited areas, out of some 130 that

---

[31] T. S. Palmer, "The Danger of Introducing Noxious Animals and Birds," in U.S. Department of Agriculture, *Yearbook of the United States Department of Agriculture for 1898* (Washington, D.C.: U.S. Government Printing Office, 1899), 87–110; Lever, *Naturalized Birds*, 305.

[32] Gordon Inglis, *Sport and Pastime in Australia* (London: Methuen, 1912), 73; Arthur Bentley, *An Introduction to the Deer of Australia with Special Reference to Victoria* (Melbourne: Koetong Trust Service Fund, Forests Commission Victoria, 1978), 30; Lever, *Introduced Mammals of the World.*

[33] Graeme Caughley, *The Deer Wars* (Auckland: Heinemann, 1983), 2–10. On New Zealand hunting see T. E. Donne, *The Game Animals of New Zealand* (London: John Murray, 1924); on the current state of introductions King, *Handbook of New Zealand Mammals.*

[34] Fascination with exotics, particularly exotic game, continues, but twentieth-century controls are more stringent. On the modern movement see Elizabeth Cary Mungall and William J. Sheffield, *Exotics on the Range* (College Station: Texas A & M University Press, 1994).

had been tried.[35] Some of the successes were even more discouraging. In the early 1860s, after many attempts, the Australian settlers established the European rabbit; twenty years later the Intercolonial Royal Commission offered a 25,000-pound reward for a way to get rid of it. In the United States the English sparrow and the starling furnished more modest arguments against the practice. By the end of the century people were warning of "the danger of introducing noxious animals and birds," and settler legislatures were passing quarantine laws.[36] Finally, immigrant memories faded with the years, experience in the new land gave it memories and associations, and a new generation grew up. By 1880 the majority of Anglo-Australians were native-born, and for them Shelly's skylark and England's hedgerows were as alien as the woolen school uniforms they wore in Sydney's heat. Their childhood memories were the bellbird's chime and the kookaburra's laugh, the ragged silhouette of eucalyptus trees, the crunch and smell of their leaves underfoot.[37]

There was, though, no putting the genie back in the bottle. What Guthrie-Smith called the "overthrow of the old world and the slow reestablishment of a new equilibrium" was becoming the great biological story of the southern countries. It was most visible in the early years and in the recurrent rabbit irruptions, but it still goes on. Baldwin Spencer, collecting in the Australian interior in 1894, found many jerboas (native marsupials) and no rabbits. Ten years later rabbits were common and in another twenty they had almost entirely displaced the jerboas. Surveying the "Red Centre" around 1930, H. H. Finlayson found much evidence that the "old Australia" was passing away, and almost thirty years later, retracing Finlayson's travels, a government ecologist, Alan Newsome, found further change.[38] A rabbit disease, myxomatosis, introduced in 1950, killed rabbits across the continent and brought a further round of change. In 1995 rabbit calicivirus disease, accidentally released in South Australia, began another round. There are reports of native plants recovering as rabbit densities drop again.[39]

[35] Thomson, *Naturalisation of Animals and Plants*, 25. See also Kazimierz Wodzicki, "Status of Some Exotic Vertebrates in the Ecology of New Zealand," in H. G. Baker and G. L. Stebbins (editors), *The Genetics of Colonizing Species* (New York: Academic Press, 1965), 432–5.

[36] Palmer, "Danger of Introducing Noxious Animals and Birds."

[37] On the early part of this process see Alan Frost, "Going Away, Coming Home," in Hardy and Frost, *Studies from Terra Australis to Australia*, 219–31. More generally, see Richard White, *Inventing Australia* (Sydney: Allen & Unwin, 1981), 85–105.

[38] Guthrie-Smith, *Tutira*; Baldwin Spencer, *Wanderings in Wild Australia* (London: Macmillan, 1928), 1: 153; H. H. Finlayson, *The Red Centre* (Sydney: Angus & Robertson, 1935), 16; Alan Newsome, author's interviews, 28 May and 4 June 1992.

[39] News bulletins of the Commonwealth Scientific and Industrial Research Organization, taken off the Internet.

Besides the landscapes, ideas changed. The skylark, singing as it soars into the sky above a Tasmanian field, was for the Victorians a reminder of home in an alien land. Now it may seem a monument to their arrogance, but to anyone raised in the country it is also part of the landscape of childhood. Walk through the stubble of fields across much of North America. The gaudy pheasant that rockets off from under your feet in a whir of wings is there because nineteenth-century hunters wanted something new to shoot. But if you were raised in farm country, what is it? Alien, certainly, if you are ecologically sophisticated, but not wholly a stranger. When, and under what conditions, do we come to see pheasants as part of the life of the land? When do we become native, and how?

Sport Hunting and Nature

There is a period in the history of the individual, as of the race, when the hunters are the "best men," as the Algonquins called them.

Henry David Thoreau, 1854[40]

Because hunting has always involved life and death people have employed ritual and symbol to control and direct the activity. On the rocks of Arnhem Land the Aborigines scratched and painted the outlines of their quarry fifty thousand, perhaps as many as a hundred thousand, years ago. Paleolithic Europeans drew great bulls and horses and bears on the walls of caves deep underground. American Indian tribes had rituals for the hunt, the kill, and the use of the animal's parts.[41] Civilized, literate Europeans were no different. The medieval hunt was pageant, spectacle, and demonstration of power, and it had as many taboos as any hunter-gather society – though the ritual was less directed toward the quarry or the spirits than the support of a social hierarchy. Modern hunting has its own rites and symbols. In the 1980s an anthropologist found plenty of grist for his academic mill among his North Carolina neighbors.[42]

---

[40] Henry David Thoreau, *Walden* (1854; reprinted New York: Modern Library, 1937), 191.

[41] Calvin Martin, *Keepers of the Game* (Berkeley: University of California Press, 1978), is a useful introduction to this point. On Aboriginal occupation of Australia and the general impact of early peoples in Australasia see Timothy Flannery, *The Future Eaters* (London: Secker & Warburg, 1994), Part 2.

[42] On medieval hunting and ritual in general see Matt Cartmill, *From a View to a Death in the Morning* (Cambridge, Mass.: Harvard University Press, 1993). The anthropology of North Carolina appears in Stuart Marks, *Southern Hunting in Black and White* (Princeton, N.J.: Princeton University Press, 1991).

The Victorian hunt was, like others, a product of the culture, in this case the society and technology of early industrial Britain. The rise of commerce and manufacturing in the late eighteenth century created a class of rich people who did not depend on the land for their wealth. They purchased land to show their wealth and establish their status and turned their new estates into game farms, complete with keepers to kill the vermin and ward off poachers. Increasingly savage laws and harsh private action reserved game for the landowning class. Rustics might have a "bit of sport" by flushing rabbits from their warrens with ferrets, but even this was frowned on. Hunting was an exercise in conspicuous consumption and social stratification. Grouse shoots were elaborately organized affairs, with gangs of beaters to flush the birds past lines of gunners, who shot them as they flew overhead. Foxes were hunted on horseback behind a pack of hounds. On Scottish estates hunters stalked red deer. As the Empire grew and weapons became more powerful, the pursuit of African and Indian "big game" became the ultimate hunt and, in the colonies of empire, a demonstration of the conquerors' power.[43]

The central element distinguishing the Victorian hunt from other ways of killing animals was its emphasis on self-imposed limitations that made for a "fair chase" and gave the quarry a "sporting chance." Quotation marks are essential here, not for irony but because technology defined fairness. Only when the invention of the percussion cap and then fixed ammunition made it possible to shoot birds on the wing did that become part of sport and, in reaction, shooting a "sitting bird" became a synonym for "unsporting" conduct. As firearms became powerful enough that a single shot would kill large game, the ideal of the one-shot kill took hold. Before that the hunter fired as many bullets into the animal as he could and waited for it to die.

Sport hunting was also a way to approach nature. There was nothing incongruous about the hunter-naturalist. Darwin hunted, and many British naturalists "relished field sports. Men like Yarrell, who, though one the finest field shots of the day, foreswore the gun for ever on being converted to natural history in 1826, were for long exceptional."[44] "We cannot but pity the boy who has never fired a gun," said Thoreau; "he is no more humane, while his education has been sadly neglected." Hunting is "oftenest the young man's introduction to the forest, and the

---

[43] On the British system see Harry Hopkins, *The Long Affray: The Poaching Wars, 1760–1914* (London: Secker & Warburg, 1985); on empire, John M. MacKenzie, *The Empire of Nature* (New York: Manchester University Press, 1988); idem, "Hunting and the Natural World in Juvenile Literature," in Jeffrey Richards (editor), *Imperialism and Juvenile Literature* (New York: Manchester University Press, 1989), 144–72; Harriet Ritvo, *The Animal Estate* (Cambridge, Mass.: Harvard University Press, 1987), 243–88.

[44] David Allen, *The Naturalist in Britain* (London: Allen Lane, 1976), 177.

most original part of himself. He goes thither at first as a hunter and fisher, until at last, if he has the seeds of a better life in him, he distinguishes his proper objects, as poet or naturalist it may be, and leaves the gun and fish-pole behind."[45] Thoreau himself laid the gun aside but kept the fish-pole. Theodore Roosevelt, who lauded the ideal of the hunter-naturalist, shot everything in sight and could identify many birds by their songs. It was only the rise of humane sentiment and the impossibility of mass hunting among a large industrial population that split hunting decisively from nature appreciation.[46]

The settlers adopted sport hunting, as they did other elements of British culture, but they had to adapt it. Social circumstances and biological realities reshaped it and gave it new meaning. There was no elite monopolizing access to land. Indeed, the great attraction and boast of these nations were of land for all. In North America hunting was already well established but not as a sport. It was a way to make money or secure food (elements opposed to sportsmanship and fair play). The great apostle of sport hunting in North America was a British immigrant, William Henry Herbert, who began preaching the ideal in the 1830s under the pen name of Frank Forester. He had an uphill pull. Americans, he lamented, regarded hunting as a way to get meat or fur or to get rid of "varmints," and they took game any way they could. Where hunting was a recreation it had no status; the respectable portion of the community looked on those who spent their idle hours in the woods as loafers. Against this Herbert preached hunting as a gentleman's pursuit, a way of making and testing character. Success was measured not by the weight of the game bag at the end of the day but by the manner in which it was obtained, by the "vigor, science, and manhood displayed." The hunter's every action should be marked by the "true spirit, the style, the dash, the handsome way of doing what is to be done, and above all, . . . the unalterable *love of fair play*" that was the mark of a true sportsman.[47]

Herbert found some converts in the Anglophile upper class in eastern cities, but hunting become a popular recreation only in the 1870s, and when it did it was a mixture of British sentiment and American history. The chase itself was governed by sporting standards. There were, as in Britain, proper ways to hunt and improper ones and a hierarchy of

---

[45] Thoreau, Walden, 191.

[46] Paul Russell Cutright, *Theodore Roosevelt: The Making of a Conservationist* (Urbana: University of Illinois Press, 1985). John Reiger, *American Sportsmen and the Origins of Conservation* (1975; reprinted Norman: University of Oklahoma Press, 1986), points out that hunting was an almost universal male recreation in this period.

[47] William Henry Herbert (Frank Forester), *Complete Manual for Young Sportsman* (1856; reprinted New York: Arno Press, 1974), 359.

game, species arranged on a scale roughly measured by the time, money, and skill the hunt required. American hunting stressed, though, as British hunting conspicuously did not, woodsmanship. The hunter's model was the pioneer. When Theodore Roosevelt and George Bird Grinnell (editor of *Forest and Stream*) organized an exclusive club for American sportsmen in 1887, they could think of no better name than the Boone and Crockett Club.[48] American hunting included technology in a way the British version did not.[49] American hunters treated firearms much as their grandsons would their cars, and for many of the same reasons. Both were industrial products that could be tinkered with, improved, and used to extend the individual's power. The syndrome was full-blown by the 1880s. Testing rifles, *Forest and Stream* warned its readers that any modern weapon was so accurate that the limiting factor was the hunter's ability to aim it. It then reported results as if from a machine shop.[50] The report on range conditions included temperature, wet bulb temperature, wind velocity and direction, barometric pressure, dew point, and humidity. For each rifle, the editors gave the weight and length of barrel, the width and depth of the rifling (the spiral grooves in the barrel that made the bullet spin), the number of grooves and their twist, and the weight of three powder charges and bullets measured to 0.1 grains. (There are 7,000 grains to the avoirdupois pound.)[51] This was what the readers wanted, for the back pages of sportsmen's magazines were filled with ads for new sights, loads, chokes for shotguns, patent bullets, and other devices that would compensate for lack of practice or skill.[52]

Hunting's social functions were distinctly American. Hunting was explicitly a way of transmitting the virtues and standards of pioneer America to a generation growing up in cities as the frontier was closing. Teddy Roosevelt, posing for a studio photograph in a buckskin outfit, holding a rifle, now looks like a parody of the mighty hunter, but in the 1880s he epitomized the appeal of the wilderness and strenuous life, the union of modern America and the pioneer past in the vigorous life of the field. Sporting and even general-circulation magazines told and retold the tale of a youth brought into manhood and responsibility

---

[48] James B. Trefethen, *An American Crusade for Wildlife* (New York: Winchester, 1975). A picture of Teddy in this pose is in the photo section of Lisa Mighetto's *Wild Animals and American Environmental Ethics* (Tucson: University of Arizona Press, 1991).

[49] See *Forest and Stream*, 2 (29 January 1874), 388–9.

[50] Ibid., 25 (29 October 1885), 261.

[51] Ibid., 344, 387, 406, 429, 447, 467.

[52] On gadgets see the magazines. One jaundiced comment on gadgetry in hunting, from a later generation, is Aldo Leopold's "Wildlife in American Culture," in *A Sand County Almanac* (1949; reprinted New York: Ballantine, 1966), 214–16.

through hunting. It begins with a boy in trouble or without a suitable adult male role model. (Often some act of waywardness sets the story in motion.) An older man, wise in the lore of the woods, takes on the child and guides his steps afield. Tests of character and skill end with a moral crisis in which the young man proves his maturity.[53]

Perhaps the most distinctive feature of American hunting was its use of magazines to spread the gospel. Specialized magazines were a feature of American life in the late nineteenth century, the product of cheap printing, regular mail delivery, and consumer advertising. *Forest and Stream* (founded 1873) was the first one devoted to the field and the best example of the breed. It had three themes: sportsmanship, woodsmanship, and game conservation. The first was Frank Forester's ideals, slightly adapted. There were, the magazine said, ways to hunt that were "sporting" and ones that were not. It preached against shooting sitting ducks, hogging shots, and other violations of the "sportsman's code," and it sought bans on "unsporting" methods (things like "shining" or "jack-lighting" deer – shooting them as they stood dazzled by the glare of fires or torches – or driving them into the water with hounds and cutting their throats). It argued against "pot hunting," the pursuit of anything edible; the true sportsman shot only proper game. As weapons became more accurate and powerful, it stressed the sportsman's responsibility to make a clean kill. Automatic and pump shotguns, which allowed many rapid shots, came under scrutiny. Were these sporting?

Woodsmanship was a distinctly Romantic idea. An early issue declared that the journal's purpose was to make the reader "familiar with the living intelligences that people the woods and the fountains, . . . [to] teach you those secrets which necessity compelled the savages to learn . . . first principles [out of which] our civilization grew but [of which] we are ignorant."[54] These were also the virtues of the hardy pioneers. An article titled "How Sportsmen Originated" said that early hunters had plunged into the wilderness alone, seeking no neighbors closer than a thousand miles, braving the panther and the Indian with only a rifle and a few crude necessities. Modern hunters had fine clothes, slept on cots with blankets that would be comfortable at home year-round, and carried cooking equipment that would suit a French chef. They did not strike out alone but huddled with their fellows.[55]

Game conservation, the third part of the program, was a response to

---

[53] Stewart Edward White, *The Adventures of Bobby Orde* (New York: Doubleday, 1901), is a pure example, but these stories were common in mass-market magazines like the *Saturday Evening Post* into the late 1940s.

[54] *Forest and Stream*, *1* (1 August 1873), 3.

[55] Ibid., *47*, (1 August 1896), 85.

American conditions. Estate hunting was not pioneer hunting, and it ran against the democratic ethos. The government, therefore, had to do what the landed class and the legal system together did in Britain – enforce rules to preserve the sport and the game – and do it on public land so all could participate. As hunting became more popular and game dwindled, the magazines demanded that states cut bag limits and shorten seasons. By 1900 they were calling for state fish and game agencies, game wardens, and an end to market hunting.

This made for a rich and contradictory mixture. Hunting was open to all on the basis of skill and character, but there was a distinct hierarchy, running from potting squirrels and rabbits in the fields behind the house through deer hunting (the democratic big game field sport) to shooting quail on plantations maintained for the birds and on to collecting trophy heads of elk and grizzly bears on long trips into the wilderness. The sport itself was steeped in the rhetoric of rural American pioneering, part of the bulwark "old stock" Americans erected against the new immigrants and the culture of the cities. Respectable hunters, that is to say white, northern Europeans, condemned things like the bird lime and nets used by "negro and Italian bird-killers." Eating songbirds, which these people did, was beyond the pale. "No white man calling himself a sportsman ever indulges in such a low pastime as killing [songbirds] for food."[56] Hunting appealed to the virtues of the primitive life and emphasized woodcraft, but the weapons were modern and hunters learned primitive wisdom from magazines run off on high-speed presses.[57]

Canadian hunting was much like American, for the land was the same and society similar. Landownership was widespread and so was the tradition of hunting for meat and money; buffalo, for instance, went as quickly north of the border, and people made the same belated efforts to save the few that survived. Sportsmanship was central, and woodsmanship and the pioneer virtues important. Canadians spoke less of pioneers, but said more about the fur trade, and the mythic figure

---

[56] William Hornaday, *Wildlife Conservation in Theory and Practice* (1914; reprinted New York: Arno, 1974), 106. On restrictions on sporting methods see Palmer, *Chronology and Index*. Social expectations with regard to wildlife use are still a factor. Vincent Serventy speaks of the Australian situation, circa 1966, in *A Continent in Danger* (Worcester: Raynell, 1966), 44. In 1990 Fish and Wildlife Service representatives in California had to educate Southeast Asian immigrants about the protection provided by wildlife refuges (discussions during author's tour as part of the Defenders of Wildlife Commission on the National Wildlife Refuge System, July 1991).

[57] On the relationship between magazines and their readers see the letter columns, articles written by readers, and the twenty-year retrospective in *Forest and Stream*, 40 (29 June 1893), 559–60.

of the Indian loomed much larger. Whatever the reality, no tale of Canadian hunting was complete without the taciturn Indian guide – skilled with a canoe, able to make camp in any conditions, peerless in finding game. The sportsmen's political program, though, did not cross the border. With fewer people and more wilderness there was less need, or apparent need, for regulation in Canada. Americans banned market hunting just before World War I; Canadians debated the issue through the interwar period. Subsistence hunting also continued, and north of farm country, where Indians, Inuit, and resident whites depended on game for food, there was very little regulation beyond what was needed to get money from outside sportsmen.[58]

Many Australians and New Zealanders were willing or eager to play the sporty English squire, and there was no tradition of meat hunting to compete with the model of gentlemanly recreation. The problem was game. Both countries had ducks and pigeons, but these were hardly sufficient, and New Zealand had no mammals and Australia's were not suitable. In the 1840s Melbourne sportsmen, mounted and in the proper attire, rode to hounds in pursuit of kangaroos, emus, and dingoes. Everyone agreed, though, that this was not "quite quite," and they brought in foxes.[59] What, you might ask, was wrong with the native animals? Dingoes could run as well as foxes, emus better than either, and in 1864 the Victorian Acclimatisation Society suggested to its British counterpart that kangaroos would be valuable and interesting additions to the English parks "and from their speed they might furnish a valuable addition to objects of sport."[60] The Australians, though, did not pursue them, at least as "game."

Australian animals were not "sporting" because they lacked the cultural associations that would embed the act of killing them in a familiar context. To see the point, try transposing Sir Edwin Landseer's enormously popular hunting canvases into colonial keys. His most famous piece, *The Monarch of the Glen*, shows a red deer stag, with arching antlers, posing in Scotch gorse. Misty Highland peaks fill the background. We

---

[58] C. Gordon Hewitt, *The Conservation of the Wild Life of Canada* (New York: Scribner's, 1921), 258–75; Janet Foster, *Working for Wildlife* (Toronto: University of Toronto Press, 1978); Robert McCandless, *Yukon Wildlife: A Social History* (Edmonton: University of Alberta Press, 1985). For a federal view and an overview from the early twentieth century see Acts and Legislation, WLU 10-2, Records of the Canadian Wildlife Service, RG 109, Public Archives Canada, Ottawa.

[59] On early hunting see Inglis, *Sport and Pastime in Australia*. Rolls, *They All Ran Wild*, chap. 14, provides an excellent introduction to these episodes. Americans hunted coyotes in the English style in the twentieth century, but with tongue in cheek. See "Life Goes to the Arapahoe Hunt," *Life*, 26 (30 May 1949), 106–8.

[60] Lever, *They Dined on Eland*, 75.

can make it North American with no trouble at all. The elk (wapiti) is the same species as the red deer, and the Rockies anywhere from Jasper to Jackson would serve as background. In Australia we would have a kangaroo standing in for the stag, old man saltbush for the gorse, and the Great Dividing Range for the Scottish hills. This will not do. The stag looks noble because centuries of myth, story, and association have made it so. To the settlers, if not the Aborigines, the kangaroo just looked odd. What would be the antipodean equivalent of *The Stag at Bay*? *The Emu at Bay*? What about *The Otter Speared*, in which a pack swirls around a hunter with an otter on the end of his upraised spear. What if the animal were a platypus?[61] If Australian animals had been fierce, shooting them would have fallen into a familiar category – the hunt as a test of skill and nerves in the face of mortal danger – but they were not. The Anglos may, over generations, incorporate Australian animals into their culture and find in hunting them a connection to the land, but at the time they could only import deer and look to the aristocratic model of gentlemanly sport.

These differences shaped wildlife law. In North America regulation began to allocate access to a transient source of food and money and evolved into the defense of a democratic, character-building recreation. Because game had to be imported and established in the southern countries, wildlife law began to protect imported species. Early laws alluded to the expense incurred by public-spirited gentlemen in bringing in game for the improvement of the colony and gave the landowner property in imported species on his land. Some even split the fines for poaching with him.[62] Legislators put seasons on native birds to save them from "wanton destruction," but fines were lower, seasons were long, and there were neither bag limits nor wardens. Enforcement depended upon someone's bringing a complaint, another bias toward exotics. The laws caused a certain amount of friction. It is only Australian legend that makes every other convict in Botany Bay a convicted poacher, but poachers were sent to Australia, and the first game laws brought angry cries of tyranny and heated declarations of the people's right to hunt as

---

[61] Reproductions of Landseer's most popular paintings are common in the Anglo world. (The *Monarch of the Glen* furnished the logo for the Hartford Insurance Company.) On his work and career see Campbell Lennie, *Landseer: The Victorian Paragon* (London: Hamish Hamilton, 1976), and Richard Ormond, *Sir Edwin Landseer* (New York: Rizzoli, 1981).

[62] New South Wales's first law was the Game Protection Act, 1866, No. 22; South Australia's was An Act to Prevent the Wanton Destruction of Certain Wild and Acclimatized Animals, 1864, No. 23. Tasmania's first statute was passed in 1865 (29 Vic. 22), Victoria's in 1862 (25 Vic. 161), Queensland's in 1863 (No. 6); and Western Australia's in 1874 (38 Vic. 4).

they pleased. A review of South Australian wildlife laws – and South Australia was a free colony, not a convict settlement – found that the debate over the 1912 act was the first in which the legislature did not discuss the English game laws when debating local ones.[63] There was, though, no repetition of the British poaching wars that had helped stock Australia with Anglos. There was land for many, the population was scattered, and pretensions to aristocratic country life soon died out. Fox hunting became an excuse for betting, coursing hares with greyhounds turned into dog racing. The imported species died out or took hold only in limited areas. The great exception was the European rabbit, but that species quickly ceased to be a game animal. It became a pest and in many areas a natural disaster. There were waterfowl and some upland birds, but hunting them seems to have been what American hunting was before Frank Forester – a pleasant way to waste a day outdoors. The colonies banned "unsporting" methods of taking game but did little more, and license laws were more a way of keeping track of firearms than of controlling hunting. Wildlife agencies were small and had few powers, and Australians did not discuss restricting market hunting until well into the twentieth century.[64]

In New Zealand deer took hold and so – for a generation – did hunting, but this story is an odd one. It stretches from the 1860s to the present; here we will review the first phase. The landscape seems to have suggested deer. In 1847 Colonel William Wakefield said that the "sport of hunting them would be highly attractive and would conduce to the breed of horses, and afford a manly amusement to the young colonists, fitting them for the more serious life of stock-keeping and wool-growing." Fourteen years later another colonist urged the importation of the animals on the grounds that they would furnish good sport and that relaxation afield would be better than "city dissipations and the laps of ballet-girls." (Poor statistics on the performing arts in early New Zealand make it difficult to assess this danger.)[65] The animals were the

---

[63] B. C. Newland, "From Game Laws to Fauna Protection Acts in South Australia: The Evolution of an Attitude," *South Australian Ornithologist*, 23 (March 1961), 52–63. On the background of this sentiment see Hopkins, *The Long Affray*.

[64] Rolls, *They All Ran Wild*, is the best source on hunting. An assessment after the founding generation is Donald Macdonald's "The Sportsman in Australia with Rod and Gun," *Australia Today* (1911), a special number of the *Australian Traveller, 1* (November 1910), 143–50. The collected state statute books in the National Library, Canberra, show the shift in laws and the lack of change. On market hunting see *Conference of Authorities on Australian Fauna and Flora* (Hobart: Government Printer, 1949). See also F. I. Norman and C. A. D. Young, "Short-Sighted and Doubly Short-Sighted Are They," *Journal of Australian Studies*, 7 (1980), 2–24, and Newland, "From Game Laws to Fauna Protection Acts in South Australia."

[65] Donne, *Game Animals of New Zealand*; quotes on xii and xiii.

acclimatization societies' most common project. A biologist found records of several hundred introductions into or relocations within the islands.[66] The Protection of Animals Act (1873) created a local landed gentry, though one more open to newcomers than the British one, by providing for the registration of acclimatization societies within districts and declaring that "the property in all animals and birds in the possession or under the control of any registered acclimatisation society shall be deemed to be absolutely vested in such society."[67]

The animals, principally the red deer of Europe, reproduced and waxed fat, and for thirty years there was British deerstalking in the South Seas. The land, though, had its own imperatives. The deer, grazing forests and fields that had not been touched since the last moa died, grew quickly and sported enormous antlers. As they ate up the accumulated surplus, the food supply dwindled and so did they. Hunters muttered about inbreeding and brought in stags from distant areas, but to no avail. Then farmers began to complain about deer eating their crops, and government officials worried about damage to plantations of imported pine. Articles like "The Effect of Deer on the New Zealand Bush: A Plea for the Protection of Our Forest Reserves" (1893) raised an argument that would become a refrain in the next generation: the native forest, since it had evolved without the grazing pressure of ruminant animals, could not survive the exotics. The deer hunter might be purchasing "his own gratification by the destruction of that forest which is the glory of his country and the birthright of the community at large."[68]

Deerstalking's decline as deer became pests paralleled a larger, social failure of hunting. Even in North America, where game and a place to hunt remained in good supply, hunting did not become a way to nature. That people were moving to the cities and the mass of immigrants had no experience with or interest in the sport was part of the problem, but the sport, with its industrial tools of killing – high-powered rifles and shotguns – and its crowds of hunters discouraged an intense engagement with the world. *Forest and Stream* might preach woodsmanship, but daily life, repeating rifles, and a weekend afield with a mass of sportsmen offered little chance for that. Hunting remained a consumer activity, focused on gadgets that were sold, tinkered with, and argued over even when there was little chance to hunt, but only in rural areas was it

---

[66] Caughley, *Deer Wars*, 1–2.

[67] New Zealand Statutes, Turnbull Library, Wellington.

[68] The description of the spread of deer is from Caughley, *Deer Wars*; P. Walsh, "The Effect of Deer," in *Transactions of the New Zealand Institute*, 25, 435–39, quoted in Wodzicki, *Introduced Mammals of New Zealand*, 192.

a significant social ritual. It appealed, though, to tradition, masculinity, and power. It did not lead into nature or give it value for the individual or the culture.

Hunting's important, permanent legacy was a legal framework that would became the basis for the management of wildlife in the twentieth century. Like hunting itself, this framework varied across the Anglo world. Laws and agencies were strongest in the United States, but even there they varied by region, strongest in the Northeast and upper Midwest (where game management would develop in the interwar period),[69] weaker in the South and West. Canada followed the same pattern. Ontario, where hunting pressure was strongest and the dollars of American tourist hunters most important, enacted the most comprehensive system. In the West little was done, and north of farming country hunting continued to be a matter of subsistence, hardly regulated in any real sense. The Australian states passed laws (but they were general), hired few wardens, and had weak agencies. In New Zealand the acclimatization societies gradually became more effective management authorities, but critics have, with reason, continued to see the system as little more than a patchwork.

Conclusion

The great conquest established basic features that still mark our physical and mental landscapes. Maps from the early twentieth century show the cities, agricultural regions, even much of the infrastructure of transportation we depend on. Speeches, books, and public policies show the national mythology that developed around the process of pioneering. There is little from this period that speaks of adaptation, for that meant failure to impose our will on nature (a strain that also remains part of our societies). Still, it was there, for nothing quite worked out as the settlers expected. They sought to live a familiar life in a familiar countryside, or to live and work in the manner or on the manor to which they wished to become accustomed. But while European farming triumphed everywhere, it did so in new forms. The Australians raised sheep and sold wool but under very different conditions than those in Britain, and even the animals had been bred for the new land. European animals and birds came to occupy at least the domestic and urban landscape, but they did not transform it, and in most places the settlers would, a

---

[69] Palmer, "Chronology and Index," and annual summaries in U.S. Department of Agriculture bulletins. The Pittman–Robertson Act of 1938, channeling federal research funds through state-level agencies funded by license fees, were a major force for new and more effective laws.

generation later, see these introductions as pests. There was countryside, but the sporty squire on his acres was a dream and the bold pioneer a fading reality. But if this was not to be England, or a New England in the American West, or even a continually expanding frontier, what was it? Who were the settlers if not triumphant conquerors?

SECTION TWO

BEYOND CONQUEST

The nineteenth century ended in a flurry of self-congratulation – the Chicago World's Fair and Victoria's Diamond Jubilee – so bright the lights almost concealed the shadows. The scattered settlements of 1800 had become prosperous nations. The opportunities that had made them beacons of economic hope and political equality, though, were fading. The good land, it seemed, had been taken up, and in places as far apart as South Australia and Saskatchewan the frontier had been thrown back. Less obviously but of equal importance, ideas had reached their limits as well. Conquest had given purpose to settlement. What would replace it? What value did the land have for the culture now? Knowledge was also changing. Natural history was at its architectural peak and in its conceptual grave. The great museums were marble monuments to an intellectual division of the world that was vanishing even as their foundations were being laid. The first generation of professionally trained scientists was dividing natural history's bones among them and shutting out the public as well.

The next three chapters deal with limits. The first takes up the physical ones, the failures of agricultural expansion and threatened shortages that culminated in the conservation movements early in the twentieth century. The next deals with the use of nature for nationalism, the search for a value beyond conquest. The settlers had, particularly in the United States, thought about this, but at the end of the century it became a greater and more general concern. Thus began an exploration of nature's place in the new societies that continues. The final chapter takes up another important current – the movement of authoritative nature knowledge from the general culture and the greater separation of experts from the rest of the population. This is the shift from natural history to ecology. The perspective that ecologists developed would become the intellectual foundation of the environmental revolution, but authoritative knowledge and popular vision would not come together again for more than half a century.

3

REACHING LIMITS, 1850–1900

Nature, to be commanded, must be obeyed.

Francis Bacon, *Novum Organum*, 1620

The settlers spoke of conquest and command, but they learned to obey. In places as far apart as South Australia and Canada's Prairie Provinces they found, at considerable cost, the limits of their agriculture. They discovered as well the bounds of their technologies. Despite poison, traps, and guns, predators persisted. Their own ingenuity turned on them. They could establish the rabbit, for instance, but not control it. The record of the species in Australia and New Zealand became a cautionary tale. The new lands were to provide for all, but shortages of resources – timber and water in particular – loomed larger and came more frequently as the century wore on. Natural history had carried the settlers onto and across their lands, mapping the country, unveiling resources, describing plants, animals, and soil, and organizing all this knowledge. It was much less successful in dealing with the problems of farming in marginal areas, making pastures safe from "vermin," and restoring abundance.

The common remedies were folk-knowledge, wishful thinking, and optimism. Farmers ignored discouraging reports about new country or suggested that settlement, plowing, or tree planting would increase the rain or moderate the winter weather. When crops failed they spoke of unusual conditions. When drought or frost persisted they muttered about grit and hanging on. Against pests they reasoned from visible events to invisible processes. Killing individuals reduced their numbers; killing enough of them would get rid of the problem. When this did not work they tried more of the same and new laws to dictate closer attention to the problem, or they searched for a "magic bullet." Against local resource shortages, mainly of wood and water, they deployed rhetoric about sturdy pioneers and greater effort. When that failed they spoke of conservation. In the extreme they turned to experts and formed new agencies. Even here there was less than met the eye. Despite pious appeals and Gifford Pinchot's fervent claims, there was little in conservation beyond common sense and sound management.[1] Only much later, during the drought of the 1930s and with the new intellectual tools of ecology, would a new perspective and new solutions begin to emerge.

Climate as a Part of Nature

Climate is what you expect, weather is what you get.

Old saying

The settlers thought of climate in several ways. Appealing to ideas that went back to classic Greece, they saw it as an influence on humans. Climate shaped temperament and culture. Thus, Anglo-Australians fretted first about the moral and physical development of children being raised in their warm, sunny land, then a century later about the problems of settling their tropical North with white people. Canadians saw their cold climate as the source of national virtue and a bulwark against corruption from the South.[2] Climate was also a physical and economic reality. These were the settler lands because they would support European agriculture, and variations in each country's rainfall and temperature range made regions and even nations rich or poor. Everywhere the settlers discovered the local knowledge they needed to survive by trial and error, using the folk-knowledge they brought from Britain to guide their first trials.[3] The first generation in each land relied on its experience in Britain, and the record of early years of settlements from

---

[1] On the development of conservation, particularly the role of colonial administrators in developing ideas of conservation in the eighteenth century, see Richard Grove, *Green Imperialism* (Cambridge University Press, 1995); on conservation as an organized movement in the United States, Samuel P. Hays, *Conservation and the Gospel of Efficiency* (Cambridge, Mass.: Harvard University Press, 1959).

[2] Clarence Glacken, *Traces on the Rhodian Shore* (Berkeley: University of California Press, 1967), chaps. 2, 9, 12, and 13; Griffith Taylor, *The Control of Settlement by Temperature and Humidity*, Bulletin 14, Commonwealth Meteorological Service (Melbourne: Victorian State Printer, 1916). See also idem, *The Australian Environment, Especially as Controlled by Rainfall*, Advisory Council of Science and Industry, Memoir No. 1 (Melbourne: Government Printer, 1918); I. Clunies Ross, "Blanks on the Map," in J. C. G. Kevin (editor), *Some Australians Take Stock* (London: Longmans, 1939), 83; Carl Berger, "The True North, Strong and Free," in Peter Russell (editor), *Nationalism in Canada* (Toronto: McGraw-Hill, 1966), 3–26.

[3] On the pragmatic definition of climate in these years, see Thomas R. Dunlap, "Agriculture and the Concept of Climate," *Agricultural History*, *63* (Spring 1989), 152–61.

Jamestown to Sydney were filled with bad crops and starvation. Knowledge gained in these footholds became the base for judging land farther out. When the new was much like the old this was often enough, but when farmers went a long way or crossed invisible thresholds they risked disaster.

This was often the case in the nineteenth century, for the pace of settlement was rapid, the dream of riches from enormous acres mesmerized people from North Dakota to South Australia, and belief in humans' power over the world was strong. Canadians, Americans, and Australians spoke of changing the climate by cutting trees or planting them and believed that rain would follow the plow. Canadian farmers (in a country where optimism is not usually considered a national trait) dreamed of farming the sub-Arctic. The belief that the Canadian climate "would be moderated by clearing and cultivating the virgin forest," said one historian, "died hard."[4] In the 1860s South Australian wheat farmers pushed across Goyder's Line, dividing the colony's agricultural land from pastures. They scoffed at the idea that the land was too dry for wheat. Americans had much the same reaction a decade later to John Wesley Powell's concerns about the "Arid Region." With rain seeming to increase each year, American and Canadian farmers pushed deeper into the rain shadow of the Rockies in the 1880s.[5]

Settlement of the interior of North America and Australia produced strong but inconclusive debates about climate – strong because riches seemed so near for so many, inconclusive because no one knew much about climate. Expert opinion and the ideas of "practical men" rested on the same foundations, and the experts usually knew no more than the farmers. Only late in the century could statistical data begin to compete with experience in most of the regions, and the numbers were, in themselves, a poor guide. This meant that expertise could as easily be called in to ratify settler dreams as to shatter them; isotherms on a map and statistics of annual rainfall in government publications were presented as ballast for the boosters' airy claims. The more fundamental problem was that climate was a general, and long-term, guide that dealt in averages, while farmers lived in the short term and by events. It took only a single very late frost in the spring, a critical week of too much or too little rain during the growing season, or bad weather at harvest and the entire season was gone. In kindly climates variations were an

---

[4] Suzanne Zeller, *Inventing Canada* (Toronto: University of Toronto Press, 1987), 122. For an extended discussion of Canada see Doug Owram, *Promise of Eden* (Toronto: University of Toronto Press, 1980). On optimism a dated but useful account is Wallace Stegner's *Beyond the Hundredth Meridian* (Boston: Houghton Mifflin, 1953).

[5] William Waiser, *The Field Naturalist* (Toronto: University of Toronto Press, 1989), 24–53, contains a good account of these developments in the Peace River District.

element, but after 1850 farmers increasingly moved into areas where they were critical. Movement west or north on the Great Plains or north on the Yorke Peninsula toward the head of Spencer Gulf in South Australia took farmers into areas where the weather might, one year, allow a bumper crop and the next one or few, nothing at all. There were no obvious divisions, and the situation was further complicated by farmers' ability to change the limits of farming by breeding crops for their new region. Like the acclimatizers, farmers could point to what had been done as proof of what could be, and there was little to refute them.

Settlement of the Prairie Provinces in the 1870s shows the state of knowledge, the degree of hope, and the elasticity of the evidence. In the 1850s Canadian expansionists began to see the region as essential to the country's future, and "[b]etween 1856 and 1869 the image of the West was transformed in Canadian writings from a semi-arctic wilderness to a fertile garden well adapted to agricultural pursuits."[6] The land was the same, and there was no new information. People simply applied hope to information. Thomas Devine created a map, using American sources, and adorned it with comments describing the region in glowing terms. With slightly more science, Loren Blodgett challenged the use of latitude as a way of determining climate. He relied instead on isotherms, one of Humboldt's techniques (though he was less rigorous than the great naturalist might have wished). He simply pushed the lines well north to show a temperate prairie.[7] Official reports showed an infertile region, "Palliser's Triangle," reaching north from the border and a fertile area beyond that. Enthusiasts gave these reports as favorable an interpretation as they could, and there was much talk of a healthy northern climate and of the cold producing a virtuous people. In 1872 Jesse Hurlbert argued that the Canadian prairies were uniquely favored. Farther south, in the United States, heat and sun produced tropical vegetation where there was water, and desert where there was not. Over generations this would make Americans tend toward the physical type of desert dweller, the "Bedouin of Arabia." Canadians, he thought, could grow grasses, barley, and oats as far north as the Mackenzie Delta, wheat as far north as Great Slave Lake (and presumably retain their Anglo-Saxon physical and cultural heritage). There was talk about the miasmas farther south and even claims that the northern part of the United States was colder than the southern Prairie Provinces.[8]

Wet years in the late 1870s encouraged optimism, and new surveys nibbled away at Palliser's Triangle, suggesting the fertile region was larger than had been thought. One expert, John Macoun, abolished it

---

[6] Owram, *Promise of Eden*, 3.   [7] Ibid., 66.
[8] Ibid., 158. The full discussion is on 151–67; see also 116–18.

altogether. A field botanist (which suggests something about expertise in climatology in these years), he was also a great booster; his "enthusiasm knew no limits except the boundaries of the region itself."[9] He made five surveys for the Geological Survey of Canada between 1872 and 1881, and they were all the boosters might have wished. His first report on the Peace River District was so positive that the Survey's director, A. R. C. Selwyn, rejected it. The land, he thought, could not be that good. Perhaps not, but who could tell? There were no special techniques for judging the land – both Macoun and the farmers relied on vegetation and successful crops – and therefore no good basis for challenging his reports. A combination of circumstances slowed settlement for a generation – the Winnipeg land boom in 1884, the decline in rainfall later in the decade, and the depression of the 1890s – but in later, better years farmers again pushed into the region.

Only gradually, and well after Macoun's day, did it become clear that the entire discussion, complete with Humboldtian isotherms, was beside the point. The critical factor was not average temperature but the length of the growing season. One of Macoun's informants had suggested that, saying the problem was late spring or early fall frosts, but Macoun had dismissed the idea.[10] Anyone might have, for accepting it required reconsidering what climate was. It was seen as a property of the region and thought of as the weather over time. It had to be taken instead as the relation between any given species and the land and in terms of the probability of key events. Farmers and agricultural scientists would gradually come to this view, though usually not consciously, as they learned more about the land and their crops. It is evident in Canadian breeders' concentration on breeding for a very short growing season and tilling to hold soil moisture. Farther south it shows up in American farmers' work in raising subtropical crops. By taking account of minute variations caused by slopes and lakes, and providing protection from winds, citrus farmers in Florida, California, and Texas managed in areas where killing frosts were a regular reality. They looked not at climate but at microclimates.[11]

The failures of settlement preserved in popular memory involve not frost but drought, and are fixed in American and Australian images of dust, wind, and abandoned farms. The great testing grounds of hope and knowledge were South Australia and the American West, where experts and ordinary citizens clashed over farming in the dry lands. Because, in the long run, the experts were right and the boosters wrong, these incidents have often been viewed as morality plays of good,

[9] Waiser, *Field Naturalist*, 16. [10] Ibid., 32.
[11] Dunlap, "Agriculture and the Concept of Climate."

prescient scientists struggling against ignorant or venal boosters. There is something to that, but not everything. There was hope and guesswork as well as science on both sides, and the debates were less about facts than about what powers humans had over nature, how nature was to be seen, and who had the authority to construct authoritative pictures of it.

South Australia came first. The early settlers on the Yorke Peninsula on Australia's southern coast, who began farming in the 1840s, lived in the path of the winds off the Great Australian Bight (the indentation in the south coast). Rainfall, from ten to twenty inches – 250 to 500 millimeters – a year, was enough for wheat, particularly since it came in the winter, and soon a boom was on.[12] Wagons and railroads brought the crop from ever-expanding fields to docks on Spencer Gulf, where it was loaded for Europe.[13] As the settlers pushed north they moved into areas not swept by sea winds. It was obvious that rain fell off as one went north, and in 1865 the colony's government sent the surveyor general, George M. Goyder, out in advance of settlement to decide what areas were suitable for pastoral leases and what could be sold for farms. The settlers observed the division, which became known as Goyder's Line, until farms began to press against it. Then they began ridiculing the idea of a line of rainfall and attacking the grazing policy as giving kingdoms to a few while denying farms to the many.[14] The government yielded and the settlers moved north. Disaster followed. The first dry year, 1881, was a shock. The next two, which were very dry, began a retreat that turned into a rout. Towns lost population and businesses failed. The railroads and ports fell silent. Entire districts were abandoned to sheep, and for the next thirty years agricultural settlement stabilized very close to Goyder's Line.[15] These events combined common settler sentiments and unique Australian realities. The arguments for expansion that filled the colony's papers could have come from country journals anywhere in the Anglo world. South Australian farmers' confidence was much like that of American and Canadian farmers who had ventured into the "Great American Desert" and were moving into the rain shadow of the Rockies. The Australian climate, though, was much less forgiving. Rainfall fell off very quickly from the coast, and in the interior it was not only low but varied much more widely than in most other areas of the world. One modern ecologist described conditions in the interior as chaotic, with

[12] On rainfall see Bureau of Meteorology, *Climate of Australia* (Canberra: Australian Government Publishing Service, 1989), 5.

[13] This account is drawn largely from D. W. Meinig, *On the Margins of Good Earth* (Chicago: Association of American Geographers, 1962), 45–92.

[14] Ibid., 45, 53.   [15] Ibid., 78–92.

no correlation from month to month or between the same month in different years. It would be another century, though, before that was understood (see Chapter 6).[16]

John Wesley Powell's *Report on the Lands of the Arid Region* (1878) was the American equivalent of Goyder's Line. Initially a report to Congress, then a publication of the Geographical and Geological Survey of the Rocky Mountain Region, it became one of the most controversial and cited land documents in American history.[17] It painted a bleak picture of the West's agricultural future, one deeply at odds with the optimistic predictions of boosters and settlers. Most of the West, it said, could never be farmed, for rainfall was too meager for anything but irrigation, and stream flow was sufficient for only a small percentage of the land. It did not counsel against settlement, though – Powell was not that radical. It proposed saving the great settler dream of independence on a family farm by changing land and water laws to suit Western conditions. It recommended abandoning the rectangular survey and the 160-acre homestead, the cornerstones of land policy since the Confederation, as well as the assumption behind them, that all land could be farmed. For areas that could be, surveying should be based on watersheds, not the compass, and homesteads given as small farms with attached irrigation rights or large ranches with water rights for stock and household use only.[18] Debate over the document and its recommendations was ostensibly about facts and figures, but at bottom it was over humans' ability to shape the land to their desires. In the boosters' eyes Powell's great offense was his sin against the gospel of unlimited opportunity and human ingenuity. That he enlisted in the service of his heresy the gods of science (potent and respected deities in their own right) only made matters worse, and it was no mitigation that he saw obedience to nature as the path to command. No one wanted to hear that nature set limits humans had to honor. It is not facetious to see the controversy in these terms. The conquest of nature and the advance of civilization were an

[16] On climate see "Physical Geography and Climate of Australia," 202–56, in Australian Bureau of Statistics, *Year Book Australia, 1988* (Canberra: Australian Bureau of Statistics, 1988); on the interior pastures, Graeme Caughley, "Ecological Relationships," in Graeme Caughley, Neil Shepherd, and Jeff Short (editors), *Kangaroos: Their Ecology and Management in the Sheep Rangelands of Australia* (Cambridge University Press, 1987), 159–87.

[17] John Wesley Powell, *Report on the Lands of the Arid Region*, U.S. Congress, House of Representatives, 45th Congress, 2 Session, H.R. Exec. Doc. 73 (Washington, D.C.: U.S. Government Printing Office, 1878); it was reprinted in 1879.

[18] Stegner, *Beyond the Hundredth Meridian.* On the larger issues of water and development see Donald Pisani, *To Reclaim a Divided West* (Albuquerque: University of New Mexico Press, 1992).

American faith, one that defined national identity and supported hopes for individual economic opportunity. Powell dissented from the faith, and he met the usual fate of heretics.

Powell was right and his critics wrong about the amount of water in the Great Basin (as current figures on irrigated land show), but this was not a contest between knowledge and ignorance. The optimists' case was stronger than it now appears and Powell's weaker. The boosters had on their side the experience of the past fifty years. Pike, Long, and others had condemned the treeless prairies as the "Great American Desert." Farmers had shown that the eastern part was rich farmland, then that Iowa and Missouri were even richer. They moved out onto the plains against expert opinion, and rain seemed to increase with settlement. Now we know this was coincidence – Anglo settlement coincided with the wet years of a generation-long cycle of high and low rainfall – but they did not. Nor was Powell's knowledge overwhelming. Rainfall records went back twenty to forty years, but they were spotty, and Powell had no theory to explain what was happening or why. This is not to say that he was lucky or guessing. He did see further than his contemporaries, perhaps because he was less emotionally committed to fashionable ideas. He knew more, though, than he could prove.

The experience of the next generation tempered enthusiasm. Rainfall on the Great Plains dipped to its cyclic low, and in Australia the "king drought" of 1895–1903 ruined pastures from Queensland to Victoria. Many North American farmers abandoned the land in the rain shadow of the Rockies, and whole towns vanished. In Australia graziers made a similar retreat to the coast. It was not just the land; the droughts coincided with a worldwide depression that ruined farmers, industries, and businesses. The rains returned and prices rose in the early twentieth century, and World War I renewed enthusiasm, but the optimism that came with the 1920s had a different form. No one planned for independent citizens working small farms; people spoke of vast, capital-intensive projects, of irrigation that would farm the Centre and the American Southwest, of plant and animal breeding that would extend Canadian agriculture to the Arctic Ocean (see Chapter 6).

Pest Animals and the Balance of Nature

It is not only in expansion that we see limited knowledge and unlimited optimism. The same characteristics appear in the settlers' drive to rid their pastures and fields of "vermin," the animals that preyed on livestock. Landowners everywhere in the Anglo world – indeed, everywhere people produced stock for a world market – deployed steel traps, repeating firearms, and poison to "wipe out" these "pests." Despite the arsenal of industrial technologies, these campaigns were based on a simple view of nature. It had parts (the various species), any of which might be removed without changing the whole. Removal could be accomplished by killing off individuals of the offending species. The consequences of this kind of action would not become apparent until later, but even then it was obvious there were flaws in the logic. Some pests, such as the coyote in North America and the dingo in Australia, survived intensive campaigns against them. Others, most conspicuously the European rabbit in Australia and New Zealand, gained ground, and at an alarming rate. The rabbit's increase in the late nineteenth century was one of the most spectacular wildlife irruptions in modern history, and not only could the settlers not halt it, they had no way to explain it.

Establishing rabbits had been part of the Anglo settlement of Australia since a shipment arrived with the First Fleet, which brought convicts to Botany Bay, but this and many subsequent introductions died out. Then in 1860 a shipment for Thomas Austin, "an English tenant-farmer who had made a fortune in Australia and wished to play the part of a sporty squire," took hold. At first they were protected; in the 1850s a man was fined ten pounds for killing one on the property of a Mr. John Robertson. "A few years later Robertson's son William began to spend 5,000 pounds a year in an attempt at rabbit control."[19] By the 1870s a ton and a half of rabbit meat was being auctioned each day in the Melbourne market. A rabbit canning industry sprang up, and hatters began to use rabbit felt (still the mainstay of the business in Australia). Later a canned-pet-food industry came to rely on rabbit meat. (It suffered a setback in the 1950s, when myxomatosis destroyed most of the rabbits; kangaroo meat was used instead).[20] None of these uses, or wholesale lots of poison, traps, or fencing, reduced the number of rabbits or slowed their march.

They spread with astonishing speed. An area 70 by 100 miles in the mallee area of western Victoria was occupied in only twelve years, and rabbits moved up the Murray–Darling River system at the rate of 125 kilometers a year, reaching the Queensland border by 1866. Introduced into South Australia in 1870, they covered the 600 kilometers to Lake Eyre in sixteen years, arrived in Western Australia in 1894 and the west coast in 1906.[21] They were good colonizers, but they had considerable

[19] Eric Rolls, *They All Ran Wild* (Sydney: Angus & Robertson, 1979), 20, 21.
[20] Ibid., 71–9.
[21] Christopher Lever, *Naturalized Mammals of the World* (London: Longmans, 1985), 401–6; K. Myers, I. Parer, D. Wood, and B. D. Cooke, "The Rabbit in Australia," in Harry V. Thompson and Carolyn M. King (editors), *The European Rabbit* (Oxford: Oxford University Press, 1994), 108.

help. Part came from humans carrying them into new areas, but more important were the changes the settlers made in the land. Cutting down the trees, killing dingoes, and introducing sheep (which grazed down much of the higher vegetation), they created a habitat. On the eastern slopes of the Dividing Range and in the highlands of New South Wales, making bush into farmland caused a series of irruptions among native and introduced herbivores. Kangaroos, then rat kangaroos, then hares and wallabies, then pademelons and rabbits went through enormous population swings. Rabbits were unique in persisting at high densities.[22]

Reactions to the animals went through several phases. Those living in areas that were not yet affected usually thought no action would be necessary. When the "gray blanket" arrived, people sought to protect their interests. "Squatters," Australia's equivalent of the American West's cattle barons, looked on rabbits as things that ate the grass God had intended for sheep, and they demanded that the legislature take action. Small farmers were often less concerned, seeing rabbits as just another kind of animal that promised a bit of sport and meat. Where climate or lack of suitable soil limited the species, or profits were high enough to make the cost of control bearable, this attitude might survive the invasion. Swagmen and hired hands took an even more benign view. Rabbits were easy meat for the pot and their skins a source of quick cash. They were the poor man's friend.

Because sheep were the basis of the Australian economy, legislators in every colony enacted laws to control rabbits. The first recourse was familiar remedies. The first statutes, passed in each colony when rabbits reached epidemic proportions, relied on the traditions of British law. Landowners had the responsibility of clearing their own acres. That was a "non-starter." Australian grazing was almost always a matter of putting sheep on many acres and getting a low return from each acre. Stockmen could not afford the poison to kill rabbits or the fences to keep new ones from coming to their pastures. Besides, until the government took action to control rabbits on Crown land, private action was often useless. Legislators, though, were understandably reluctant to commit so much tax money to what soon seemed an endless and impossible task. New South Wales turned to another established method, bounties, and quickly turned away. Between 1882 and 1887 it paid out almost a million pounds for dead rabbits with no obvious result except to make a number

---

[22] Myers et al., "The Rabbit in Australia," 110. On the rate of spread and the relation to human aid (Flux discounts the effect of direct transport) see John E. C. Flux, "World Distribution," in Thompson and King, *The European Rabbit*, 8–21.

of professional "rabbiters" rich.[23] Fences enjoyed a vogue. If rabbits could not be controlled, they might at least be confined to already "infected" areas. (Metaphors of disease were as common as those of war.) In 1886 Queensland and New South Wales began building a fence on their border; the next year Victoria and South Australia started another; and a maze of shorter barriers soon covered the Southeast. Western Australia erected the wire-mesh equivalent of the Great Wall of China; Fence No. 1 stretched 1,100 miles from the south coast to the Indian Ocean. Enthusiasm quickly waned. Swagmen were often accused of carrying rabbits across the barriers, but rabbits hardly needed them. Floods undermined fences; wind made bridges of sand or brush over them; large animals tore holes. Then there were the inevitable flaws in design or construction: slanted support posts that allowed rabbits to climb over, carelessness or warped timber that left gaps at gates, loose soil or sloppy work that allowed rabbits to dig under.[24] By the time World War I diverted all available manpower, fences were in disrepute. By the end of the war, they were in disrepair.

As conventional remedies failed, Australians searched for unconventional ones. Regularly, the colonial, later state, governments convened rabbit control conferences and commissions to investigate new methods, and the newspapers reported new ones every year.[25] One man suggested importing from South Africa a species of carnivorous ant that would eat the young rabbits while they were still in the nest.[26] Another thought it would be possible to spread liver flukes among the rabbits by feeding the carcasses of diseased rabbits to dogs. Their contaminated feces

---

[23] This is a general picture based on an analysis of the collected colonial (later state) statutes in the national libraries of Australia and New Zealand. The complexities of the situation deserve a book of their own. On the rabbiters see James Matthams, *The Rabbit Pest in Australia* (Melbourne: Specialty Press, 1921), 25.

[24] A good analysis can be found in David Stead, *The Rabbit Menace in New South Wales* (Sydney: Government Printer, 1928). More extensive and candid is Stead's five-volume unpublished "Report on the Rabbit Menace in New South Wales (1925–1926)," in the Mitchell Library, Sydney.

[25] New South Wales, for instance, had conferences in 1885, in 1886, and again in 1895, the last to review the findings of the Royal Commission appointed in 1888. "The Rabbit Conference," Report of the Proceedings of a Conference Respecting the Rabbit Pest in NSW, presented to the Parliament (Sydney: Government Printer, 1895). The conferences received coverage in the agricultural and local presses as far away as New Zealand. See, e.g., "Destruction of Rabbits," *New Zealand Country Journal, 13* (May 1889), 209–13. This, with the control boards, was an enduring phenomenon; there was a conference in 1964, and the current resurgence and the accidental release of rabbit calcivirus disease in 1995 may produce another.

[26] "Rabbit Destruction," *Journal of the Department of Agriculture of Western Australia, 14* (20 October 1906), 281–4.

would spread the disease back to the rabbits. Matters could be hurried along by feeding the dogs a purgative of areca nut.[27] Neither of these attracted much interest, but the "Rodier system" did. In 1889 William Rodier reported that rabbits could be controlled if they were trapped, the does killed, and the bucks released. As the proportion of males to females rose, the bucks would eventually kill the remaining does or harass them so much they could not produce young. As late as 1925 David Stead, head of the New South Wales Rabbit Menace Commission, found it necessary to point out forcefully that Rodier had not been able even to clear his own property using this scheme.[28] An even more durable enthusiasm was piping poisonous gas into the rabbits' warrens. In the 1870s carbon dioxide was the favored agent. Later materials included carbon disulfide, various cyanide compounds, and truck engine exhaust.[29]

The Intercolonial Rabbit Commission's offer, in 1887, of a £25,000 reward for an effective method of rabbit control brought the greatest number of suggestions, more than 1,400. They ranged from the prosaic to the fantastic; some were vague, others quite specific.[30] Disease was popular: 115 people from nineteen countries recommended it. The idea was plausible, and Australians would use it to deadly effect in the 1950s (Chapter 9), but then no one knew how to turn this idea into a program. Some of the correspondents did not try; they just suggested spreading a disease without naming one. Others proposed a disease without any proof that it would kill rabbits. Some of the diseases they suggested were known to infect humans, which probably reduced the Commission's interest. Only one scheme had any scientific backing. Louis Pasteur suggested using chicken cholera and sent one of his assistants to Australia with a culture. Quarantine and health authorities turned him and the germs back on the grounds that the disease was too hazardous.

These were legitimate concerns, but there was a larger problem no one spoke of – how to propagate the disease. No one said anything because the ecology of disease had hardly been thought of, and certainly not by that name. Germ theory was in its infancy, and while it was known that some diseases could be transmitted from person to person or animal to animal, no one knew how it was done. When, for example, Pasteur

[27] Rolls, *They All Ran Wild*, 163.

[28] "Rabbit Extermination," printed in *Webster's Tasmanian Agricultural and Machinery Gazette,* reprinted in *New Zealand Country Journal, 13* (March 1889), 119–20. Stead's counterblast is in *The Rabbit Menace in New South Wales.*

[29] T. E. Mahoney, *The Rabbit in Australia* (Ararat: privately printed, n.d.); David Stead, *The Rabbit in Australia* (Sydney: Winn, 1935).

[30] New South Wales Royal Commission (Rabbits), 1889, *Progress Report*, xxxi.

tried chicken cholera on rabbits on an isolated estate on South Island, New Zealand, he could not say why the disease did not produce an outbreak. Under these conditions, any trial would have been hit-or-miss, and Australian authorities were understandably reluctant to try, particularly since a remedy might backfire, causing illness or death among stock or people.[31]

Conventional remedies could have been made to work – New Zealanders used them with success after World War II – but the colonies lacked the resources or the political will that would have been required. It was not simply that authority was divided among the colonies (states after 1901), rabbits were in an unusually favorable position and graziers an unusually bad one. The ocean voyage had filtered out many of the diseases and parasites that held rabbits in check in Europe. The climate was ideal over most of the continent and the soil good for burrowing. There was only a small suite of native predators that could adapt to preying on them, and the rabbits' rapid reproduction meant that this was not, under normal circumstances, an effective check on their population. (After the myxomatosis epidemic of the 1950s, populations were small enough that for a number of years it was). On much of the land, agriculture and stock raising were extensive, and farmers or graziers could not afford measures to control their vast acres. There was in addition much vacant land that returned no profit at all but on which rabbits had to be killed, lest they spread to field and pasture.[32] The states fell back on the requirement that landowners clear their own fields and pastures but district boards, in charge of the work, always included graziers, and they saw little reason to bankrupt themselves or their neighbors. The laws that remained on the books were a dead letter in most districts. By the 1920s rabbits seemed established as a continent-wide, permanent check on agricultural production.[33]

Events in New Zealand followed a similar course. There were multiple introductions of tame and wild rabbits from the 1820s – there were traders on shore before the influx of settlers in the 1840s – and sometime in the 1850s the species became established. As in Australia, humans had prepared the land, and the rabbits spread over it with

---

[31] Ibid., xxx; Rolls, *They All Ran Wild*, 152–5. On a later trial of chicken cholera, cut short by official concern about side effects, see Lech Paszkowski, "Dr. Jan Danycz and the Rabbits of Australia," *Australian Zoologist*, 15 (August 1969), 109–20.

[32] On this see Harry V. Thompson, "The Rabbit in Britain," 88–89, and Myers et al., "The Rabbit in Australia," 131, both in Thompson and King, *The European Rabbit.*

[33] Australian state statutes, National Library, Canberra. For overviews see Stead, "Report on the Rabbit Menace," and Rolls, *They All Ran Wild.* The economics are in Francis Ratcliffe, *The Rabbit Problem* (Melbourne: Commonwealth Scientific and Industrial Research Organization, 1951), 3–4. Stead, *The Rabbit Menace*, discusses the 1920s at length.

astonishing speed, reaching incredible numbers. Between 1877 and 1884 graziers in the province of Otago abandoned seventy-seven sheep runs, a total of 627,935 hectares (roughly 1,500,000 acres) to the invaders.[34] Here too the introducers, in this case the acclimatization societies, took full credit until rabbits became a pest. Then they dodged the limelight. Legislation and remedies followed Australian practice. New Zealand's first law, the Rabbit Nuisance Act (No. 63, 1876), was modeled on Tasmanian legislation of 1871, and New Zealanders resorted as well to poisons and fences. Like the Australians they sought to turn the pest into a cash crop. By 1881 rabbit products were the colony's seventh largest export.[35]

The islanders, though, tried something the Australians had not: predators. In the 1870s graziers, reasoning from the fact that weasels, ferrets, and stoats ate rabbits to the conclusion that they would keep the rabbit population down, began agitating for their introduction. There were immediate objections, both to the program and to the reasoning behind it. Alfred Newton of Cambridge said that "[i]t would be ridiculous to suppose" that the predators "will confine their destructive powers to rabbits and there will be an end not only of your brevipennate birds, but of many of the other native species."[36] Witnesses before the Select Committee Inquiring into the Rabbit Nuisance said much the same. Mr. William Smith thought the animals would "merely go into the bush after the small birds and leave the rabbits alone." The Honorable Mr. Robinson, "having experienced rabbits in Australia," did not think the weasels would prey on them.[37] In 1882, however, the legislature bowed to the graziers and hundreds of these animals were imported. The Hawkes Bay Acclimatization Society sought to protect native birds by offering a guinea a head for these "vermin dead or alive," but this kind of opposition did not check interest or the weasels.[38] The animals took hold. They did not have any visible effect on the rabbit population, and as New Zealanders placed more value on their native wildlife the experi-

---

[34] John A. Gibb and J. Morgan Williams, "The Rabbit in New Zealand," in Thompson and King, *The European Rabbit*, 162.

[35] Andrew Hill Clark, *The Invasion of New Zealand* (New Brunswick, N.J.: Rutgers University Press, 1949), 267–71; Kazimierz Wodzicki, *Introduced Mammals of New Zealand* (Wellington: Department of Scientific and Industrial Research, 1950), 127–9; George M. Thomson, *The Naturalisation of Animals and Plants in New Zealand* (Cambridge University Press, 1922), 85–95.

[36] Quoted in Ross Galbreath, "Colonisation, Science, and Conservation" (unpublished Ph.D. dissertation, University of Waikato, 1989), 131.

[37] Gibb and Williams, "The Rabbit in New Zealand," 174.

[38] H. Guthrie-Smith, *Tutira: The Story of a New Zealand Sheep Station* (London: Blackwood, 1921), 334.

ment was looked on with even less favor. In 1922 George Thomson, reviewing the record of introductions into the islands, said that "[n]othing in connection with the naturalisation of wild animals into New Zealand has caused so much heart-burning and controversy as the introduction of these bloodthirsty creatures." A year earlier the naturalist and sheep raiser H. Guthrie-Smith had rendered a blunter judgment. The project was an "attempt to correct a blunder by a crime."[39] By then the government had restricted further importations, and in 1936 it removed all protection from weasels and their relatives. That did no good.[40]

All these measures rested on a view of nature so well accepted that the settlers did not even discuss, much less debate, it. They saw the world around them as made up of separable parts – the various species – interacting on the apparently neutral backdrop of the land by processes that could be understood by observation. It was clear, for example, that predators killed other animals. From these examples of individual behavior the settlers reasoned that the predator species in some way controlled the population of its prey. Understanding, though, was schematic. It was vaguely assumed that each species had some natural level of population to which it was held by the "balance of nature." No one knew, though, how it worked, and when introductions failed or their populations exploded out of control, the only intellectual tools the settlers had were observation and common sense – which had already proved futile. It would be another half-century before there would be an effective way to think about these situations. In the meantime, legislatures held to conventional measures even when it was clear they were as useful as baling the ocean with a teaspoon, and everyone grasped at any straw.

That the problem was not a lack of sophistication or experience in the southern countries is clear from the record of coyote control in North America. Ranchers and legislatures in the United States and Canada approached that in the same fashion and had much the same success. The coyote was a late addition to the category of vermin, attacked only when wolves, bears, and mountain lions had been reduced to remnant populations. Then ranchers turned on it the same weapons and logic. Control was a matter of killing some individual pest animals. Eradication a matter of killing more. This approach was as effective as

---

[39] Thomson, *Naturalisation of Plants and Animals*, 70; Guthrie-Smith, *Tutira*, 334. On the introductions see relevant chapters in Carolyn King (editor), *Handbook of New Zealand Mammals* (Auckland: Oxford University Press, 1990).

[40] Carolyn King, *Immigrant Killers* (Auckland: Oxford University Press, 1984), and King, *Handbook of New Zealand Mammals*.

it was against rabbits. Despite the widespread use of poison (one could buy strychnine over the counter in South Dakota and Montana into the 1920s without even signing a poison book), traps, and guns, coyotes lived and moved into new territory. Here, no more than in Australia or New Zealand, did this prompt new thoughts. Ranchers continued to demand new poisons, new ways to deliver them, and more intensive campaigns. Even the Biological Survey, coordinating the work and doing research on coyotes and their control, stuck to the old ways.[41]

Conservation: Organization as Science

The same intellectual limits are visible in programs to conserve and manage what the settlers were coming to see as dwindling, or at least finite, resources. From the 1860s there was a counterpoint to the expansive faith in nature's bounty that had been a great theme of settlement literature, expressed in debates about dwindling forests, water shortages in cities, and the need for irrigation to develop new land. Like pest control work, though, these programs were stopgaps, concerned not so much with how humans affected the earth or how nature worked as with economic opportunities for the next generation.[42] Like hunting, conservation was less important for what it did at the time than for the legal and administrative foundation it laid. Its precedents and agencies became important in land and resource management in the twentieth century, part of the structures the environmental movement would use, transform, and battle.

Conservation was, also like hunting, an idea from the metropolis adapted to the conditions on the various peripheries. Europeans had been dealing for hundreds of years with dwindling resources at home, and in the European overseas empires people developed and applied ambitious programs of use and management.[43] The experts they trained were central to conservation in the settler countries. German foresters staffed the South Indian Forest Service, and went on to advise the Australians. Carl Schurz, the U.S. Secretary of the Interior, who in the

---

Stanley Paul Young and Hartley H. T. Jackson, *The Clever Coyote* (1951; reprinted Lincoln: University of Nebraska Press, 1978), 171–226. On the species' modern range see Ron Nowak, *Walker's Mammals of the World* (5th ed., Baltimore: Johns Hopkins University Press, 1991), 2: 1070.

The classic on the conservation movement in the United States is Hays, *Conservation and the Gospel of Efficiency.* On the antecedents of conservation see Grove, *Green Imperialism.* An example of the deficiencies of expert knowledge in this period and in the core area of forestry is Nancy Langston's *Forest Dreams, Forest Nightmares* (Seattle: University of Washington Press, 1995).

Grove, *Green Imperialism,* deals with imperial programs.

1870s warned against the waste of forests, was a German immigrant. So was the first U.S. government forester, Bernard Fernow, trained at Prussia's forestry school. Gifford Pinchot, the American apostle of conservation, learned his gospel at the French National Forestry School at Nancy and from British foresters. He based the U.S. Forest Service's administrative structure on the South Indian Forest Service, and the imperial model was written all over the American movement. It was largely national policy applied to a region, the West, which had only recently been divided into states (and some areas were still territories). Western discontent with these policies was, not surprisingly, couched in the rhetoric of colonialism and outside domination.[44]

Americans, though, led as well as followed, an indication of their growing importance in the Anglo world. George Perkins Marsh's *Man and Nature, or, Physical Geography as Modified by Human Action* (1864), generated one of the first discussions of the problem. The author, fittingly for a New Englander, went against the current. In the midst of the great expansion, he warned of the dangers of conquest. Humans, he said, were capable of destroying the earth. The ancient civilizations had destroyed the land around the Mediterranean. North Africa, once the granary of the Roman Empire, was a desert, and the rocky hills of Italy and Greece the bare bones of once-rich lands. Modern civilization was repeating these follies at a faster pace and on a larger scale. We are "breaking up the floor and wainscoting . . . of our dwelling, for fuel to warm our bodies and seethe our pottage." The crisis was upon us, and "the world cannot afford to wait until the slow and sure progress of exact science has taught it a better economy." We must appeal to the "[m]any practical lessons [that] have been learned by the common observation of unschooled men."[45] In 1865 Australian newspapers cited Marsh in articles about the contributions forests made to civilization and the need to preserve them.[46] New Zealand legislators appealed to his book in debates over forestry bills in 1868 and 1873, and W. T. L. Travers, a member of Parliament, drew heavily on Marsh for a national lecture series.[47]

---

[44] On American conservation see Hays, *Conservation and the Gospel of Efficiency*. Joseph Powell treats Australian conservation in *Environmental Management in Australia* and two state studies, *Watering the Garden State* (Sydney: Allen & Unwin, 1989) and *Plains of Promise, Rivers of Destiny* (Brisbane: Boolarong, 1991). Canada and New Zealand have been less thoroughly treated in the historical literature.

[45] George Perkins Marsh, *Man and Nature, or Physical Geography as Modified by Human Action* (1864; reprinted Cambridge, Mass.: Harvard University Press, 1965), 52.

[46] Powell, *Watering the Garden State*, 67.

[47] David Lowenthal, Introduction to Marsh's *Man and Nature*, xxi–xxiii. Galbreath, "Colonisation, Science, and Conservation," 186–192; Powell, *Watering the Garden State*, 67–70.

The experience of the next generation added weight to Marsh's indictment. Concern was greatest in the United States, where settlement was most rapid and thorough. In 1860 the frontier line was hardly across the Mississippi, the lumber industry was centered in the Great Lakes region, the plains were filled with buffalo, and there seemed land for all. The census of 1890 showed no frontier line; the lumber barons were on the West Coast; the buffalo were bones; and westerners were calling for federal irrigation projects to make farmland from the desert. Conservation might have remained a matter of discussion or the impetus for some disconnected initiatives but for Gifford Pinchot. One of the first Americans trained in forestry (in Europe), he made conservation his cause and, as head of the U.S. Forest Service, made it a national crusade. With the help of his friend Theodore Roosevelt, who became president in 1901, he expanded the national forests and sent his rangers out to manage the land and spread the gospel. To do that, he tied the work to American dreams. Conservation, he said, would preserve economic opportunity and democracy in an era of large corporations. It would rationalize the use of resources, guaranteeing continued economic expansion for the nation and opportunities for individuals.[48] The rhetorical peak came in 1908, at a national conservation conference held at the White House and attended by the governors of the American states and delegates from Canada, Mexico, and Newfoundland. Pinchot fell out with William Howard Taft, Roosevelt's successor, and was fired, but the Forest Service and conservation remained, as did the schools and professional societies dedicated to scientific management.[49]

The other countries considered and discussed American actions, but none did as much. Their programs were less ambitious and their agencies, which were smaller and had fewer resources and less authority, focused more on public education than on administration.[50] One reason was that the sense of crisis that drove the Americans was lacking. Another was that conservation programs required more money and expertise than could easily be found. Finally, the United States had the advantages of a more centralized government and an extensive federal public domain. Australia was a set of independent colonies until 1901, and its

---

[48] Gifford Pinchot, *The Fight for Conservation* (1910; reprinted Seattle: University of Washington, 1967), gives a good sample of Pinchot's rhetoric and ideas.

[49] On the persistence of conservation two older but still useful volumes are Elmo Richardson, *The Politics of Conservation, 1897–1913* (Berkeley: University of California Press, 1962), and Donald C. Swain, *Federal Conservation Policy, 1921–1933* (Berkeley: University of California Press, 1963).

[50] For examples, see J. M. Powell, *An Historical Geography of Modern Australia* (Cambridge University Press, 1988), 151, 161.

new constitution left Crown land and natural resources to the states. In Canada federal powers were almost as limited, provincial ones almost as expansive.[51] New Zealand had a central government but few people and fewer resources for ambitious programs, particularly when there seemed no great problems.

The influence of American ideas, though, was evident. Canada had seen a few, disconnected conservation initiatives in the early 1900s, but it formed its Commission on Conservation only in the wake of the White House conference, which had recommended such action. The agency was the creation and tool of Clifford Sifton, minister of the interior from 1896 to 1905, who lobbied for it and headed it until he retired in 1921 (it was disbanded three years later). It showed, though, the limits of Canadian concern. It had no policy powers and worked mainly to reduce waste, expand forest reserves, and raise the professional level of the dominion and provincial forest services. The public showed little enthusiasm for more, even when Sifton invoked the specter of Americans buying up Canadian resources and stripping the country.[52]

In Australia, as in Canada, the belief that there were still land and resources for the taking discouraged conservation. The states established forestry commissions, but they were small, poorly staffed, and had few powers.[53] Only Victoria, a compact state with more intensively farmed land than the others, developed a comprehensive program. It was developing close ties with California in this period, and it looked across the Pacific for its inspiration.[54] Alfred Deakin, a prominent state, then national, politician, was, as head of the 1884 Royal Commission on Water Supply, the Moses of Victorian water development (as an editorial cartoon depicted him). He toured the United States and not only took ideas home but compiled *Irrigation in Western America*, which "became a standard work in the United States as well as Australia."[55] When the state's ambitious plans proved too much for local engineers, the Victorian government hired an American, Elwood Mead. He headed the state's Rivers and Water Supply Commission from 1907 to 1915,

[51] Bruce W. Hodgins et al., *Federalism in Canada and Australia* (Peterborough: Frost Centre, 1989). Any textbook on environmental or natural resource law will give a basic discussion.

[52] D. J. Hall, *Clifford Sifton* (Vancouver: University of British Columbia Press, 1985), 236–52; Commission on Conservation, Canada, *Annual Report* (Toronto: various printers, annual).

[53] John Dargavel, *Fashioning Australia's Forests* (Melbourne: Oxford University Press, 1995), 61–7, provides an overview.

[54] On this see Ian Tyrrell, *True Gardens of the Gods: California–Australian Environmental Exchanges, 1860–1930's* (Berkeley: University of California Press, 1999).

[55] Powell, *Watering the Garden State*, 104–5.

supervising the construction of Victoria's water system, establishing his agency as a professional bureau, training the first generation of Australian water experts, and shaping water management and administration throughout the country.[56] The only hints of a national program were suggestions that the Murray–Darling River system be developed, and two generations would pass before that began.

Conservation in New Zealand was even more a matter of debate than action. Discussing Premier Vogel's forestry bills in the 1870s, some legislators pointed to Marsh's *Man and Nature* and warned of coming shortages. Others spoke of the inexhaustibility of the "bush" or thought it, like the Maori, destined to fade before white civilization. There were complaints that conservation meant "locking up" resources (a charge also heard in Washington and Ottawa). The post of conservator of forests was established, allowed to lapse when Vogel left office, and renewed only when he returned in 1884. There was, as in Australia, considerable interest in exotic trees, but this was plantation forestry, not reforestation.[57]

In its view of nature – as resources – conservation looked back to the era of conquest. In other ways it looked forward, if not always with a clear view. Its concern for multiple use implied connections, consequences, and the balancing of various goods. Unlike earlier programs of development, it sought not immediate gain but sustained production, and it saw social goods as something more than a consequence of enriching individuals. It also called for expert knowledge in a way that expansion had not. Rather than just finding resources that would then be developed, it would guide the pace and direction of development. Conservation, though, was not environmentalism. It did connect nature to society, but through common sense, and its goal was primarily more (if more careful) development. Environmentalism appeals to biological science and is concerned with changing our lives and ways of living to bring them into line with nature's imperatives.

The Structures of Knowledge

Even as natural history's perspective was reaching its limits in helping the settlers understand nature, it was reaching its institutional and orga-

---

[56] Ibid., shows various influences in Victoria; on Mead, 150–67. See also Powell's "Protracted Reconciliation: Society and the Environment," in Roy MacLeod (editor), *The Commonwealth of Science: ANZAAS and the Scientific Enterprise in Australia, 1888–1898* (Melbourne: Oxford University Press, 1988), which offers a description of the parallels between North American and Australasian conservation.

[57] Galbreath, "Colonisation, Science, and Conservation," 188–9; David Thom, *Heritage: The Parks of the People* (Auckland: Lansdowne, 1987), 77–9.

nizational apogee. In the second half of the nineteenth century the infrastructure of national science matured rapidly, except in Canada, where the much larger U.S. institutions remained the focus of organization and information for another century. In 1851 the Americans had followed the British example by forming a national umbrella group, the American Association for the Advancement of Science.[58] The Australians and New Zealanders followed, working on a smaller scale. In 1844 Tasmania formed the first of the new "Royal" societies. Victoria followed in 1859, New South Wales in 1866, South Australia in 1880, and Queensland in 1884. Except in Western Australia, whose organization was not granted the title "Royal" until 1914, and in New South Wales, unaccountably slow off the mark, the dates reflected settlement and population growth. In New Zealand the institutions established during the first generation dwindled or died, but in 1876 the New Zealand Institute was established on a permanent basis. It was authorized to conduct a geological survey, published its proceedings, and became the Royal Society of New Zealand in 1933. The combined scientific societies of the southern colonies formed, in 1888, the Australasian Association for the Advancement of Science.[59] This was also the time of the great public natural history museums, which sprouted in every major settler city.[60] In the United States the Smithsonian Institution (1846) and the American Museum of Natural History in New York (which dates as a museum from the 1870s) were the flagships. Sydney's Colonial Museum, founded 1827, became the Australian Museum. Queensland established a museum in 1855, Victoria and South Australia in 1856. Only Tasmania, 1891, and Western Australia, 1892, lagged.[61] In Wellington the Colonial Museum, established in 1865, became part of the New Zealand Institute two years later. Four other cities founded museums in the next decade

[58] On its roles, particularly as a center of international science, see Debra Lindsay, *Science in the Subarctic: Trappers, Traders, and the Smithsonian Institution* (Washington, D.C.: Smithsonian Institution Press, 1993); Carl Berger, *Science, God, and Nature in Victorian Canada* (Toronto: University of Toronto Press, 1983), 5, 23–7.

[59] Colin Finney, *Paradise Revealed* (Melbourne: Museum of Victoria, 1993); C. A. Fleming, *Science, Settlers and Scholars* (Wellington: Royal Society of New Zealand, 1987); Roy Macleod, "Organizing Science under the Southern Cross," in Roy Macleod (editor), *Commonwealth of Science* (Melbourne: Oxford University Press, 1988), 19–39.

[60] Susan Sheets-Pyenson, *Cathedrals of Science* (Kingston: McGill-Queens University Press, 1988). See also Sally Gregory Kohlstedt, "International Exchange and National Style: A View of Natural History Museums in the United States, 1850–1900," in Nathan Reingold and Marc Rothenberg (editors), *Scientific Colonialism* (Washington, D.C.: Smithsonian Institution Press, 1987), 167–90.

[61] Anne Moyal, *"A Bright and Savage Land": Scientists in Colonial Australia* (Sydney: Collins, 1986), 87–105. There is a full account of Queensland's Museum in Patricia Mather, *A Time for a Museum* (South Brisbane: Queensland Museum, 1986).

– Christchurch, Nelson, Auckland, and Dunedin. Canada had major museums in Montreal, Quebec, and Ottawa.[62]

Each building was a "cathedral of science" and a monument to the country's (or colony's) cultural progress, but they showed even in their designs the growing divide between the public and science, and the science they proclaimed was dissolving. Their visible justification was the vast exhibition halls that were packed with specimens, skeletons, "native" artifacts, and dioramas showing nature past and present, at home and abroad. These drew appreciative crowds and hordes of schoolchildren on tour.[63] Behind these halls, inaccessible to the public, were cases upon cases of preserved birds, animals, fish, and plants, the specimens that were the museums' scientific contribution. The scientists, though, were becoming less interested. At least for the larger forms, the great adventure of taxonomy had become a matter of filing and a few minor disputes about classification. Natural history itself was dissolving into specialized fields as the accumulation of information required specialization and the development of new theories and research methods encouraged it. Geology was the first to establish its independence, followed by ornithology, entomology, piscatology, herpetology, mammalogy, and a number of other even more obscure fields.[64] Observation of organisms in the field was yielding to the study of their internal processes by chemical and physical methods and by observation at the microscopic level, and even in the field, attention was shifting from organisms to their interaction, from observation to statistical and experimental studies.

This changed the public's relation to the culture's knowledge, and it affected the ability of the smaller settler societies to contribute to our understanding of nature. Natural history had been accessible even to a person with little education. Wallace had made himself into a more than competent naturalist, had indeed reached to theoretical heights, without formal training. Darwin's ideas were accessible, if not always welcome. Even the smaller countries had supported amateur societies,

---

[62] On the museum movement see Sheets-Pyenson, *Cathedrals of Science*, 8–23.

[63] On Australian collections see P. J. Stanbury, "The Discovery of the Australian Fauna and the Establishment of Collections," in Bureau of Flora and Fauna, *Fauna of Australia* (Canberra: Australian Government Publishing Service, 1987), 1A: 202–6; Sybil Jack, "Cultural Transmission: Science and Society to 1850," and Ian Inkster and Jan Todd, "Support for the Scientific Enterprise, 1850–1900," both in R. W. Home (editor), *Australian Science in the Making* (Cambridge University Press, 1988), 45–66 and 102–32. On the general process see Sheets-Pyenson, *Cathedrals of Science*.

[64] Ornithology is a good example. See Paul Lawrence Farber, *The Emergence of Ornithology as a Scientific Discipline, 1760–1850* (Boston: Reidel, 1982). See also Mark V. Barrow, Jr., *A Passion for Birds* (Princeton, N.J.: Princeton University Press, 1998).

then more professional ones, then museums. The new fields dealt with nature on a level people could not see or easily imagine, required specialized training and laboratory equipment, and used concepts that were far from popular understanding. This made careers like Wallace's or Walter Buller's impossible. More important, the new fields cost much more than natural history.[65] Of the group, only the United States had the resources and population to support the expensive infrastructure that was required.

In some ways the United States had advantages over even some of the European countries. Science, field studies included, was moving into the universities, and the United States had and was building an extensive system. Many institutions modeled or remodeled themselves on German research schools. There was also the unique set of land-grant colleges and their associated research stations (which supported field biology on a scale unequaled elsewhere). The country's booming industrial economy and a developing system of private philanthropy created other academic niches. The Carnegie Institution, for instance, established one of the first ecological research stations in the world. The University of Chicago, which became a center for ecological research, was largely the creation of John D. Rockefeller. Miss Annie Alexander, amateur collector and heir to a Hawaiian sugar fortune, almost single-handedly funded what became the premier institution of field studies on the West Coast, the University of California's Museum of Vertebrate Zoology. Americans had been an important influence in conservation. In ecology they would be at the center.

Toward the Nature of the New Lands

As the settlers tried to understand their lands they also asked what place they had in them and in their societies, and as the century ended this question came to the fore. The land had been a place to conquer and transform, and as that process ended the settlers came to see their lands as having formed them. The identification of settlers with the country was less obvious in the United States, where it had been going on since independence, than in the Commonwealth countries, but it was a general movement. Plants, animals, landscape, even climate came to be seen as part of a complex that identified and shaped the settler society.

---

[65] The process of professionalization, though, even in established fields, went on at different speeds in the different societies. A useful study here is Ray Wallace, "A Body in Twain? The Royal Australasian Ornithological Union and the Events of 1969" (unpublished M. A. thesis, University of Melbourne, 1990).

People also turned to nature – outdoor recreation became more popular – and learned about it in nature literature, elementary school classes, and new or revived amateur natural history groups. Governments established national parks to preserve scenery or provide open space. This was the counterpoint to the disappointments of conquest – a new direction for thinking about the land, people's place in it, and its place in them.

4

NATIONAL NATURE, 1880–1920

During the great expansion the settlers took possession of the land and converted it to their own ends. Now they began to examine more consciously what effect the land had on them, to incorporate the land into their culture, even to call themselves "natives" and find other names for the people they had dispossessed.[1] Conspicuous or emblematic plants and animals became national symbols; the kiwi in New Zealand and the kangaroo in Australia took center stage. Landscape paintings, especially in Australia and Canada, became important elements of a national vision. The settler's nature literature, too, focused more closely on what was around them. Outdoor recreation, in forms ranging from hiking to bird-watching, became part of urban, middle-class leisure. Nature education became more popular, informally in youth organizations like the Boy Scouts but also in the schools. Governments took new steps to protect nature, setting aside rural or wilderness areas for public recreation and providing some protection for at least a few favored species of plants and animals.

People had always forged ties to their lands, but the settlers were doing it under new conditions. The Australian Aborigines and the American Indians had had many generations to explore and settle their lands – even the Maori in New Zealand had had a few centuries. The Anglos had done most of their conquering in the span of a single lifetime, and even the oldest of their settlements was, comparatively speaking, still very young. The earlier peoples had been largely isolated once they arrived; the Anglos had easy communication with the cultural hearth and each other, and it became easier as the century went on. They also transformed the land far more thoroughly and obviously than

---

[1] On Australia as an example see J. W. Gregory, *Australia* (Cambridge University Press, 1916), 45, and Edward Jenks, *A History of the Australasian Colonies* (Cambridge University Press, 1912), 16. Language was changing in the other nations as well.

earlier cultures. They faced it, too, with a new kind of knowledge, a formal framework that tied their local and immediate experience to the Anglo and European worlds.

Nature and Nationalism

Europeans had long associated the new lands with their plants and animals. The beaver was on the armorial bearings Charles I granted Sir William Alexander, baronet of Nova Scotia, in 1633, and the kangaroo and the emu were on the coats of arms of virtually all the Australian colonies.[2] In the late nineteenth century, though, native species more clearly became national symbols. The shift was particularly apparent in Australia and New Zealand. The Revolution had given the United States a political identity, and Canadians' reaction against that change gave Anglo Canada one of its own. (Into the mid-twentieth century, at least, Canadian schoolchildren learned about invasions from the United States and were taught to "Remember Lundy's Lane," a slogan from one of these expeditions against their country.) The southern nations lacked a political focus or a military tradition, and as they came to see themselves as something more than British plantations on a far shore, they found in native nature something to distinguish them within the Empire. In Australia "wattle" and "gum" (acacia and the dominant eucalyptus) became prominent; the Victorian schools established, along with Bird Day and Arbor Day, "Wattle Day."[3] The kangaroo, which had shared top billing with the emu, came to the fore (in part because it was easy to anthropomorphize). Editorial cartoons showed stern 'roos with rifles and slouch hats marching off to defend the Empire. They also appeared, less martially, in advertisements, where they pounded pianos and drank beer, brewed up (tea in a billy can), and washed up.[4]

In New Zealand native birds took pride of place. Images from Walter Buller's monumental volumes were printed, reprinted, and pirated. The huia, a bird emblematic of Maori nobility, was popular, and it received a boost in 1890 when the son of Lord Onslow, the governor-general, received the name Huia in a Maori ceremony. This, as historian Ross Galbreath noted, was great empire copy: "the fair child of a noble

[2] Jim Cameron, *The Canadian Beaver Book* (Burnstown, Ontario, Canada: General Store Publishing, 1991), 17–18. On Australia see R. M. Younger, *Kangaroo Images through the Ages* (Melbourne: Hutchinson, 1988).

[3] Tom Griffiths, "'The Natural History of Melbourne': The culture of Nature Writing in Victoria, 1880–1945," *Australian Historical Studies*, 23 (October 1989), 359; Richard White, *Inventing Australia* (Sydney: Allen & Unwin, 1981), 117–19.

[4] Younger, *Kangaroo Images*, 129–53. On Australian nationalism of the period see White, *Inventing Australia*, 85–119.

English house taking his place at the head of a dusky tribe, amid curious native customs."[5] The huia, though, was vanishing, and by the early
twentieth century was extinct. This was not an insuperable obstacle to
national fame. The moa, which had been extinct before the Anglos
arrived, was a popular emblem; it and the kiwi often appeared on entries
for a national coat of arms. Still, it did not help. In any event, the prize
committee rejected the native birds, choosing the Southern Cross, the
silver fern, and a heraldic fleece. That was only official sentiment.[6]
Across the nation the kiwi swept the field. This was not because people
knew it by sight. A small, secretive bird, living on the forest floor and
active mainly at night, it was heard far more often than seen. It was,
however, distinctive, lived only in New Zealand, and was easy to draw.
Like the kangaroo it came to represent the country in editorial cartoons,
advertisements, and commercial logos. (The most famous use, by an
Australian shoe polish firm, was the result of one company director's
being married to a New Zealander.) Today the national rugby team, the
All-Blacks, puts the silver fern on its jerseys, but New Zealanders refer
to themselves as "kiwis," financial reporters on television report the
movement of "the kiwi" against the Aussie dollar and the greenback, and
the roundel of the Royal New Zealand Air Force has a kiwi silhouette in
the center. (The oddity seems not to bother New Zealanders.)[7]

The settlers also pressed landscape and climate into use, and in
Australia and Canada these became central to national myths. Australia
had entered European consciousness as a topsy-turvy land, full of
strange marvels, and the experience of living on the land and being
changed by it had been a theme in Australian literature and stories
about Australia since the early nineteenth century. In children's stories,
where it was common, the protagonists had at first been British. Toward
the end of the century, "that degenerate native product," the cousin
from Sydney, replaced the "Pom," but the moral remained the same: the
"bush" either broke the "new chum" or made a man of him.[8] Between
1880 and 1900 a group of writers, mainly Sydney journalists, used the
record and the stories of three generations to make life in the bush the
theme of what became the national popular literature. In prose and
poetry men like A. B. "Banjo" Paterson and Henry Lawson created the

---

[5] Ross Galbreath, "Colonisation, Science, and Conservation" (unpublished Ph.D. dissertation, University of Waikato, 1989), 347; idem, *Walter Buller: The Reluctant Conservationist*
(Wellington: GP Books, 1989), 159–66.

[6] Galbreath, "Colonisation," 223.

[7] Galbreath, "Colonisation," and the author's observations.

[8] Brenda Niall, *Australia Through the Looking-Glass: Children's Fiction, 1830–1980* (Melbourne: Melbourne University Press), 6, 2. On the general use of nature for nationalism in Australia in this period see White, *Inventing Australia*, 85–105.

"Australian legend." The outback was the "real" Australia, and the cities (where the bush writers held their jobs) but poor and trifling places. Those who lived and worked in the "backblocks" – the drovers, swagmen, small farmers, even squatters – were the true Australians. "Mateship," the ties between men who depended on each other in a harsh land, they made the national creed, endurance the national virtue. Paterson's "Waltzing Matilda," set to music, became an unofficial national anthem, and poems like "The Man from Snowy River" set pieces for recitation. (I heard Paterson's "Clancy of the Overflow" among the canned music on an interstate bus in 1992.) The most famous literary piece, though, was not their doing. That was Dorothea Mackellar's poem "My Country," which became a classroom staple, memorized by generations of school-children. It was an aesthetic declaration of independence. The first verse describes the landscape some prefer, a well-watered English scene. The "second verse, beginning 'I love a sunburnt country,' is perhaps the most quoted piece in Australian poetry." It describes the author's delight, the Australian countryside, and ends: "the wide brown land for me." The legend took on new dimensions as Australian troops went off to fight for the Empire; after the Boer War and World War I, the bush was hailed as the school of Australia's soldiers.[9]

It was the paintings of the Heidelberg school, though, that most shaped the popular Australian view of the land. The group had come together in Melbourne in the 1880s, taking their name from a summer camp they had staked out near the city. Some had been born in the country, others had immigrated, some had been trained in Europe, others in Australia, but they were united in an artistic project: to use the fashionable techniques of French Impressionism and painting *en plein air* to depict the Australian bush. Showing the land by the conventions of European art was not a new project; every generation from the convicts on had taken its turn. Nor did these artists have a unique ability to "see" the country or more technical skill in portraying it. After the experiments and adjustments of the first generation, landscape images were recognizably Australian, not distortions of European scenes. The Heidelberg artists, though, worked at a time when Australians were

[9] Russel Ward, *The Australian Legend* (1958; 2d ed., Melbourne: Oxford University Press, 1978), is the standard. On the city and the bush legend see Graeme Davison, "Sydney and the Bush: An Urban Context for the Australian Legend," in John Carroll (editor), *Intruders in the Bush* (Melbourne: Oxford University Press, 1982), 109–30. On literary nationalism and the Australian myth see John Barnes (editor), *The Writer in Australia* (Melbourne: Oxford University Press, 1969). Elizabeth Ward, "A Child's Reading in Australia," *Washington Post, Book World*, 4 November 1990, 17, 22, comments on the ubiquity and influence of the poetry into the 1950s. Quote on Mackellar from Sarah Dawson (editor), *The Penguin Australian Encyclopedia* (Melbourne: Penguin, 1990), 313.

looking for beauty in the land, and their canvases did catch Australian light, convey the feel of the country, and show the outback life as the foundation of the nation and the national character. Their paintings became icons, "mediating the relation to the bush of most people growing up in Australia. . . . Perhaps no other local imagery is so much a part of an Australian consciousness and ideological make-up."[10] Their popularity peaked in the 1920s, but their images are still part of the national consciousness. The originals hang in major national collections and cheap reproductions in hotel dining rooms. Still cruder versions decorate the covers of paperback copies of Lawson's stories and Paterson's verse.

How they did it and what came of it is a study in the adaptation of European ideas to local conditions. The French Impressionists had worked in the soft light of dawn and twilight, or at most the sunshine of northern European summer. The Heidelberg painters had to deal with the harsh light of noon in a sun-filled land. Instead of the verdant colors of a European spring or summer, they had to capture the eucalyptus's dusty, washed-out greens and grays, the soil's harsh yellows and reds. They did. Paintings like Arthur Streeton's *The Purple Noon's Transparent Might* and *Road Cutting on a Hot Day* or David Davies's *Golden Summer* showed a land shimmering under an overpowering sun. They were celebrations of light, heat, and space as elements of the country.[11] They also took a new stance. Paintings like Eugene von Guerard's *Mount Kosciusko Seen from the Victorian Border* (1866) and W. C. Piguenit's *Mt. Wellington from New Town Bay* (1879) looked out on the country from a height or spread it before the viewer as a "scene." Some of their paintings had this point of view, but most, and most of the best known, did not. They were immersed in it. There was no "magisterial gaze." Vegetation fills the canvas.[12] Tom Roberts's *Wood Splitters* (*the Charcoal Burners*) (1885), for instance, shows three men in a clearing splitting and stacking wood. Behind, the bush closes in. The darkness in these paintings

---

[10] Ian Burn, "Beating about the Bush: The Landscapes of the Heidelberg School," in Anthony Bradley and Terry Smith (editors), *Australian Art and Architecture* (Melbourne: Oxford University Press), 83.

[11] Almost all the paintings mentioned here can be found in William Splatt and Susan Bruce, *Australian Landscape Painting* (Melbourne: Viking, O'Neil, 1989), and William Splatt and Barbara Burton, *Masterpieces of Australian Painting* (Melbourne: Viking, O'Neil, 1987). See also Curry O'Neil, *Classic Australian Paintings* (Melbourne: Curry O'Neil Ross Pty, 1983). *Road Cutting on a Hot Day* is in the collection of the National Museum in Canberra.

[12] Albert Boime, *The Magisterial Gaze* (Washington, D.C.: Smithsonian Institution Press, 1991). A longer discussion of landscape paintings and nature can be found in Barbara Novak, *Nature and Culture* (1980; rev. ed., New York: Oxford University Press, 1995).

is emotional as well as physical. Frederick McCubbin's *On the Wallaby Track* (the title is a euphemism for looking for a job) shows a man, his wife, and their child camping in the bush as they trek from one station to the next. *Bush Burial* depicts a family's loss and grief. *Down on His Luck* is a study of a tired swagman brewing up at the end of the day. *The Pioneer* presents this life in triptych. On the left a young couple camp in the forest as they prepare the homestead. In the center is the young family, the husband sitting on a log talking to his wife, who holds an infant. Smoke curls up from the cabin in the clearing behind them. The right-hand panel shows a man clearing the encroaching bush from their graves. The scene opens to a view over a bay to a city in the distance, but that is the only indication of progress.

The dangers of the bush, and the feeling of enclosure, even entrapment, are most intense in paintings of lost children, images that mixed Victorian (period) sentimentality, with the reality of the colony of Victoria. The population was widely scattered, inexperienced people had ready access to the bush, and it was easy to get lost; newspapers of the period are full of stories of children lost and found, or lost forever. The best-known painting in this genre is Frederick McCubbin's *Lost* (1886). We look through a thin screen of brush at a young girl, dressed for a picnic, standing in tawny grass and wiping her eyes. All around is a chaos of vegetation. Only in one corner can we see the sky. There is a suggestion of a happy ending – a broken twig in the foreground that trackers can use (Aborigines were sometimes used in these searches) – but only a suggestion.

The painters' bush was, like any representation, selected from a larger reality. It was not the whole of Australia, but the coast where the Anglos lived and the inland pastures where their sheep grazed. The Centre had entered national mythology with the Burke and Wills expedition of 1864, becoming the place where, as a school geography book of 1903 put it, was recorded "our forefathers' deeds of courage and self-sacrifice."[13] The Heidelberg artists, however, did not look at the rock and sand deserts and the salt pans that lay beyond the inland pastures. Nor, though they depicted hardship, did they show the two great realities of interior settlement, fire and drought. Australia is the only country where paintings of great fires hang in national galleries, and a title like *Gippsland, Sunday Night, Feb. 20th, 1898* was immediately understandable – and for many in that generation evocative.[14] The great dry spells that

[13] Charles R. Long, *Stories of Australian Exploration* (Melbourne: Whitcombe & Tombs, 1903), Preface. Explorers continue to occupy a large part in the Australian pantheon. Cook appears on the ten-dollar bill, Douglas Mawson, who explored Antarctica, on the hundred.
[14] Stephen Pyne, *Burning Bush* (New York: Henry Holt, 1991), deals with the conditions that created these fires. Tom Griffiths, *Secrets of the Forest* (Sydney: Allen & Unwin, 1992),

shriveled the bush, killed the sheep, and left tons of dead rabbits piled against barrier fences are only suggested, in some pictures of wind-blown settlers, and these are not the canvases that have become popular favorites.

Their subjects and treatment of them were much like that of their American western contemporaries, men like Frederic Remington and Charles Russell. All used the accepted techniques of the period, and an art historian can trace changing fashions in at least some of the painters' work. All, and this includes the Canadians discussed later, show the rejection of science as a guide to art. Church's mid-century concern with life-zones and accuracy in the painting of vegetation had been exiled to scientific illustrations. There were some clear parallels in the focus on outdoor work and frontier life. Tom Roberts's *The Breakaway*, showing a rider turning a flock of sheep, is much like American cowboy paintings, and his *Bailed Up*, depicting an Australian stagecoach robbery, only needs different trees and "Wells Fargo" on the coach to pass for American. The Australians, though, celebrated labor and laborers in a way the Americans did not. Paintings like *Shearing the Rams* showed honest toil, while the Americans focused on danger – stampedes, accidents, and bucking broncos. The "West" was a place of conflict, of nature and the Indians conquered. The "bush" was loved and feared but not "conquered." Indians were central to many Western canvases, Aborigines conspicuous by their absence. In Australia animals were symbols, not subjects, while American art abounded with deer, elk, grizzly bears, and buffalo. On the other hand the Australians made plants central to many of their works, and the Heidelberg school made the eucalyptus "a dominant symbol of the Australian bush."[15]

"Everybody," Canadian historian Carl Berger said, "talks about the weather and the climate; seldom have these been exalted as major attributes of nationality."[16] In Canada they were. Appealing to the classical tradition linking climate and temperament, Anglo-Canadians saw their harsh climate as a guarantor of civic virtue. They made, Berger noted, a close association between snow and civil liberty.[17] This is hardly

---

45–57, provides an overview of the meaning of fire in Australia and an analysis of the Victorian fires of 1939; this kind of natural disaster did not vanish with the pioneers' death.

[15] Smith, "The Artist's Vision of Australia," 163.

[16] Carl Berger, "The True North, Strong and Free," in Peter Russell (editor), *Nationalism in Canada* (Toronto: McGraw-Hill, 1966), 4. This analysis draws on Berger's book, as well as Cole Harris, "The Myth of the Land in Canadian Nationalism," in Russell, *Nationalism in Canada*, 27–43, and David Heinimann, "Latitude Rising: Historical Continuity in Canadian Nordicity," *Journal of Canadian Studies, 28* (1993–4), 134–40.

[17] Berger, "True North," 22. For a recent statement of this see Robertson Davies, *The Merry Heart* (New York: Viking, 1997), 180. On climate, temperament, and virtue in Western

surprising. Canada's climate had shaped settlement in obvious ways. The country's true frontier was the North, the Arctic, which the annals of nineteenth-century exploration had fixed in popular memory as a place of death and disaster.[18] Nature was not a refuge or place of spiritual enlightenment. One scholar described the nation from its literature as a people living in a fortress, huddling together in fear of the wilderness. Another, Margaret Atwood, found the theme of death in nature so pervasive she titled her survey of Canadian literature *Survival*.[19]

As in Australia, landscape painting was an important vehicle for national sentiment. The search for a national Canadian art was apparent in the 1890s, at the time the Heidelberg school was forming half a world away, and just before and during World War I a group came together around Tom Thomson, who painted hardwood and spruce forests, lakes, and the northern light in Algonquin Provincial Park in northern Ontario. In 1913 J. E. H. McDonald and Lawren Harris found another inspiration, in an exhibition of modern Scandinavian landscape art in Buffalo, New York. There they saw how the techniques of French Naturalism and Impressionism could be used to capture northern light and the look of northern wilderness. Thomson drowned in a canoeing accident in the park in 1917, but by then the techniques and ideas "or more properly the cult of Canadianism, symbolized by the Precambrian northern shield, and a sense of common purpose and kinship were generated which would hold these Toronto painters together."[20] Out

civilization see Clarence Glacken, *Traces on the Rhodian Shore* (Berkeley: University of California Press, 1967), 80–115.

[18] The centerpiece was the search for Sir John Franklin, his two ships, and more than a hundred crewmen, who vanished on an expedition to the Northwest Passage in 1845. Expeditions from several countries combed the Canadian Arctic for a decade, but established only that all had been lost. Sir John Franklin was lieutenant-governor of Van Diemen's Land, and his statue stands today in downtown Hobart, half a world away from his Arctic grave. An accessible account of the lure of the Arctic is Pierre Berton, *The Arctic Grail* (1988; reprinted New York: Penguin, 1988). See also Trevor H. Levere, *Science and the Canadian Arctic: A Century of Exploration, 1818–1918* (Cambridge University Press, 1993), 190–234.

[19] Gaile McGregor, *The Wacousta Syndrome* (Toronto: University of Toronto Press, 1985); Margaret Atwood, *Survival* (Toronto: Anansi, 1972).

[20] J. Russell Harper, *Painting in Canada* (Toronto: University of Toronto Press, 1977), 273. Recent assessments are those of Michael Tooby (editor), *The True North: Canadian Landscape Painting, 1896–1939* (London: Barbican Art Gallery, 1991), written to accompany the exhibition at the Barbican Art Gallery, which has extensive reproductions, and critiques of the Gallery of Ontario Exhibit, "The Group of Seven: Art for a Nation," Margaret Rodgers, "Exploring the Quintessential," and Matthew Brower, "Framed by History," the last two in *Journal of Canadian Studies, 31* (Summer 1996), 174–8 and 178–82.

"of trees, rocks, and Lakes ... [they] established the basic symbols of national identity."[21] In 1920 they formed the Group of Seven.

Like the Heidelberg painters, they used a particular landscape to symbolize a national ideal. It was not the landscape of settlement, however, but a part of their land that Canadians had not settled or tried to settle – the Shield. From Hudson's Bay to the Great Lakes and from western Ontario to the edge of the prairies, the foundation rock of the continent lies on the surface. There is little soil, and the lack of drainage means that the land is laced with lakes, streams, and bogs. The forest consists of spruce – short and short-lived trees, twisted by the wind and winter. It is not like the rest of the country, but the Group's paintings show it as a psychological landscape Canadians could identify with. They emphasize the stunted trees struggling for a foothold amid enormous boulders, the cold, and the snow. There are neither settlers nor Indians in their canvases. This is the "unconquerable North." The Group's paintings now have pride of place in the National Gallery in Ottawa and adorn the covers of paperback collections of Canadian literature.[22]

Nature into Culture

Nature literature was also part of the search for national identity. In North America the classic works of nature writing belong to these years, the works of John Burroughs and John Muir – John o' the Birds and John o' the Mountains – the one writing of rural nature in New York's Catskills, the other of the western mountains. In the late 1890s two Canadians, Ernest Thompson Seton and Charles G. D. Roberts, made North Americans' first distinctive addition to the development of the genre, the "realistic" animal story, and became popular figures on both sides of the border.[23] Nature stories and essays were then not set apart. They appeared in general-circulation magazines, and Muir and Burroughs became the first nature celebrities. Burroughs went camping and tramping with President Roosevelt and wrote about the experience for a national audience. His house, "Slabsides," became a place of pilgrim-

---

[21] Roald Nasgaard, *The Mystic North* (Toronto: University of Toronto Press, 1984), 158.

[22] Gaile McGregor finds the closing in of the landscape, in contrast to the more open and elevated view in American paintings, a consistent theme from the late eighteenth century. Gaile McGregor, *EcCentric Visions* (Waterloo: Wilfrid Laurier University Press, 1994), 289–91. On the landscape and the Canadian Indians, as well as the school's relation to Canadian nature, see Jonathan Bordo, "Jack Pine – Wilderness Sublime or the Erasure of the Aboriginal Presence from the Landscape?" *Journal of Canadian Studies*, 27 (Winter 1992–3), 98–128.

[23] A list of Anglo classics can be found in Edwin Way Teale, "The Great Companions of Nature Literature," *Bird-Lore, 46* (November–December 1944), 363–6. It reflects that period, but is a good start for a long view of the genre.

age. Muir accompanied Roosevelt through Yosemite, and his Sierra Club attracted San Francisco's elite. In Australia a smaller movement developed around a Melbourne journalist, Donald Macdonald, and there were a number of children's nature stories treating native nature, mainly by women (these are still being reprinted).[24] In New Zealand, which had fewer people, and consequently fewer writers and a much smaller market, the principal outlet was newspaper nature notes, which were collected for school nature study texts. In the early 1920s two classic studies (discussed in Chapter 7) appeared, but they were not part of a national stream.

North American nature literature depicted a realm that was outside human society but one humans needed and hungered for. Nature was both a refuge and the gateway to transcendence. Wilderness, vaguely thought of as areas of "untouched" Romantic scenery, was its center and Muir its high priest. "Thousands of tired, nerve-shaken, over-civilized people," he wrote, "are beginning to find out that going to the mountains is going home; that wildness is a necessity; and that mountain parks and reservations are useful not only as fountains of timber and irrigating rivers, but as fountains of life."[25] Defending the Hetch-Hetchy valley against San Francisco's plans to use it for a reservoir, he spoke of it as a cathedral, its destruction by the dam not a bad policy decision but desecration. Nature was Eden; Muir and Burroughs both averted their eyes from things that smacked of chaos and cruelty – predation, starvation, and the enormous production of life cut short by violent death.[26] Even the "realistic" animal story, which commonly dealt with struggle and death, did not seek to alienate the reader from nature. Seton said that his stories taught a lesson "as old as Scripture – we and the beasts are kin."[27] The modern animal story, Roberts said, "helps us to return to nature. . . . It leads us back to the old kinship of earth."[28] Their work spoke to the classic goal of nature literature – including awe and wonder at the marvels of nature. Their great achievement was to do this in a Darwinian world, to show that the endless struggle and multiple deaths produced strength and skill.[29]

---

[24] See Naill, *Australia Through the Looking-Glass*, 4, on the situation in children's literature.

[25] John Muir, "The Wild Parks and Forest Reservations of the West," *Atlantic Monthly*, January 1898, 15, reproduced in William Cronon (editor), *Muir: Nature Writings* (New York: Library of America, 1997), 721.

[26] This can be seen in their works. For an examination of Muir see Lisa Mighetto (editor), *Muir Among the Animals* (San Francisco: Sierra Club Books, 1986).

[27] Ernest Thomson Seton, *Wild Animals I have Known* (New York: Scribner's, 1898), 11–12.

[28] Charles G. D. Roberts, *The Kindred of the Wild* (Boston: Page, 1902), 29.

[29] Thomas Dunlap, *Saving America's Wildlife* (Princeton N.J.: Princeton University Press, 1988), 18–33; idem, "'The Old Kinship of Earth,'" *Journal of Canadian Studies*, 22

The separation of nature and society was more marked than it had been. From Walden Pond, Thoreau had criticized the values his fellows put on work, money, and property. For Muir and Burroughs nature was a place in which to become refreshed for further toil, and one of the goals of Muir's Sierra Club was to make the mountains more accessible. (The club would drop this after World War II, recognizing that there were now many, or too many, roads.) They drew back even from the implications of their own defense of nature. Muir railed against the greed and short-sightedness that would destroy "nature's temples," but not the society and economy that produced and served this greed. He sought to save areas like Yosemite, but as islands around which progress might flow. These sentiments are characteristically North American. Thoreau had his admirers in Australia – one was the Melbourne nature writer Charles Barrett – but none wrote as if he wished "to speak a word for . . . absolute freedom and wildness, as contrasted with a freedom and culture merely civil, – to regard man as an inhabitant, or a part and parcel of Nature, rather than a member of society."[30] None echoed Muir's appeal to nature as a source of ultimate meaning or religious experience. A generation later the New Zealand ecologist Leonard Cockayne would wax poetic about the "sacred" reserves and the need to guard them "religiously," but his passion was for plants, not self-knowledge.[31] The realistic animal story did not travel well either. With the exception of a late Australian example (see Chapter 7), it remained a part of North American literature.

Australian nature writing had other goals and models, the result of settler experience and continuing ties with the metropolis. It did not seek to bring people into union with nature but emphasized the ties that made the land their home. Like most beginnings, the early works showed their origins as much as a new direction. Donald Macdonald, whose *Gum Boughs and Wattle Bloom* was the first major work of the Melbourne school, followed the lead of W. H. Hudson and Richard Jeffries, the popular British nature writers of that generation.[32] He spoke of the English birds and plants in Australia as reminders of "home" and depicted the village of his youth (then an agricultural town of 600, now a suburb of Melbourne) as a rural Arcadia. He drew his title, though,

---

(Spring 1987), 104–20. On the natural history aspects see Ralph Lutz, *The Nature Fakers* (Golden, Colo.: Fulcrum, 1990).

[30] "Walking," in Henry David Thoreau, *Walden* (1854; reprinted New York: Modern Library, 1937), 597.

[31] Leonard Cockayne, "Report on a Botanical Survey of Stewart Island," Department of Lands and Survey (Wellington: Government Printer, 1909), 41, 42.

[32] Donald Macdonald, *Gum Boughs and Wattle Bloom* (Melbourne: Cassell, 1888). This analysis draws heavily on Griffiths, "'Natural History of Melbourne,'" 339–65.

from the characteristic plants of the country, eucalyptus and acacia, and he saw the inhabitants of the village putting down roots in the land, as memory and experience gave them personal associations with the rhythms of nature in Australia.

Children's nature literature shows the same mix of Anglo form and Australian subject. The first of the classic children's stories was Ethel Pedley's *Dot and the Kangaroo* (1899), which tells of Little Dot, lost in the bush, who is saved and restored to her family by a mother kangaroo. The animals are humans in animal suits; there is a good deal of moral instruction; and kindness to animals is emphasized. Here, though, the animals are Australian, and the book is dedicated to "[t]he children of Australia, in the hope of enlisting their sympathies for the many beautiful, amiable, and frolicsome creatures of their fair land; whose extinction, through ruthless destruction, is being surely accomplished."[33] The most popular works came toward the end of our period; May Gibb's stories of the gumnut babies appeared as *Tales of Snugglepot and Cuddlepie* (1918), *Little Ragged Blossom* (1920), and *Little Obelia* (1921).[34] They are in the nursery book tradition of Victorian Britain but self-consciously and accurately Australian in their detail. The gumnut babies are drawn from gumnuts, the villains – the "bad Banskia men" – from that genus's fruiting cob, and botanists can identify the species Gibb used as models.[35] Like Pedley, she preached the preservation of native nature. Besides the use of Australian material there is the frontispiece of *Snugglepot and Cuddlepie*, which shows a gumnut baby writing on a leaf, "Humans, Please be kind to all Bush Creatures and don't pull flowers up by the roots."[36]

The settlers did not just find new meanings in nature; as they ceased to work in nature and live in the country, they played there with more intensity and attached more meaning to it. Middle-class people began

---

[33] Ethel M. Pedley, *Dot and the Kangaroo* (London: Thomas Burleigh, 1899). On an earlier writer, whose career provides a useful comparison, see Vivienne Rea Ellis, *Louisa Anne Meredith* (Sandy Bay: Blubber Head Press, 1979), 196–8. A North American counterpart is Thornton W. Burgess.

[34] May Gibb, *Tales of Snugglepot and Cuddlepie* (Sydney: Angus & Robertson, 1918). *Little Ragged Blossom* (1920) and *Little Obelia* (1921) were published by the same house. All are in print in cheap and deluxe editions, and in 1988 the Royal Australian Ballet Company put on a one-act ballet, *Snugglepot and Cuddlepie* (the costumes were on display in the National Gallery of Victoria, Melbourne, in 1990).

[35] Peter Bernhardt, "Of Blossoms and Bugs: Natural History in May Gibb's Art," in Royal Botanic Gardens, *Gumnut Town: Botanic Fact and Bushland Fantasy* (Sydney: Royal Botanic Gardens, 1992), 5–21.

[36] Gibb, *Snugglepot and Cuddlepie*. On nature protection through fantasy see Robert Holden, "Gumnut Town: Fact, Fantasy & Folklore," in Royal Botanic Gardens, *Gumnut Town*, 29–47.

taking vacations at the shore or in the mountains and sent their children to the country for the summer. They climbed mountains, hiked, skied, and swam. This was in part a matter of seeking a change from the daily round, but it had other purposes. Mountain climbing, like sport hunting, was an upper-class European activity that formed and tested character. Hiking, canoeing, and camping gave people experience of the pioneer life. Amateur nature study, which enjoyed a revival in these years, was self-improvement. This generation was probably no more entangled in its culture than earlier ones, but the dissonance between rhetoric and social context was more striking than usual. People learned about simple virtues and nature from magazines, joined clubs to celebrate the solitary life of the wild, took the trolley out to hike, fish, or walk in the woods.

The settlers paid little attention to these incongruities. They wanted a relationship with nature in the context of their lives, and if that required modern methods to find a primitive experience, so be it. The results are apparent in Australian hiking, or bushwalking, as it came to be called. Australia was even then an urban nation – critics bemoaned the "swollen" cities and called for more dispersed settlement – but Sydney was only fifty miles from the Blue Mountains, Melbourne about as close to the Ash Range. The light rail systems built to take people from suburbs to city center began boosting the new enthusiasm for the outdoors as a way of filling cars that would otherwise be idle on the weekend. They ran excursions to the bush and published guides to country walks and day trips. Soon bushwalking was thriving on every level, from Sunday strolls down country lanes to camping and tramping in the mountains. People began by walking from point to point, finding food and beds at the end. The more adventurous moved off the roads and trails, carrying their gear as the old-time bush workers had: in a "swag" (a roll of blankets and gear held together with a canvas cover). For some it was a solitary pleasure, but others organized clubs that became, and remained until the rise of environmentalism, the backbone of the nature preservation and national park movement.[37]

Another aspect of the enthusiasm for nature was a renewed interest in natural history. The Field Naturalist's Club of Victoria, established 1880, was the forerunner of a new group of societies. Earlier ones had

---

[37] Myles Dunphy, New South Wales's leading figure in bushwalking, was of the next generation, but his papers are a useful source. On ideas and philosophy see Myles Dunphy, "Some Reminiscences," 73, File 16, Box MLK 3304, Myles Dunphy Papers, Mitchell Library, Sydney. A more accessible source of Dunphy's ideas is Patrick Thompson (editor), *Myles Dunphy: Selected Writings* (Sydney: Ballagirin, 1986). On the movement in Victoria see Tom Griffiths, *Secrets of the Forest* (Sydney: Allen & Unwin, 1992), 76–84.

been part of the apparatus of science. These were often associated with the mechanics' institutes bringing education to the "working classes." The Natural History Association of New South Wales, for instance, said it sought to help people "who would like to know something of the natural objects around them, and as a means of encouraging students and assisting them by advice or technical information in the pursuit of their studies and by naming specimens forwarded."[38] The relationship, though, was not quite as simple as that of experts assisting amateurs. Natural history had incorporated a mass of field observations by everyone from amateurs to experts, and even as the study of visible nature became a specialized set of disciplines in field biology (including ecology) it continued to rely on amateur observations. This was especially true in countries like Australia, where there were few professionals or resources for research. It was not possible to rely on experts alone or even to form the closed society of self-conscious professionals that was the disciplinary ideal. American ornithologists, for example, began excluding amateurs in the late nineteenth century. But they were full members of the Royal Australian Ornithological Union until well after World War II.[39]

The Australian market was too small to support specialized magazines and books, and newspaper columns featuring nature notes played a larger role in amateur natural history and informal education than they did in North America. The genre, common to the Anglo world, was a way of filling newspaper space with an attractive feature but it had important consequences. It was an introduction to formal nature knowledge for many people, and since editors answered letters that did not make it into the column, it served as a correspondence course for the more serious students in the hinterlands. The columns' most important function was to tie clubs and individuals to a larger community by making them aware of others with a common interest. In Australia they have been a fixture since the start of the twentieth century and have lasted much longer than their North American counterparts. Donald Macdonald edited one from 1904 until his death in 1934, and from 1908 a second one, "Notes for Boys." A younger nature writer, Crosbie Morrison, assisted him in the early 1930s and took over the columns briefly after Macdonald's death. The better-known Alec Chisholm, who

[38] Announcement, Record of the Field Naturalists Society of New South Wales, Collection 8090, National Library of Australia, Canberra. On the movement see Colin Finney, *Paradise Revealed* (Melbourne: Museum of Victoria, 1993), 124–7.

[39] On this issue see Ray Wallace, "A Body in Twain? The Royal Australian Ornithological Union and the Events of 1969" (unpublished M.A. thesis, University of Melbourne, 1990). More generally see Roy MacLeod (editor), *The Commonwealth of Science* (Melbourne: Oxford University Press, 1988).

had begun as one of Macdonald's young rural readers, took over. After World War II Vincent Serventy, a naturalist from Western Australia, carried on the tradition.[40]

Like outdoor recreation, nature study took on forms appropriate to mass participation by city dwellers. The study of birds, for instance, was not what it had been even a generation before. Then it consisted of shooting birds and making their skins into specimens and collecting clutches of eggs. By 1900 the Audubon Society was discouraging egg collecting because it threatened the popular and vulnerable species. (One measure of the difference in national populations is that Australia's Gould Leagues and New Zealand naturalists did not take this stand for another generation). The increase in humane sentiment meant that many people wanted to study birds but not shoot them.[41] They needed a book to help them identify what they saw. (One thing that did not change was people's desire to connect their knowledge to the culture's by putting the "right" name on their specimens.) The great productions of an earlier generation, Gould's or Audubon's volumes, were intended for the library, not the field. There were smaller volumes, but they told about birds, not how to tell one from the other.[42] The scientific literature was intended for identification, but the books were too bulky to carry around. Besides, they aimed to help identify birds in the hand, not in the bush. Descriptions went from head to toe and gave all characteristics in order, including many that could not be seen from a distance. Elliott Coues's description of the North American robin in *Key to North American Birds* (1872) is a good example:

> Adult male, in summer: Upper parts slate-color, with a shade of olive. Head black; eyelids and spot before eye white; throat streaked with white. Quills of the wings dusky, edged with hoary-ash, and with color of back. Tail blackish; outer feather usually tipped with white. Under parts to vent, including under wing-coverts, chesnut. Under tail-coverts and tibiae white, showing more or less plumbeous. Bill yellow, often

---

[40] Griffiths, "'Natural History of Melbourne'"; Graham Pizzey, *Crosbie Morrison* (Melbourne: Victoria Press, 1992). Vincent Serventy Papers, National Library, Canberra, are a rich source of information on this genre and in particular on Serventy's readers.

[41] See collection of State Gould League publications, Mitchell Library, Sydney, and File 248, Ms. Papers, 444, Royal Forest and Bird Protection Society Papers, Turnbull Library, Wellington. David Allen, *The Naturalist in Britain* (London: Allen Lane, 1976), 230–1, thinks that bird-feeding and photography, not binoculars, which developed in the 1850s, were behind the growing interest in birds. This may understate the need for inexpensive and readily available equipment.

[42] Thomas Nuttall's *Ornithology* (Cambridge, Mass.: Hilliard & Brown, 1832; reprinted New York: Arno, 1974) is a good example of the type.

with a dusky tip; mouth yellow; eyes dark brown; feet blackish; soles yellowish.[43]

Among those who stepped into the gap was Frank Chapman, a curator at the American Museum of Natural History in New York. He described his *Color Key to North American Birds* (1903) as "the natural outcome of the recent remarkable interest in the study of birds which, fostered by Audubon Societies and nature study teachers, has assumed an ethical and educational importance of the first magnitude." There were many people who did not want to shoot birds, but did want to identify them, and ornithologists "should make some special effort to meet their peculiar wants."[44] *The Color Key* did. It was of a size to fit into a pocket or day pack, gave only the information necessary for identification, and was organized for the amateur. It did not, for instance, put the robin in its taxonomic place, with the other thrushes, but in a section on "perching birds marked with chesnut or reddish brown." The description was not systematic but concentrated on things that could easily be seen: "Adult male. Breast and belly rich rust-brown; above dark-slaty, head and spots in back black."[45] In *What Bird Is That?* (1920), subtitled *A Pocket Museum of the Land Birds of the Eastern United States Arranged According to Season*, he departed even further from the scientific texts, categorizing 301 species as permanent residents, north and south, winter visitors, and spring migrants.[46]

In Australia J. A. Leach's *An Australian Bird Book: A Pocket Book for Field Use* (1911) did much the same thing for the birds of Victoria, though Leach's background produced some differences. Chapman was a museum curator seeking a popular audience, and the *Color Key* was geared to informal education. Leach, whom the Melbourne nature writer Charles Barrett described as "the genius of Nature Study in Victorian state schools," wrote for a curriculum. His book grew out of a list of Victorian birds published, with short descriptions, as a supplement to

---

[43] Elliott Coues, *Key to North American Birds* (1872; reprinted New York: Arno, 1974), 249. This same technique may be seen in the literature of the other settler colonies. It is used even when the object is to allow amateurs to contribute to science. See Frederick Wollaston Hutton, *Catalogue of the Birds of New Zealand* (Wellington: Government Printer, 1871). This is a publication of the Geological Survey of New Zealand, another indication of the lack of specialization in this small country.

[44] Frank M. Chapman, *Color Key to North American Birds* (1903; 2d ed., New York: Appleton, 1912, same text), vii.

[45] Ibid., 207.

[46] Frank Chapman, *What Bird Is That?* (New York: Appleton, 1920). Chapman began with *Handbook of the Birds of Eastern North America* (New York: Appleton, 1895) and continued to publish. He had a number of imitators and successors, with Chester Reed's line of pocket-sized guides being, perhaps, the most popular.

the *Education Gazette and Teacher's Aid.*[47] It had a lengthy section on Australian birds and their relation to the world avifauna, and the descriptions had more the flavor of taxonomic language. Like the American, though, Leach was concerned with field identification by basic markings. He urged the reader, confronted with an unfamiliar bird, to pick out "one or two prominent markings and the size."[48]

The new guides straddled the separating worlds of science and popular knowledge (as would the second generation, which appeared in the 1930s and is discussed in Chapter 7). Living in cities or suburbs, on a daily round confined to pavement, people lacked not only woods lore but acquaintance with people who did have it. The books, though, were not an exact substitute for the old settler. Documents written for a mass audience, they provided neither feedback nor local knowledge. On the other hand they did not pass on the misinformation that often went with folk-knowledge, and pointed the reader to a wider world by showing related species over a large area on the same page and giving scientific names and organization.[49] Leach's method, the cornerstone of modern practice as well, is an appeal to folkbiology. People automatically sort birds by size and shape, then by smaller marks, behavior, and habitat. Over much of North America, for instance, anyone seeing a bird confidently hopping across a lawn, chest out and head up, will identify it as a robin even without seeing the red breast. Australians will as quickly tag a small bird on a post as "Willie Wagtail" by the way it vigorously switches its tail. Anyone can test this out (and verify the soundness of Leach's approach) by looking at a group of unfamiliar birds. North Americans in Australia, for instance, will quickly turn to the proper section of the modern Australian guide if they see a new hawk, duck, or gull. Confronted with a drongo, a honey-eater, or a cuckoo-shrike, they will find the proper section just as the bird flies off.

Nature Education

"[N]ature-study" . . . stands everywhere for the opening of the mind directly to the common phenomena of nature.

Liberty Hyde Bailey, 1904[50]

---

[47] J. A. Leach, *An Australian Bird Book: A Pocket Book for Field Use* (Melbourne: Whitcombe & Tombs, 1911); Charles Barrett, "The Doctor," a memoir of Leach, in 8th ed. of *An Australian Bird Book* (Melbourne: Whitcombe & Tombs, 1939), 3–8; J. A. Leach, *Nature Study: A Descriptive List of the Birds Native to Victoria* (Melbourne: State Printer, 1908).

[48] Leach, *Australian Bird Book,* 9.

[49] Anyone who grew up in backwoods America with an interest in nature knows how erratic folklore is. Others may consult the autobiography of almost any American naturalist.

[50] Liberty Hyde Bailey, "What Is Nature Study?" in Anonymous, *Cornell Nature-Study Leaflets* (Albany, N.Y.: J. B. Lyon, 1904), 16.

Childhood education was another element of enthusiasm for nature. Elementary schools and the new organized children's groups taught nature studies and urged children to explore nature. Both the focus on nature and the children's organizations were responses to what were seen as problems caused by an industrial, urban society. There was, for instance, the "boy problem." Young men growing up in cities, without healthy outdoor work and in a "feminized" society, would not, it was feared, be sufficiently masculine. Child psychology, seeing individual development as the recapitulation of social progress, reinforced the need for nature. Children, born little better than animals, went through stages of barbarism and savagery before becoming fully civilized adults. Adolescent boys were in particular danger, for they were thought to need physical activity, adventure, and the chance to compete and cooperate with their peers, forming the tribes appropriate to their stage of development.[51] Of the groups that sprung up around the turn of the century to allow boys this freedom – under adult supervision, of course – the most successful was the Boy Scouts.

The movement began in Britain, where it was closely associated with the Empire and the military. It acquired other overtones and associations as it spread to the settler countries. In the United States the Scouts emphasized the pioneers and the Indians, particularly the Indians. Much of scouting was built on the body of "lore" borrowed from a competitor it overwhelmed, Ernest Thompson Seton's Woodcraft Indians. The focus on the forests was a reflection of Romantic nations about the forest primeval, the frontier, and the "ancient wisdom of the Red Man," but also of reality. Many North Americans lived inland, and there was a large rural and small-town population for whom forested areas were an escape from fields and towns. In Australia Donald Macdonald's *The Bush Boy's Book* (1911), which Australian historian Tom Griffiths described as the scouting movement's "ideal handbook," concentrated on the shore and the local bush.[52] Macdonald mentioned the Australian pioneers but

---

[51] A general treatment of recapitulation in child psychology of the period can be found in Stephen Jay Gould, *Ontogeny and Phylogeny* (Cambridge, Mass.: Harvard University Press, 1977), 135–47. On Seton and recapitulation in the Woodcraft Indians, a Boy Scout precursor, see H. Allen Anderson, *The Chief* (College Station: Texas A&M University Press, 1986), 129–50.

[52] Griffiths, "'Natural History of Melbourne,'" 356; Donald Macdonald, *The Bush Boy's Book* (Melbourne: Endacott, 1911); Ernest Thompson Seton, *The Book of Woodcraft and Indian Lore* (New York: Doubleday, 1912). On Seton, see Anderson, *The Chief*, 151–75; on themes and interests, Dan Beard, *The American Boys' Handbook of Camplore and Woodcraft* (Philadelphia: Lippincott, 1920). The chapter titled "The Snake Trouble" points even more clearly to a national difference. North America had many poisonous snakes, and outdoor manuals from that day to this treat the problem, but Macdonald's devoting a chapter to it is a reflection of the continent's uniquely varied, abundant, and deadly suite of reptiles.

also looked to North America. Planking fish and cooking with hot stones are "borrowed from the Indian." Stalking rabbits rather than using ferrets "is in a small way the 'Still Hunting' of the American backwoods." He paid little attention to the Aborigines.[53]

Nature education was in these years added to the elementary school curriculum.[54] It drew, in part, on the nostalgia for rural and small-town life that grew with industrialization and urbanization and the fear that city children would grow up not knowing even the names of common plants and animals or where their food came from. It was also a way to inculcate desirable values, particularly in children of the "lower orders." This was very much the case in the United States, where "old-stock" Americans fretted about the effects immigrants from southern and eastern Europe might have on the nation's values. An article in *Bird-Lore*, for instance, gently warned schoolteachers to remember that the children in their classes often came from "races with instincts concerning what are called the lower animals, quite beyond the comprehension of the animal-loving Anglo-Saxon."[55] Contemporary theories of learning, holding that children learned most quickly and effectively when they were directly involved with the subject matter, justified making nature a focus. It was ideal – close to every school, at least in some form, important, and already attractive to most children.

The movement to add nature to the school curriculum began in Britain among teachers in the grammar schools formed under the 1870 Education Act. By 1902 it was popular enough to be the subject of an international conference in London, and it was part of a 1908 conference on education in the British Empire.[56] In the colonial countries, state or provincial education agencies introduced it. In the United States, where authority was more dispersed, private organizations played larger roles. At Cornell University a group interested in rural life produced much of the basic teaching material. Anna Botsford Comstock wrote a series of nature study pamphlets and lessons with contributions from her husband, John H. Comstock, an eminent entomologist, and Liberty Hyde Bailey, a horticulturist. In 1904 the New York State Department of Agriculture published a selection in a book, "Cornell Nature

[53] Macdonald, *Bush Boy's Handbook*, 43, 73, 17.

[54] A recent useful study of the movement is Sally Gregory Kohlstedt, "Nature Study in North America and Australasia, 1890–1945," in R. W. Home (editor), *The Scientific Savant in Nineteenth Century Australia*, Historical Studies of Australian Science, 11, No. 3 (Canberra: Australian Academy of Science, 1997).

[55] Mabel Osgood Wright, "The Law and the Bird," *Bird-Lore*, 1 (December 1899), 203–4.

[56] Allen, *Naturalist in Britain*, 203. He considered the whole thing a mistake. Nature acquired a "fatal association with lessons – and, even worse, lessons for infants" (204). See also *Official Report of the Federal Conference on Education Convened by the League of the Empire* (London: League of the Empire, 1908), 296–324.

Study Leaflets." The Audubon Society sponsored school bird clubs and essay contests and provided pamphlets, lessons, posters for classroom use, and plans for birdhouses (which could be built as school projects and for Audubon contests). *Bird-Lore* printed club information and had a children's section.[57]

Nature education in Canada was much like that in the United States, since the plants and animals were the same, and Canadians used the Cornell material, but government officials (the Wildlife Office of the Dominions Parks Branch of the federal Department of the Interior), not a private organization, produced instructional materials.[58] The Australians also looked to the United States. The first Australian nature study text, *Nature Studies in Australia,* followed the American plans. In the introduction the authors paid homage to the mother house at Cornell.[59] In 1910 South Australia proclaimed a school bird day, "backed by a recommendation from the United States." A decade later one enthusiast estimated that at least 20,000 children in Victoria, New South Wales, South Australia, and Queensland had been enrolled in nature leagues, and two "of these states specialize in bird clubs, on lines successfully followed in the United States."[60] The genesis of these clubs, state organizations called the Gould League of Bird Lovers, was the suggestion, in 1908, by Miss Jessie McMichael, a teacher in the Victorian education department, that something "equivalent to the Junior Audubon Society of the United States" be established. The name came from John Gould, the London-based naturalist and publisher who had produced the great nineteenth-century volumes on the continent's plants and animals. The idea quickly took hold. The Melbourne nature writers took up the cause, and Prime Minister Alfred Deakin agreed to become the first president of the club. The next year two teachers in Wellington, New South Wales, copied the

---

[57] Anonymous, *Cornell Nature-Study Leaflets;* Frank Graham, Jr., *The Audubon Ark* (New York: Knopf, 1990), 83–7.

[58] Harrison Lewis, "Lively: A History of the Canadian Wildlife Service," unpublished manuscript, RG 109, Records of the Canadian Wildlife Service, Public Archives Canada, Ottawa, 20–1, 56–7. There is a list of early pamphlets in Box 1, Acts and Legislation, RG 109.

[59] William Gillies and Robert Hall, *Nature Studies in Australia* (Melbourne: Whitcombe & Tombs, 1903; 2d ed. is undated). On the subject in Victorian schools see Charles R. Long (editor), *Review of the State Schools Exhibition* (Melbourne: Government Printer, 1908), 47–56. See also *The Geelong Naturalist, 11* (September 1905), especially Frank Tate, "Introduction," i–ii. A more general treatment is Alfred G. Edquist's *Nature Studies in Australasia* (Melbourne: Lothian, 1916), which also cites Liberty Hyde Bailey, ix.

[60] Alec H. Chisholm, *Mateship with Birds* (Melbourne: Whitcombe & Tombs, 1922), 70, 73.

Victorian constitution and formed a local society, then a state organization.[61]

The New Zealand movement was on a smaller scale, with fewer materials and less input from the United States. The first school text, *Nature in New Zealand*, began as a series of newspaper articles, "Rambles in a Museum," written for the *Lyttelton Times* in 1901. A publisher, "knowing the want of an elementary book on the Natural History of New Zealand," got one of the authors to modify it for school use. In 1905 a second appeared, *The Animals of New Zealand*, the result of a series of articles that combined "popular information with the purely scientific" and were to serve "naturalists and at the same time be interesting to the general public."[62] Nature study was added to the curriculum in 1904, with *Nature in New Zealand* as the recommended text.[63] In the early years there was no organized outside support, but in the 1920s a new organization, the Forest and Bird Protection Society, began helping the schools with lessons and posters.[64]

The two southern countries show the extremes in using nature and nature education for nationalism. *Nature in New Zealand* said of the English skylark that its song "appealed to the feelings of the early colonists" and perhaps allowed them to "let their minds wander back to days gone by, and to imagine that once more, hand-in-hand with friends of their childhood, they roamed the fields and pastures of Merry England."[65] *Animals of New Zealand* judged that "on the whole, the results of acclimatisation in New Zealand must be considered favourable, although unfortunate mistakes have been made." It hinted at a value for native nature, but only timidly, saying that "the kiwi is the most notable living bird of New Zealand, and should be classed among the colony's treasured possessions."[66] In Australia, on the other hand, nature

[61] Neville Caley, *What Bird Is That?* (Sydney: Angus & Robertson, 1931), xv–xvi; Alec Chisholm, preface to the 1958 edition (same publisher), v–vii. The clubs were part of the school system, while the Audubon Society was a private association, but both promoted nature education, and the Australian organization evolved into an adult nature appreciation club – which is where Audubon started. Australian nature writer Vincent Serventy described the separate state Gould Leagues as "basically a nature conservation society working through both primary and secondary schools." Folder 101, Box 13, Serventy Papers, National Library, Canberra.

[62] James Drummond, *Nature in New Zealand* (Christchurch: Whitcombe & Tombs, n.d.), Introduction; F. W. Hutton and James Drummond, *The Animals of New Zealand* (Christchurch: Whitcombe & Tombs, 1905), 13.

[63] Galbreath, "Colonisation," 336 and 489n.

[64] File 56, Papers of the Royal Forest and Bird Protection Society, Turnbull Library, Wellington.

[65] Drummond, *Nature in New Zealand*, 69.

[66] Hutton and Drummond, *Animals of New Zealand*, 28, 308.

education was consciously and sometimes enthusiastically nationalistic. Introducing Leach's guide, Frank Tate, head of Victoria's education department, rejoiced that the children of Australia were casting off English associations and coming to appreciate the beauties of Australian nature. Nature study would promote a love of country and identification with it. Leach agreed. Too many names, he complained, were not really Australian; they had come from the United States. "Thus the Gum-tree (Eucalypt) is not a Gum, the 'Possum is not the carnivorous Opossum of America, the Goanna is not the equivalent of the vegetarian American Iguanna; the 'Wild Cat' is not a Cat, nor is the 'Native Bear' a Bear, nor even remotely related to one, nor is the Kestrel a Sparrowhawk."[67] (Tom Griffiths noted, "A high point of the confluence of interest between nature writing and Australianism was A. H. Chisholm's eloquently titled *Mateship with Birds* (1922)."[68] Though Leach complained about American names, the critical influence was British, and it continued well into the twentieth century and went far beyond nature study. Novelist Thomas Keneally, born in 1935, said that his school books were full of things and places he had never seen and about which he and his classmates knew nothing. "We were educated to be exiles." Others, writing of the 1950s and even later, echoed this sentiment.[69]

Protecting Visible Nature

Interest in nature fueled interest in protecting it. The settlers passed laws, set aside land for recreation and beauty, and established agencies to administer them. The most important contribution to nature protection in these years was the idea, or ideas, of a "national park," which provided precedent as well as legal and social definition for preserving land for noneconomic uses. This now seems an American idea; park histories throughout the Anglo world appeal to the act establishing Yellowstone National Park in 1872 as the fountainhead of the movement. This is hindsight and bad history. Canada's first park, Banff Hot Springs (1885), was modeled on Yellowstone, but Canadian legislators were working under the same conditions and knew it, carving out a tourist attraction from public domain in the still-unsettled western part of the country. In Canada, though, federal power was weaker, and so was

---

[67] Leach, *Australian Bird Book* (1911), 1, 72–3, 74. This is still a problem. See Vincent Serventy to E. M. Russell, 20 January 1970, Folder 15, Box 2, Serventy Papers, National Library, Canberra.
[68] Griffiths, "'Natural History of Melbourne,'" 352.
[69] Peter Quartermaine, *Thomas Keneally* (London: Edward Arnold, 1991), 5; Jill Ker Conway, *The Road From Coorain* (New York: Knopf, 1990), 98–9; Tim Flannery, *The Future Eaters* (London: Secker & Warburg, 1994), 13–14.

the perceived need to set aside wild areas. Parks could do little to save nature if provincial authorities wanted development. New Zealand parks were like American ones in preserving monumental scenery, but this was less a matter of national heritage than of money; from the first, the parks were intended to lure overseas tourists. Australian parks owed more to the British idea of green spaces for urban dwellers than any ideal of scenic beauty.[70] They all moved toward what we now see as Yellowstone's model – large natural areas with the scenery, plants, and wildlife in a "natural" state – but everywhere local culture was as important as foreign example.

Setting aside areas for national parks was a step toward making recreation and scenery socially acceptable uses of the land, but hardly a bold one. American park historian Alfred Runte has argued that American parks were "worthless lands," created only where no one wanted the land. This may be debated at the edges, but the core is surely sound, and it holds for the other countries as well.[71] In Canada the provinces used their power not only to block parks but to draw their boundaries to exclude timber and mining lands.[72] In New Zealand the commission recommending additions to Tongariro National Park in 1908 put first among the principles "that no land shall be included which is of economic value, either for agriculture or as having forests containing milling-timber."[73] About Australian parks one commentator noted that until well after World War II parks were selected on the principle that "if the scenery is good and nobody else wants it, then it could be a park."[74]

[70] Alfred Runte, *National Parks: The American Experience* (2d ed., Lincoln: University of Nebraska Press, 1987); W. F. Lothian, *A History of Canada's National Parks* (Ottawa: Environment Canada and predecessor agencies, 1977–87), 4 vols.; David Thom, *Heritage: The Parks of the People* (Auckland: Lansdowne Press, 1987). On common goals in parks see John Shultis, "Improving Wilderness," *Forest and Conservation History*, 39 (July 1995), 121–9. Ross Galbreath, "Colonisation," 205, sees New Zealand parks as part of a common development of the settler colonies going beyond English precedents.

[71] Runte, *National Parks*, 48–64. See also Colin Michael Hall, "The 'Worthless Lands' Hypothesis' and Australia's National Parks and Reserves," in Kevin J. Frawley and Noel M. Semple (editors), *Australia's Ever Changing Forests* (Canberra: Department of Geography and Oceanography and Australian Defense Force Academy, 1983), 441–56.

[72] Leslie Bella, *Parks for Profit* (Montreal: Harvest House, 1987), chap. 3.

[73] Leonard Cockayne, "Report on a Botanical Survey of the Tongariro National Park," Department of Lands (Wellington: Government Printer, 1908), 33.

[74] Eric R. Guiler, *Thylacine: The Tragedy of the Tasmanian Tiger* (Melbourne: Oxford University Press, 1985), 31. See also D. F. McMichael, "New South Wales," in "Further Case Studies in Selecting and Allocating Land for Nature Conservation," in A. B. Costin and R. H. Groves (editors), *Nature Conservation in the Pacific* (Morges: International Union for the Conservation of Nature, 1973), 53. The extent to which this continues to be true can be seen in the annual reports of the Australian National Parks and Wildlife

North Americans had the clearest ideas of what a park should be – a place of Romantic scenery and rustic comfort. The ideal park had forests of large trees, free of underbrush, near lakes and streams. There were vantage points where people could look over vast chasms, splendid valleys, and enormous cliffs. In the evening, sitting on the porch of their lodge, visitors could look out over verdant meadows, past deer and elk posing nobly in the middle distance, to a vista of snow-capped peaks. In both countries park advocates appealed to national pride. Americans had been making their land a patriotic asset since the early nineteenth century, and they easily incorporated the parks. Canadians came to speak in the same accents. One legislator, urging passage of the Banff Hot Springs act, asserted that anyone who had gone there, seen the area, "and not felt himself elevated and proud that all this is part of the Dominion, cannot be a true Canadian."[75] There were warnings that developers would ruin these delightful areas. Calling for a larger park at Banff, a Canadian newspaper said, "There is lots of elbow room in the Northern Rockies as yet, and the time to take action is now, – before vested interests shall have made it difficult to enlarge this, the most beautiful playground possessed by any people."[76] In the United States, Niagara Falls, which had been turned into a tourist trap, was held up as the example to be avoided, and every new park was justified on the grounds that unspoiled land had to be saved.

North Americans may have had the clearest ideas, but they were not entirely modern ones. Early visitors did things that would now be considered entirely inappropriate. In Yellowstone they threw objects into the geysers to see how high the next eruption would carry them, and hunting was permitted in the park until 1894. Wildlife, now a major attraction, was treated as mobile scenery or a novelty on a par with hot springs and odd rock formations. People fed bears along the roads and park officials created garbage dumps to give visitors a better look.[77] The Yellowstone buffalo, survivors of the great massacres, were kept near the road and managed like so many cows – rounded up, checked for disease,

Service (now the Australian Nature Conservation Agency). A study of park policy in one state is Sarah Bardwell's "National Parks in Victoria, 1866–1956" (unpublished Ph.D. dissertation, Monash University, 1974).

[75] *Winning Free Press*, quoted in Lothian, *Canada's National Parks*, 1: 32.

[76] Runte, *National Parks*. See also Lothian, *Canada's National Parks*, 1: 32. A recent history of Ontario's park system, Gerald Killan, *Protected Places* (Toronto: Queen's Printer, 1993), chaps. 1 and 2, shows the same ideas and influences on the provincial level.

[77] See, e.g., the bear photos in Horace M. Albright with Robert Cahn, *The Birth of the National Park Service* (Salt Lake City: Howe Brothers, 1985), 182, and the photoessay in Lisa Mighetto, *Wild Animals and American Environmental Ethics* (Tucson: University of Arizona Press, 1991).

fed during hard winters, and when their numbers grew too large some were sent to the slaughterhouse. They were stampeded to give distinguished visitors a view of the "thundering herd."[78] Other animals were ignored, except those that preyed on the "nice" animals the tourists wanted to see. Park managers shot, trapped, and poisoned these "varmints," less because of policy than common sentiment. In 1912 John Brown, in charge at Waterton Lakes National Park, seemed quite surprised that Howard Douglas, the new superintendent of parks, would question his operations. Poison baits, he admitted, were killing some birds, but only "magpies and whiskey jacks [Canada or gray jays], which are very destructive to young grouse and will eat the eggs of those or any other birds." There was, he went on, no danger to other animals. He had been using poison for forty years and had found no destruction of useful wildlife except some dogs.[79] In the 1920s the elimination of predators became policy, and the Dominion Park Service bought poison from the U.S. Bureau of Biological Survey (which was then eliminating the large predators from the western United States) and got instructions on its use. Killing these animals was so well accepted that when, a decade later, U.S. and Canadian park officials decided that predators should be preserved as part of the life of the land (see Chapter 8), they faced strong protests, including those from otherwise ardent defenders of the parks who did not see that predators should have any place in them. In the 1940s Canadian park officers were still killing these animals, in violation of policies by then a decade old.[80]

---

[78] Elk were treated in the same fashion, and regulation of numbers has been a continuing problem. Douglas B. Houston, *The Northern Yellowstone Elk: Ecology and Management* (New York: Macmillan, 1982). On Canadian parks see Lothian, *Canada's National Parks*, 4: 18–34. On Canadian bear problems, Box 114, U212, v.2, Records of the Dominion Park Service, RG 84, National Archives, Ottawa; this record group, organized by subject, provides a wealth of material for a comparative study.

[79] Douglas to Brown, 3 July 1912, reply of 9 July, Brown to Douglas, 28 December 1912, Box 35, U300; Diary, Warden, Waterton Park, March 1921, and instructions from the Survey on poisoning, sent to the parks in late 1924, Box 43, W300. Policy documents are in Boxes 33–9, U300, Universal, Game Protection–General. All material in RG 84, Records of Parks Canada, Public Archives Canada, is cited hereafter as RG 84, Public Archives Canada. On U.S. policy see Adolph Murie, *Ecology of the Coyote in the Yellowstone*, Fauna Series No. 4, National Park Service, Department of the Interior (Washington, D.C.: U.S. Government Printing Office, 1940); Murie, *Ecology of the Coyote*.

[80] Thomas R. Dunlap, "Wildlife, Science, and the National Parks, 1920–1940," *Pacific Historical Review*, 59 (May 1990), 187–202; idem, "Ecology, Nature, and Canadian National Park Policy," in *To See Ourselves/To Save Ourselves*, Proceedings of the Annual Conference of the Association for Canadian Studies, 1990 (Montreal: Association for Canadian Studies, 1991), 139–47. See also Box 162, W266, RG 84, Public Archives Canada.

Parks in Australia and New Zealand drew on different traditions and responded to different circumstances, but interest came at the same time as in North America. Only two years after Yellowstone was established, former premier William Fox suggested to Premier Julius Vogel, citing the American example, that Lake Rotomahana and its volcanic wonders be set aside. Nothing came of this, however; the country's first park was established in 1887 when Te Heu Heu Tukino ceded the land around the sacred mountain Tongariro to the government for a park to forestall white settlement. Later additions – Mt. Egmont (the lovely symmetrical cone the Maori know as Taranaki), which was set aside in 1900, and Arthur's Pass, established in 1901 – were scenic wonders to lure foreign tourists. (The same motive would a decade later cause the government to bring in new animal species for hunting.)[81] Australian parks were rural retreats for city dwellers, closer to the British model of open space than North American scenic grandeur. Royal National Park, established by New South Wales in 1879, and Ferntree Gully and Tower Hill, set aside by the Victorian Parliament in 1887 and 1891, were to serve the people of Sydney and Melbourne. South Australia's first park was on the government farm at Bel Air, and Western Australia's was near Perth. They were, despite the title, not "national parks." The separate colonies created them and retained control after confederation in 1901. Even now Australia has few parks under Commonwealth control.[82]

Parks were, at first, established as independent units, but there soon developed a need for some central administration. In the early twentieth century the U.S. parks were run by a variety of agencies, seemingly whatever was at hand when Congress established the park. A cavalry troop, for example, was in charge of Yellowstone. In the other countries the parks were administered by boards, usually composed of local citizens. New Zealand was the first to establish a central park agency, under the Scenery Preservation Act (1903), but its duties were primarily to attract foreign tourists; administration was left to local boards. Canadian plans were more extensive; the 1912 act that provided for the Dominion Parks Office in the Department of the Interior set up a national authority. The provinces, though, still had a great deal to say about "their" parks. There was a Wildlife Branch – a first in park governance – but its three-man staff was responsible for all the wildlife work in the Department of the Interior, including the game laws of the Northwest Territories. The U.S. agency was the most powerful. The National Park

---

[81] On park formation see Thom, *Heritage*, 4.

[82] Ontario named its first park Algonquin National Park, and there were calls for the preservation of another area as a "public and national park." Killan, *Protected Places*, 16. Only in Australia, though, has the name lasted.

Service, established in 1916, managed all the American parks, which it staffed with its own employees, and it was legally independent of the states. It also had an effective lobby, the National Parks Association, which the first director, Stephen Mather, had encouraged as an aid in his continuing battle for congressional funds and public approval.[83]

Australia did the least. There was no national authority until the 1970s (and then it had no authority over the parks the states had set up and continued to administer), and the states had no common model. Queensland's Forests and National Parks Act (1906) lumped together economic and noneconomic uses. South Australia's had a National Pleasure Resort Act (1914) and Tasmania a Scenery Preservation Act (1915). Parks continued to be administered by individual boards, and into the 1950s these sometimes had the responsibility of raising money for visitor services. There was little statutory protection, and some parks were destroyed by mining or military activity.[84] Differences in administration were in part due to differences in government – that the United States had the strongest park agency and the strongest federal government was not coincidental – but the number of visitors was also a factor. By 1920 the pressure on the most popular American parks was such that management was necessary to keep them in anything resembling a "natural" state. Canadians could afford to do less because there were fewer visitors, but officials were acutely aware of problems south of the border.[85] Australian and New Zealand parks survived with little management simply because fewer people used them.[86]

Parks were one expression of a rising interest in nature and experience in nature. Another was wildlife. People dismissed animals and plants in the days of conquest, or saw them only in terms of dollars, but as settlement transformed the land they sought to save at least a few emblematic species that seemed in danger. This sentiment was strongest in North America, where the effects of settlement were obvious and dra-

---

[83] Robert Shankland, *Steve Mather of the National Parks* (3d ed., New York: Knopf, 1970); Donald Swain, *Wilderness Defender* (Chicago: University of Chicago Press, 1970); Albright with Cahn, *Birth of the National Park Service.*

[84] Queensland's act was No. 20, 1906; South Australia's No. 1173, 1914; and Tasmania's No. 15, 1915 (6 Geo. V.). An overview of this period can be found in J. G. Moseley, "National Parks and Equivalent Reserves in Australia," draft prepared for the Australian Conservation Foundation, 1968, Box 1824, Australian Conservation Foundation Collection, State Library of Victoria, Melbourne. On the post–World War II period the Vincent Serventy Collection, National Library, Canberra, contains a mass of papers. Though it is not well organized, it is a useful source.

[85] Dominion Park Service records, RG 84, Public Archives Canada, testify to this on a wide variety of topics.

[86] Tower Hill was lost to mining, but this was not a visitor issue.

matic. The slaughter of the buffalo herds in the 1880s was the watershed. People rallied to save the remnant, and the species became an example of heedless destruction of the natural world. It was the plight of birds, though, that brought organized and continuing action. Birds had always been popular, and when in 1886 George Bird Grinnell, editor of *Forest and Stream*, proposed in the magazine a society devoted to protecting them, named after America's foremost artist-naturalist, the response was so great that in a few years he had to abandon the project. It demanded too much time. Others took up the cause and the name, forming state societies (the first in 1896). In 1905 these came together as the National Association of Audubon Societies. *Bird-Lore*, which Frank Chapman edited from his position at the American Museum of Natural History, became the group's journal.[87] The national group formed around opposition to the use of bird feathers on women's hats and dresses, and it continued an active campaign for protection. It was part of the coalition that convinced Congress to pass the first federal wildlife statute, the Lacey Act (1900), which banned from interstate commerce wildlife killed in violation of state laws. It wrote a model state law for the protection of non–game birds and by 1910 had persuaded legislators in all but seven western states to adopt it. This was a milestone. The presumption was that wildlife could be killed unless it was protected. Now it was protected unless the legislature set a season or declared a species to be a pest. The society was also instrumental in ending market hunting of geese and ducks, and it campaigned for the Migratory Bird Treaty of 1916, the first effective international treaty on wildlife.[88]

Elsewhere threats to birds and native wildlife were fewer, or at least less obvious. Canadian officials were concerned about migratory birds, but none of the provinces found it necessary to enact strict measures for resident wildlife. The United States outlawed the sale of game before World War I; Ontario, the most thickly settled Anglo province, was still debating the issue in the 1940s.[89] The Australian Ornithologists' Union, formed in 1901, sought legal protection for non–game birds but did not

---

[87] Graham, *Audubon Ark*, 83–7.

[88] On early action see Theodore Cart, "The Struggle for Wildlife Protection in the United States, 1870–1900" (unpublished Ph.D. dissertation, University of North Carolina, 1971); Robin Doughty, *Feather Fashions and Bird Preservation* (Berkeley: University of California Press, 1975). On laws see T. S. Palmer, "Chronology and Index to the More Important Events in American Game Protection, 1776–1911," Biological Survey Bulletin 41 (Washington, D.C.: U.S. Government Printing Office, 1912). The holdouts were Nebraska, Montana, Idaho, Nevada, Utah, New Mexico, and Arizona. On the law see Michael Bean, *The Evolution of Federal Wildlife Law* (New York: Praeger, 1983).

[89] Acts and Legislation, WLU 10-2, Records of the Canadian Wildlife Service, RG 109, National Archives, Ottawa.

campaign against market hunting, specimen collecting, or amateur egg collecting – all of which the Audubon Society sought to ban.[90] Australians continued to kill kangaroos for leather or dog food, or just to keep them off sheep pasture. They hunted platypus and koala for their fur (almost wiping these species out in much of the Southeast) and killed native birds for the table or the market. Laws were few and weak. In 1893 New South Wales set a five-year closed season on some two dozen native or imported species or groups and in 1903 a two-year closed season on several native animals, including the platypus and koala. Queensland provided absolute protection for the platypus in 1906, as did Tasmania in 1907 and South Australia in 1912, but enforcement was minimal. The koala trade continued in New South Wales and Queensland into the 1920s. Legislators added species to the lists of those protected, but this was less a movement than a set of disconnected initiatives. Around 1900, for example, South Australia protected more than 100 species, while Tasmania had seasons on 152. As in the United States, there was a shift from the presumption that wildlife could be shot to making it protected unless a season was set. Queensland was first, in 1912, Western Australia followed the next year, and New South Wales in 1922.[91] In the absence of wardens, though, this meant little.

New Zealanders did no more. They shot rare native birds for specimens and abundant ones, like ducks and pigeons, for the table. In the 1890s Parliament, alarmed at the decline of native birds, acted on recommendations that dated back twenty years by buying Little Barrier Island and stocking it with rare species from the mainland, but it left the administration of this and other refuges to dedicated volunteers. It did keep adding species to those protected year-round; the Animal Protection and Game Act (No. 57, 1921), codifying earlier enactments, listed some 171 native birds, both native bats, the tuatara (a native lizard), and native frogs.[92] Since the acclimatization societies were responsible for game law enforcement and increasingly for non-game management, and they were in no position to hire enough people to do the necessary work, the laws were little more than pious wishes.

---

[90] Box 11, Royal Australian Ornithologists' Union Papers, State Library of Victoria, Melbourne.

[91] Public statutes of the Australian states. The most complete collection is located in the National Library, Canberra. Overviews are J. M. Thomson, J. L. Long, and D. R. Horton, "Human Exploitation of and Introductions to the Australian Fauna," 227–49, and George R. Wilson, "Cultural Values, Conservation and Management Legislation," 250–60, both in Bureau of Flora and Fauna, *Fauna of Australia* (Canberra: Australian Government Publishing Service, 1987), vol. 1A.

[92] Galbreath, "Colonisation," 201, 203, 219, 221, 233–4, 253, 256. New Zealand statutes, Turnbull Library, Wellington.

Conclusion

The turn toward native nature marked a new phase in the settlers' search for a place in the land. They had tried to remake the country by killing the wildlife, destroying the forests, plowing up the ground, and bringing in familiar plants and animals. Even the popular recreation of sport hunting had been infused with conquest. By the end of the century, as the limits of conquest became clear, these dreams were fading. People began to see the land as a force that shaped them and their society even as they shaped it. What this meant was not clear. For the settlers to say they were native by virtue of conquest, or that they had been formed by the trials and hardships of settlement, was stirring but vague. Worse, it made the present generation second-class citizens, inheritors of a great tradition to which they could not contribute. There was no frontier, at least one that could be settled, and people were moving to the cities or being driven from the land. Fewer and fewer got their living directly from the land or even lived in daily contact with what they could see as nature. They could not even participate in the search for knowledge about the land – that was for the experts.

Some things were clear. Whatever the outcome, the lands would be, biologically, a combination of native and exotic species seeking a new equilibrium. Mental landscapes were just as confused. The settlers celebrated conquest even as they searched for a place in the land, chose national animal symbols and shot the actual animals, set aside land but in order that people might be refreshed in nature for further toil in the industrial city. Nationalism grew, except in the United States, as part of loyalty to the British Empire, and even in the United States ties to Britain and Europe would continue to be a vital part of the culture and people's identity. Another certainty was that the settlers wanted a relation to nature. Their work had provided a connection, and they glorified this "conquest of nature" even as it faded into history. Now they sought through education, study, and play to form new ties to that world. For this they needed knowledge, and natural history, their guide for more than a century, had reached its limits, while ecology was not yet formed. It would come to shape people's views, but not for two generations. The next part of our story is the development of this new field within, and as part of, the new and more isolated infrastructure of science, a matter of professionals testing explicit theories in research programs housed in universities and research centers. It is a detour that will, half a century later, lead back into the high road.

The four maps on the following two pages suggest one of the most striking and constant elements of Anglo settler history – the connection between climate and settlement. These show population and rainfall; the reader can add temperature for North America by tracing a line from south to north. That no map of New Zealand need be included is one indication of that country's stable, temperate climate.

The two illustrations after this are from nineteenth-century American landscape painting; Thomas Cole's famous depiction of settlement, wilderness, and civilization and Frederic Church's Humboldtian vision of an ordered, majestic natural world (see Chapter 1).

Next is Edward Landseer's famous painting of a red deer stag, showing the object of the hunt as an examplar of nobility (see Chapter 2).

The three Australian paintings that follow show in subject matter and treatment classic themes in Australian understanding of the bush – the dust and heat of summer, danger in the bush, and the distinctive beauty of the eucalyptus. Hans Heysen made a career of painting these trees, as specimens and forests (see Chapter 4).

The final trio, paintings by the Canadian Group of Seven, show the distinctly different national vision of the bush in that country (see Chapter 4).

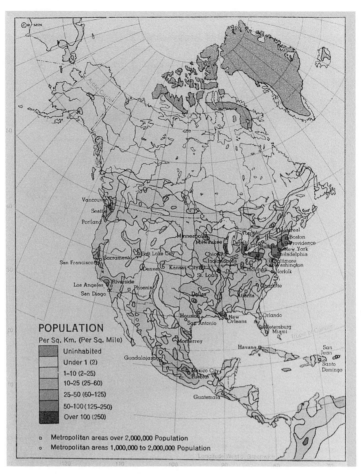

1 North American population distribution. Copyright 1998 Rand NcNally, R.L. 98-S-89.

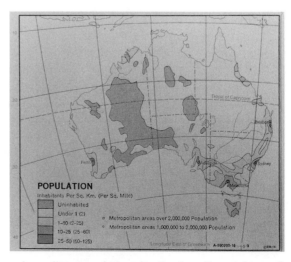

2 Australian population distribution. Copyright 1998 Rand McNally, R.L. 98-S-89.

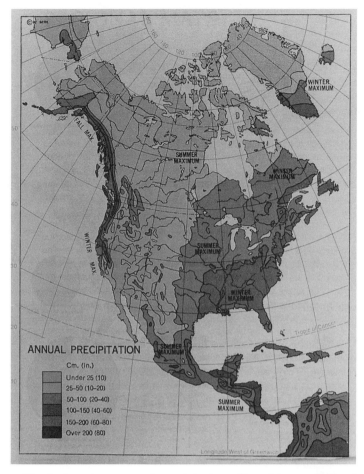

3 North American annual precipitation. Copyright 1998 Rand McNally, R.L. 98-S-89.

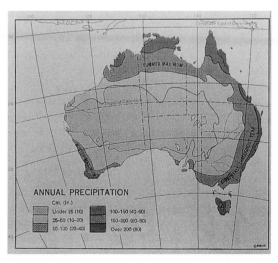

4 Australian annual precipitation. Copyright 1998 Rand McNally, R.L. 98-S-89.

5 Thomas Cole, *View from Mount Holyoke, Northhampton, Massachusetts, after a Thunderstorm (The Oxbow).* The Metropolitan Museum of Art, New York, Gift of Mrs. Russell Sage, 1908.

6 Frederic Edwin Church, *The Heart of the Andes*. The Metropolitan Museum of Art, New York, Bequest of Mrs. David Dows, 1909.

131

7 Edward Landseer, *The Monarch of the Glen.* By kind permission of United Distillers & Vintners.

8 David Davies, *A Hot Day (Golden Summer)*. National Gallery of Victoria, Melbourne.

9 Frederick McCubbin, *Lost (1886)*. National Gallery of Victoria, Melbourne.

134

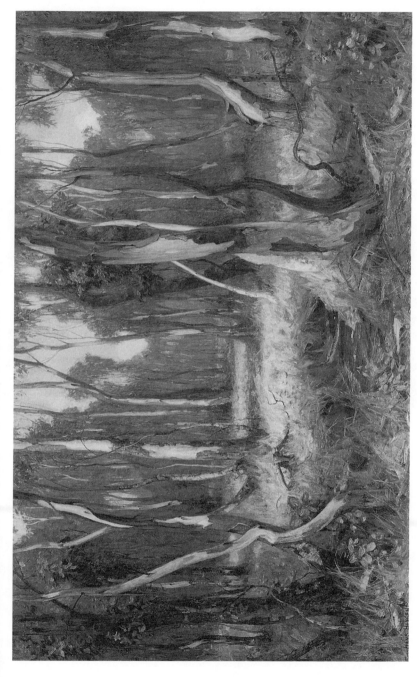

10 Hans Heysen, *Sunshine and Shadow*. National Gallery of Victoria, Melbourne and by kind permission of Mr. Christopher Heysen.

135

11 Tom Thomson, *The Jack Pine*. National Gallery of Canada, Ottawa.

2 A.Y. Jackson, *Terre Sauvage (1913)*. National Gallery of Canada, Ottawa.

13 A.Y. Jackson, *Algoma Rocks, Autumn, c. 1923.* Art Gallery of Ontario and by kind permission of Dr. Naomi Jackson Groves.

5

CHANGING SCIENCE, 1880–1930

Ecology, even most theoretical ecology, can be seen as largely a formalization of common knowledge, as institutionalized walks through the woods. This aspect of the discipline's character comes from the essential tangibility of the research material; apparently it is difficult for ecologists to look beyond phenomena that are in tune with familiar rhythms. Sophistication may lie in the methods of quantification of field data and in the formalization of these numbers into ecological models; but the underlying common-knowledge base, it could be argued, sets limits upon the level of sophistication that can be achieved.

Allen and Starr, 1982[1]

The transition from natural history to ecology is the single most important change affecting the settlers' understanding of the land and their relationship to it in the nineteenth century. It is not, however, a matter of one system of knowledge replacing another. Natural history had not replaced folkbiology and experience, but complemented them. It had to. People have, after all, "direct experience" with nature on this level, the foundation on which natural history must build. In a similar fashion ecology developed out of natural history and retained its ties with folkunderstanding. In this regard "develop" and "out" are equally important, and so are the ties to older systems. Ecology is clearly a stage beyond. The essential tangibility of the research material and the familiar rhythms of nature link theories to common knowledge, but the links are not always obvious to nonscientists, and by the time the ecologist reaches the limits of sophistication outsiders have been left far behind.

---

[1] T. F. H. Allen and Thomas B. Starr, *Hierarchy: Perspectives for Ecological Complexity* (Chicago: University of Chicago Press, 1982), 3.

On the other hand, because they have "direct experience," people are not willing to take their ideas entirely from scientists. There are as well the emotional ties people have to nature, ties that scientists, as humans, share.[2] For all these reasons, though it is part of academic science, ecology is an unusual part, and its ties and separations have affected what the public has learned, how it has used ecology, and how ecologists have used their knowledge in the public arena.

This chapter is about the interaction of popular and scientific understanding, especially how the new conditions under which authoritative knowledge was generated shaped views and popular access to those views. Ecology was not, as natural history so clearly had been, a part of high culture and many individuals' search for a place in the land. It was part of the infrastructure of modern science, a theoretically sophisticated and semidetached part of the culture. Professionals working in universities and research stations, almost always in the United States or Europe, generated and tested theories and trained students. The smaller societies, unable to support this costly research, had to work out different ways to get access to this knowledge. The rise of the United States means that much of our story takes place there; in the other countries we can see the first stirrings of interest, which serve largely to show the difficulties of developing modern scientific fields in small societies. With the field established, the chapter turns to the quirks and problems of ecology, not for their own sake, but as indications of a continuing theme, the ties between formal knowledge and passionate conviction.

New Science, New Setting

The changes in science that would bring about the new field of ecology were visible in the second half of the nineteenth century. Natural history's project of classifying the diversity of life was, at least for the larger forms, reaching the point of diminishing returns, and there was

---

[2] On the construction of nature at this level see Scott Atran, *Cognitive Foundations of Natural History* (Cambridge University Press, 1990), ix–xii, 1–13. One example of people's willingness to appropriate science on this level is the widespread acceptance of Aldo Leopold's dictum that diversity means stability. Aldo Leopold, *A Sand County Almanac* (1949; reprinted New York: Ballantine, 1967), 262. Allen and Starr, *Hierarchy*, 188, dismiss this as part of the "folklore" of ecology. See also Daniel Goodman, "The Theory of Diversity–Stability Relationships in Ecology," *Quarterly Review of Biology*, 50 (September 1975), 237–64, esp. 261, and Frank Golley, *A History of the Ecosystem Concept in Ecology* (New Haven, Conn.: Yale University Press, 1993), 100. Nevertheless, it is an article of faith in the environmental community.

so much information that no one could grasp it, even in outline. The field splintered into specialities, each with a part of natural history's subject matter, about which it asked new questions, sometimes in new ways. Ecology developed around the question of how organisms affected each other on the ground, and what impact the ground had on those processes. Think of the world we see as a set of card games. The cards (plants and animals) vary around the world, and so, less obviously, do the games (the interactions). Folkbiologies had been efforts to name the local deck and arrange the cards in suits. Natural history had sought to classify all the cards around the world and all the local decks. Ecology was concerned with the rules of the games. This is not to say people had no "ecological" information and ideas before the nineteenth century. They surely did. They knew a great deal about reproduction, distribution, associations of different plants and animals, abundance and scarcity, and how conditions affected populations, and they knew that all creatures were in some ways linked. Their knowledge, though, was operational; it helped them gather what they needed. There was no "folkecology" comparable to folktaxonomy. Nor did natural history fill this gap. There was a concept of a "balance of nature," but that was an acknowledgment that there were rules. It was not an explanation of them.[3] Ecology became possible when enough was known about the cards – what they were, where they occurred, and how they were associated – to study the rules of the various games and the overlap of rules among them, and when people had developed methods of probing relationships in nature on something more than a commonsense basis.

Natural historians laid the base. Biogeography, the study of the distribution of plants and animals across the earth, was developed by Humboldt's time, and many natural historians contributed to it as the century went on. In Britain, a bit later in the century, Edward Forbes defined zones of fauna in the sea by depth. A German, Karl Mobius, first used the word "community" in its ecological sense in describing an oyster bed and its relation to its surroundings. Another, Ernest Haeckel, coined the word "ecology."[4] By the 1890s the name was coming to designate a defined field. One historian said that ecology may be said to have begun as a self-conscious field "with the application of experimental and mathematical methods to the analysis of organism–environment relations,

---

[3] On the balance of nature see Frank Egerton, "Changing Concepts of the Balance of Nature," *Quarterly Review of Biology, 48* (June 1983), 320–50.
[4] On early work see the introductory chapter of W. C. Allee, Orlando Park, Alfred E. Emerson, Thomas Park, and Karl P. Schmidt, *Principles of Animal Ecology* (Philadelphia: Saunders, 1949).

community structure and succession, and population dynamics."[5] It began, that is, with the systematic study, using mathematical tools, of the processes of visible nature within a theoretical framework. Botanists took the lead in defining the field because the first requirement of ecological study was knowing what was there and, as one eminent animal ecologist said, the botanists finished classifying their specimens first. This was "because there are fewer species of plants than of animals, and because plants do not rush away when you try to collect them."[6] Plant ecologists could also be confident that conditions where they found their specimens were suitable for life and growth. Animal ecologists could not. There was more here, though, than names and concepts like "community," "association," and "succession." These early terms in ecology were evidence of deeper change, new methods of asking new questions.

These developed in a new institutional setting. Science was moving into the university system, and ecology developed within its structure of departments and research stations. Instead of apprentices learning for the sake of their own interest, there were graduate students working for degrees. Experts were not leisured enthusiasts but people who carried on research for a living. The field was developing its own means of communication (meetings and journals) and social structures (professional societies), and research was more and more a matter of testing ideas generated within this community. Soon the public would, as it already had with physics and chemistry, learn about the culture's authoritative knowledge from interpreters, a class of science popularizers who translated the arcane jargon of scientific concepts into terms the generally educated knew. Ecology's concepts would remain tied to popular ones (certainly far more so than those in physics), but the ties would stretch and the theories would be more remote from people's experience on the land.

Another consequence of this shift was the exclusion of the smaller countries from full participation in the search for knowledge. They had all helped catalog nature and built natural history museums, but they could not afford the universities, graduate schools, and research programs the new discipline required. All but the United States and Britain were priced out of the market. The others, though, still needed the

---

[5] Sharon E. Kingsland, "Defining Ecology as a Science," in Leslie A. Real and James H. Brown (editors), *Foundations of Ecology* (Chicago: University of Chicago Press, 1991), 1: 1–13. On the development of ecology see Robert P. McIntosh, *Background to Ecology* (Cambridge University Press, 1985), 22–68.

[6] Charles Elton, *Animal Ecology* (London: Sidgwick & Jackson, 1927; reprinted London: Methuen, 1966), 3.

results of science. Indeed, the increasing emphasis on production for the market, and for a world market, made its guidance even more important. To get it they developed, or began developing, a different set of institutions than those that were becoming the standard apparatus to train people in developing and testing theory. They built ones that would take advantage of the people and knowledge generated in the centers of Europe, Great Britain, and the United States. This seems a continuation of the tradition of the "bird of passage," the expert from "home" making a temporary or permanent living in the "colonies," a feature of science in these countries since their settlement. That, however, had been a temporary and local expedient on the way, it was always said, to national greatness. This was a response to conditions that shaped and still shapes the introduction and use of science in these countries. They had specialized agencies whose task it was to identify local problems and use the nation's ties to the metropolis to draw in experts from abroad and to identify places to send promising people for advanced training. That, however, gets ahead of our story.

The United States was a center of ecological studies from the start, one indication of the maturity of its scientific establishment. The formal beginnings were at the 1893 meeting of the Botanical Congress in Madison, Wisconsin, where the members organized an ecological section, but there was a generation of activity behind this.[7] It began in the state natural history surveys and government agencies, with the work of a generation that had begun as natural historians and then supported or trained the first generation of ecologists. Consider Stephen Forbes, whom one historian of ecology described as "perhaps closer to the 'compleat' ecologist than any [other] figure of the 19th century."[8] Born in Illinois in 1844, he had entered medical school, and the botanical studies there led him into natural history (as they had many others). He interrupted his education to serve in the Illinois cavalry during the Civil War, and abandoned medicine after the war to follow John Wesley Powell into the Illinois Natural History Society (which became the Illinois Natural History Survey in 1917). Powell went on to national fame exploring the West and advising federal authorities on land policy. Forbes remained in Illinois and pieced together a living holding several positions: curator of the society's museum, professor at the state university, and state

---

[7] Eugene Cittadino, "Ecology and the Professionalization of Botany in America, 1890–1905," *Studies in the History of Biology*, 4 (1980), 171–98.

[8] McIntosh, *Background to Ecology*, 64. The obituary in *Science*, 71 (11 April, 1930), 378–81, provides a summary of his career. A larger study is Robert Allyn Lovely's "Mastering Nature's Harmony: Stephen Forbes and the Roots of American Ecology" (unpublished Ph.D. dissertation, University of Wisconsin, 1995).

entomologist. He worked, as state entomologist, on insect control, but his interests were broader than that. A colleague said he was "probably the first entomologist in the United States to adopt the word *ecology* and to insist upon the broad applications of studies of that character in a consideration of the insect problems of agriculture."[9] He studied insect population dynamics, looked at the effect of bird predation, and was among the first to investigate diseases as a way to control insects. He was also a pioneer in limnology. He developed a program of river and lake studies in the state, and his paper "The Lake as a Microcosm" (1876) was an important early contribution to what would now be called food cycles and energy budgets in aquatic ecosystems.[10] He also worked in terrestrial ecology. Though the work was outside academe, he drew academic scientists into his projects. Major figures in plant and animal ecology served on the board of the Illinois Natural History Survey, and the agency hired, among others, Victor Shelford and Charles C. Adams, early graduates in ecology of the University of Chicago.[11]

Another agency that played a key role in the period before academic programs were well established was the U.S. Department of Agriculture's Bureau of Biological Survey. It did not seem, on the surface, a place for ecology. Its chief (from 1885 to 1911) was C. Hart Merriam, a man so identified with natural history that his biographer called him "the last of the naturalists," and the agency was established for quite practical projects.[12] Congress had established it, in 1885, in response to a petition by the American Ornithological Union Congress for an agency to study the impact of wildlife on crops. It was a one-man office, the Office of Economic Mammalogy and Ornithology, and it was part of the Department of Agriculture's Division of Entomology – a venue dedicated to helping American farmers. Merriam built the agency, changed its name to the Bureau of Biological Survey, and made it a center for field research. He never had much money (in part because of congressional dissatisfaction with his penchant for work that seemed only distantly related to agriculture) but there was always some, and the agency bridged the gap

---

[9] Leland O. Howard, *A History of Applied Entomology*, Smithsonian Miscellaneous Collections, Vol. 84 (Washington, D.C.: Smithsonian Institution Press, 1930), 28; Stephen F. Forbes, "The Ecological Foundations of Applied Entomology," *Annals of the Entomological Society of America*, 8 (1915), 1–9.

[10] *Bulletin Science Association of Peoria, Illinois*, 1887, 77–87. This is reprinted in Keir Stirling (editor), *Ecological Investigations of Stephen Alfred Forbes* (New York: Arno, 1977), which offers a fuller selection of Forbes's papers. On the development of limnology see Golley, *A History of the Ecosystem Concept*. On "The Lake as Microcosm" see Joel B. Hagen, *An Entangled Bank* (New Brunswick, N.J.: Rutgers University Press, 1992), 8–10.

[11] McIntosh, *Background to Ecology*, 47–8.

[12] Keir Sterling, *Last of the Naturalists* (New York: Arno, 1977).

between the museum era and that of academic departments. In the first decades of the century almost every major figure in mammalogy and animal ecology worked on one or another of its surveys and projects, and all depended upon its publications. Even after academic departments came to be the center of research, it continued to play a role in field biology, supporting the developing field of game management in the 1930s and, as the U.S. Fish and Wildlife Service, doing pioneering work on restoring endangered species in the 1950s and 1960s.

Merriam's own contribution was an exercise in classical biogeography. Using the Survey's work on Arizona's San Francisco Peak, he divided the continent into a set of biological regions, which he called life-zones. The idea, and the correlation he made between altitude and latitude, dated to Humboldt and were by now common wisdom. Speaking of crop introductions in this period, David Fairchild, head of the USDA's Office of Introductions, spoke of the "common delusion . . . that high altitudes in tropical mountains resemble the temperate zone."[13] Merriam's formulation, though, was more rigorous and mathematical. Temperature set the boundaries, the northern extension of plants being determined "by the sum of the positive temperature for the entire seasons of growth and reproduction, and the southward distribution . . . by the mean temperature of a brief period during the hottest part of the year."[14] A generation of ecologists used the formulation to think about the plant and animal communities around them, and its publication in a government report took it into a wider world. Several generations of American schoolchildren studied maps of the continent with variously colored regions accompanied by illustrations of an idealized mountain peak labeled "Arctic" at the top and "Tropical" at the bottom.

Agencies, though, were stopgaps, for they could not provide the stability that research programs testing theory required. University departments could, and the American system was open to the new field. It was expanding, so in many universities there were no or few established interests to block a new discipline. Science was also becoming more important, as American universities modeled themselves on the German system. There was also the developing system of land-grant schools and their associated experiment stations, which created opportunities for applied biological research on an unprecedented scale.[15] By 1900 two

[13] David Fairchild, *The World Was My Garden* (New York: Scribner's, 1938), 77.
[14] Quoted in Frederic E. Clements, *Plant Succession and Indicators: A Definitive Edition of Plant Succession and Plant Indicators* (New York: H. W. Wilson, 1928), 288.
[15] On the land-grant colleges and their contributions see A. Hunter Dupree, *Science and the Federal Government* (Cambridge, Mass.: Harvard University Press, 1957), chap. 8, and Howard, *History of Applied Entomology*, 74–7. On Britain see John Sheail, *Seventy-five Years in Ecology: The British Ecological Society* (Oxford: Blackwell, 1987), 40–1.

schools had developed programs that would dominate ecological studies in the United States for the next two generations. One was at the University of Nebraska, a "cow college," the other at the University of Chicago, a mushroom school nourished by John D. Rockefeller's philanthropy. Nebraska's grew out of Charles Bessey's graduate seminar in botany. He had come to the school in 1884 as an advocate of the "new botany," a movement to give that field a new foundation, replacing observation and qualitative judgments with quantifiable data and testable theories.[16] Taxonomy was to yield to physiology, the field to the laboratory. When Bessey formed a graduate seminar in the mid-1890s, he infused that emphasis on rigor and new methods into his students' work.

The stimulus for their ecological approach came from Oskar Drude's work, *Deutschlands Pflanzengeographie* and *Handbook of Geographical Botany*. The first introduced them to the European tradition of plant biogeography; testing the second gave them a research program. Unable to confirm Drude's division of North America into botanical regions, they began to study plant distribution and abundance on the prairies, concentrating on problems of succession. Botanists had commonly estimated plant abundance by looking at the landscape. Bessey's students, working with more rigor, found that just looking at the land was not enough. Even long field experience in a region, they warned, "without actual enumeration of individual plants" gave an erroneous picture of what was there.[17] Choosing, at random, areas of one or five square meters, called quadrats, they counted the individual plants and from a series of these plots – in the extreme, across a region or the entire state of Nebraska – developed a precise picture of what was there and how it changed in space. This seems simple, but, as historian Ronald Tobey pointed out, it involved not just a new method but an epistemological shift.[18] Nature was not seen, it was constructed, built up in the observer's mind not from sense impressions but from data gathered in a theoretical framework. This was a sharp break with popular understanding and, particularly as it was extended over time, gave a very different picture of the land.

This vision was strongest in the Nebraska school's major contribution to ecological theory, Frederic Clements's work on plant succession. It dominated American plant ecology for two generations (and gave it a distinct national cast) and trickled down to the classroom and Boy Scout

[16] Ronald Tobey, *Saving the Prairies* (Berkeley: University of California, 1981), is the basic source for this discussion of the grasslands school.
[17] Roscoe Pound and Frederic Clements, 1902, quoted in ibid., 65.
[18] Tobey, *Saving the Prairies*, 57–75.

manuals. In its principles and on a small scale Clements's work seemed not only reasonable but part of common sense. A bare area of ground, say a sandbar thrown up by a river or a burned-over field, was first colonized by species that flourished in open areas. They changed the land, creating new conditions that favored other species. These in turn prepared the ground for another set. Eventually the "climax community," as Clements called it, grew up, a group that reproduced in place, remaining until some disturbance set back the succession. The idea of plants succeeding each other, though, was beyond what people saw, and as ecologists moved from small areas to large, their view was very different from the common, commonsense view. Ordinary people saw an apparently changeless sea of grass. Ecologists saw vast associations covering entire continents, succeeding each other over centuries, their rise and fall caused by changes in rainfall and temperature. Even stranger was Clements's belief (which was never fully accepted even by his colleagues) that the associations themselves formed superorganisms which had, like conventional ones, a life and a life cycle.

All this grew out of common ideas and apparently simple research techniques. Ecologists defined plants as everyone did, and their conception of climate was familiar, if stated in strange language. It was easy, Clements said, to trace rainfall and evaporation on a map and use the lines to mark off the division between prairies and plains. "Such an assumption reverses the proper procedure, in which the associations themselves must be permitted to indicate their respective climates."[19] That was what farmers did. They saw the climate in terms of what would grow. It was Powell's method as well. The ecologists' basic technique, counting plants, was understandable. Yet all this, produced by a formal approach to common things and common activities, produced a view as strange as Darwinian evolution, that other great advance based in observation, and it reshaped the study of visible nature even more thoroughly than evolution did.

Ecology at the University of Chicago, the other major center of American work, also drew on botany and European work and concentrated on questions of community and succession. It had, though, a different base, in the newer disciplines of embryology, physiology, and animal behavior, and the program developed in animal as well as plant studies.[20] The stimulus for early work was the studies of Eugene Warming, a Danish scientist, which appeared in the United States in

---

[19] Clements, *Plant Succession and Indicators*, 246.
[20] The best study of the Chicago ecologists is Gregory Mitman, *The State of Nature* (Chicago: University of Chicago, 1992). On the foundations of the school see Kingsland, "Defining Ecology," and McIntosh, *Background to Ecology*.

1896 in German translation. Henry Chandler Cowles quickly applied them to his work on the sand dunes on the southern shore of Lake Michigan. It was a unique area that would support a generation of ecologists. The lake was continually building dunes, and by walking inland from the shore Cowles could walk back in time, examining successively older plant communities. In 1907 one of his students, Victor Shelford, used the same area and techniques to chart the distribution of species of tiger beetles on the dunes and relate them to the plants. The school continued to develop, producing students in both plant and animal ecology and developing a research tradition as influential and more far reaching than that of the grasslands ecologists of Nebraska.[21]

Both these schools produced what seemed to be, and were at times said to be, universal laws. This seems reasonable. Plants were plants, animals were animals, and sun, rain, temperature, and other physical variables that affected the globe drove all natural systems. Each land, though, was different, and those differences shaped what ecologists studied and even how they studied it. This was particularly true of plant ecology. Prairie ecologists, for instance, worked with large areas, much of it only slightly affected by Anglo settlement when they were doing their early work. Their theories emphasized natural processes and said little about the role of humans in nature. British ecologists worked under very different conditions, and they had other views about a proper study area. The "belief is sometimes met with," the English ecologist A. G. Tansley said, "that only perfectly 'natural' vegetation is a proper subject for ecological study. If this were true, the ecological field in Great Britain would be very limited indeed."[22] British ecologists developed theories and research techniques suited to a smaller scale and a finer mosaic of communities, and they took people into account. They had, consequently, little interest in Clements's ideas. The Americans' most enthusiastic disciple outside the United States was a South African, John Phillips, who worked in a country with large grasslands hardly affected by industrial farming.[23] There were also national intellectual differences. British ecology owed more to natural history and less to the continental tradition of biogeography than American ecology, and researchers tended to be less specialized (which may have been a consequence of a

---

[21] Victor E. Shelford, *Animal Communities in Temperate America* (Chicago: Geographic Society of Chicago, 1913). On Shelford's career see Robert A. Croker, *Pioneer Ecologist: The Life and Work of Victor Ernest Shelford, 1877–1968* (Washington, D.C.: Smithsonian Institution Press, 1991). More generally, see Mitman, *State of Nature.*

[22] A. G. Tansley, *Practical Plant Ecology* (London: Allen & Unwin, 1923), 22–3.

[23] On British reaction see Tobey, *Saving the Prairies*, chap. 6, and McIntosh, *Background to Ecology*, 44–7.

smaller university system and fewer jobs).[24] One of the British pioneers in this field, Julian Huxley, spoke of the period in which he trained a group of bright future ecologists ("the half-dozen years after the War") as preparing for a "renascence of general biology." His students certainly turned their hands to a wide variety of problems. They went on to careers in embryology, oceanography, ecology, entomology, genetics, and anthropology.[25] The Nebraska ecologists wound up in botany, plant ecology, and range management.

Animal ecologists faced more common conditions, at least in part because the available technology forced them to work on the same kinds of animals and ask the same kinds of questions. Had the Americans been able to study grizzlies, wolves, or mountain lions (but that would wait until the 1960s, when radio collars became available), they might have developed an animal ecology very different from Britain's. As it was, everyone worked on questions of growth, reproduction, association, and competition among and within populations, and the kinds of small animals they could study were widely scattered. They all used the same conceptual tool kit, which Charles Elton laid out in his *Animal Ecology* (1927). It was an attempt to capture essentials. Forty years later Elton recalled that he had written the book in three months, "working at strong pressures and with a feeling of clarity that I wish that I could call back now, when the complexities of the subject are more apparent."[26] He did not precisely invent, but he did make comprehensible, such concepts as food chains, food cycles, trophic levels, and niches, which seemed to offer powerful tools for studying natural processes. Who ate whom and in what quantity was certainly a promising way to sort out relationships among and pressures on populations. The ideas and techniques were applicable everywhere; all systems had plants and plant eaters, species that preyed on the plant eaters, and recyclers that took care of them all. The scheme also unified many things through a single kind of process. Though Elton did not quite put it in this fashion, what he was recommending was the study of the flow of energy through the system. (That his terms echoed those of physics, the epitome of "science," may have increased the attraction.) He also insisted on rigor, another hallmark of respectability, and suggested ways to get it. Research should "revolve around censuses, the structure of the population by age

[24] Sheail, *Seventy-Five years of Ecology*, 16–21.

[25] Julian Huxley, Introduction to Francis Ratcliffe, *Flying Fox and Drifting Sand* (1938; Australian edition, Sydney: Angus & Robertson, 1947), vii. On ecological training see Peter Crowcroft, *Elton's Ecologists: A History of the Bureau of Animal Population* (Chicago: University of Chicago Press, 1991).

[26] Elton, *Animal Ecology*, vii.

and sex, the birth-rates and death rates, movements, as well as the influence on these of outside changes and interrelations."[27] His ideas gained force from his work as a teacher. (The equivalent mechanism in natural history had been correspondents and protégés.) He founded the Bureau of Animal Population at Oxford in the early 1930s and taught there for thirty years.

The background that went into these ideas suggests some of the connections among the new science, earlier nature knowledge, and the culture at large.[28] He owed much to natural history, relying heavily on the "old-fashioned" work of mapping out populations. His respect was great enough that he named his department after the Bureau of Biological Survey, whose work he admired. Like his predecessors, he drew on studies of human populations. Both Darwin and Wallace credited Thomas Malthus's *Essay on Population* with stimulating their thinking about competition in nature. Elton used Carr-Saunders's *The Population Problem*. His geographic focus, though, was directly opposite that of natural historians. They had sought the tropics and reveled in the profusion of life and the multitude of species. He worked in the Arctic, where species were fewer and it was therefore easier to sort out interactions among them. He took part in the Oxford expeditions of 1921, 1923, and 1924 that surveyed Spitzbergen Island. He developed a continuing interest in a phenomenon of the Arctic, the almost regular rise and fall of several wildlife species. His interest in the problem came from a pioneering study by Gordon Hewitt, a British zoologist who had gone to Canada in 1909 as Dominion entomologist. (He also served as consulting zoologist and Canadian representative on the International Commission for the Protection of Nature.) In *Conservation of the Wildlife of Canada* he had used the Hudson's Bay Company fur returns to chart the almost regular changes in the abundance of snowshoe hares and lynx in northern Canada that had occurred over more than two hundred years.[29] Elton also used contemporary theory. As an undergraduate he had read Shelford's *Animal Communities of North America*.

Elton combined theory and testing theories with a concern for the application of his work, which ran against the scientific ethos of the period. Science was supposed to be "pure," but Elton worked, and

---

[27] Ibid., xi.

[28] This account relies on Crowcroft, *Elton's Ecologists*, 3–10.

[29] On Hewitt see Howard, *History of Applied Entomology*, 186–7. Elton's major work on cycles was *Voles, Mice and Lemmings: Problems in Population Dynamics* (Oxford: Clarendon Press, 1942). The subject is still of interest. A recent review is Kai Norrdahl, "Population Cycles in Northern Small Mammals," *Biological Review*, 70 (1995), 621–37. One participant's overview, stretching from the 1930s, is Dennis Chitty, *Do Lemmings Commit Suicide?* (New York: Oxford University Press, 1996).

insisted his students work, on projects that had immediate utility. In the 1930s that meant the study of agricultural pests and – with money from the Imperial Chemical Industries, manufacturers of shotgun shells – game birds. During World War II, rodent control, in everything from crops to stored grain to city sewers, came to the fore. After the war the Bureau gave short courses on applied problems in agriculture (particularly pest control), public health (the ecology of mosquitoes was a prime interest), and other subjects ranging from fisheries to forestry.

Elton developed close ties to American game managers after meeting Aldo Leopold in 1931 at the Matemak conference. This venue suggests some of the less obvious connections in the study of scientific questions. It was sponsored by an eccentric millionaire, Copley Amory, and held at his fishing camp in Canada. He was not a scientist and not particularly interested in wildlife. His fascination, perhaps obsession, was with cyclic phenomena, and he looked at everything from sunspots to the stock market. He later funded an institute for the study of cycles. (Among its findings was the information that there was a four-year cycle of mouse populations in the United States whose low coincided with presidential election years. The report said that "there is probably no connection between these phenomena.")[30] Elton and Leopold had a more concrete agenda. "Elton was laying the foundations of ecology; Leopold was attempting to apply the science even before the foundations were set."[31] Leopold was soon recommending Elton's work to his friends, and he was one of Elton's hosts on the latter's trip to the United States in 1938. Elton, in turn, acknowledged "the great stimulus which American investigators such as Leopold, Stoddard, and Errington have given me, both personally and from their writings."[32]

Science Outside the Centers

Departments and research were essential, but so were professional means of communication. Here too there was a marked difference between metropolis and periphery. In 1913 British ecologists formed the British Ecological Society and immediately began publishing the

[30] Notes from a meeting between representatives of Amory's Foundation for the Study of Cycles and the Dominion Park Service's Wildlife Office, 13 November 1941, Box 38, U300, Records of the Canadian Wildlife Service, RG 84, Public Archives Canada, Ottawa.

[31] Kurt Meine, *Aldo Leopold* (Madison: University of Wisconsin Press, 1988), 284.

[32] Charles Elton, "On the Nature of Cover," *Journal of Wildlife Management*, 2 (1939), 332. On Leopold's recommendations see, e.g., letters for 1931–2 in Correspondence – Stoddard, Aldo Leopold Papers, Department of Wildlife Ecology, University of Wisconsin Archives, Madison.

Journal of Ecology. The next year American ecologists, many of them from Nebraska and Chicago, established the Ecological Society of America and in 1918 took over *Plant World*, renaming it *Ecology*.[33] Australia, Canada, and New Zealand lacked resources and people, and until the 1950s research there was a matter of isolated researchers often looking abroad for intellectual companionship and places to publish.[34] Canadians often went to the United States. One of them was Harrison Lewis, in charge of enforcing the Migratory Bird Act in Canada. He found that institutions at home were not interested in the applied work he wanted to do. The chairman of the zoology department at McGill University indignantly rejected his proposal for a thesis on the life history of the eider duck with the objection that "this is a university, not a draper's shop."[35] Lewis went south, where he earned his degree at Cornell under the eminent American ornithologist A. A. Allen and established close professional and personal ties with American scientists.[36] So did many others. When they did not, American ideas came to them, for the continent was a biological whole. One example of this was the work done under the Migratory Bird Treaty of 1918. Requiring Canada and the United States to manage continental waterfowl populations, it almost forced professional interchange and provided opportunities that contributed to Canadians' education.[37] Only after World War II did Canada get its own wildlife research agency and only in 1960, when Ian McTaggart Cowan established a program at the University of British Columbia, could Canadians learn ecology in their own country.[38]

[33] Robert L. Burgess, *The Ecological Society of America* (Oak Ridge, Tenn.: Oak Ridge National Laboratory, n.d.), reprinted in Frank N. Egerton (editor), *History of American Ecology* (New York: Arno, 1977).

[34] George Basalla, "The Spread of Western Science," *Science, 156* (5 May 1967), 611–22, provides the conventional view, now largely rejected, of this development as the "colonial phase" of national development. Critiques can be found in R. W. Home (editor), *Australian Science in the Making* (Cambridge University Press, 1988) and Nathan Reingold and Marc Rothenberg (editors), *Scientific Colonialism: A Cross-Cultural Comparison* (Washington, D.C.: Smithsonian Institution Press, 1987).

[35] Harrison Lewis, "Lively: An Informal History of the Canadian Wildlife Service," unpublished manuscript, RG 109, Records of the Canadian Wildlife Service, Public Archives Canada, Ottawa, 108.

[36] Ibid., 128.

[37] F. Graham Cooch provided recollections of his own education here in "Canadian Connections," in U.S. Department of the Interior, *Flyways* (Washington, D.C.: U.S. Government Printing Office, 1984), 304–9.

[38] On Canadian reliance on American training see memo on training in Box 37, U300, RG 84, Public Records Canada, Ottawa. Lewis, "Lively," is an excellent source of information on this topic. Graham Cooch, author's interview, 5 June 1989, said that the development of training in wildlife management in Canada was causing students to lose their close personal connections to the larger American community, which he regarded as a major source of information and stimulation.

Australia and New Zealand had even fewer people and no well-developed and biologically identical neighbor from whom to borrow. The result was little organized work and less government support. In the early meetings of the Australia and New Zealand Association for the Advancement of Science (formed in 1889) proposals circulated for a biological survey. In 1921 a biologist proposed an Australian bureau of biological survey, on American lines, and in 1939 Crosbie Morrison, then the most prominent nature writer in Australia, made a similar suggestion.[39] Nothing came of these. This is not to say that no ecological work was done. In the interwar period an Australian, A. J. Nicholson, developed a mathematical model of population stability among insects based on density-dependent controls (pressures that became more severe as population density increased and declined as it did). This set off a long, international argument over the nature of population controls and made him "in the light of a minor deity" among Elton's students at Oxford, but it did not grow out of a research tradition nor did it establish one.[40] Nicholson operated on a shoestring, working with insects, which were small and abundant. A larger form might have yielded results that would be seen as more widely applicable, but it would have raised the cost of research considerably. The only ecological work on mammals was Francis Ratcliffe's study of flying foxes (fruit-eating bats), and Ratcliffe was an Englishman, hired by the Committee on Scientific and Industrial Research to study what was seen as a pest control problem (see Chapter 6). (Graeme Caughley, a New Zealander, said that his 1961 dissertation on kangaroos was the first ecological work done at the University of Sydney on a mammalian species.) To find a mathematically sophisticated collaborator Nicholson had to turn to a physicist, V. A. Bailey. To publish he had to send his papers to Britain.[41]

Australian museums continued to be the center of research, and the major efforts were natural history surveys. This was, in part, a response to the situation. Little systematic collecting had been done over much of the country; the first scientific surveys of what became Kakadu

---

[39] Linden Gillbank, "The Life Sciences: From Collection to Conservation," in Roy MacLeod (editor), *The Commonwealth of Science* (Melbourne: Oxford University Press, 1988), 99–129; Walter W. Froggart, "Presidential Address: A Bureau of Biological Survey," *Australian Zoologist*, 2 (January 1921), 2–8; Crosbie Morrison editorial, March 1939, in *Wildlife*, quoted in Graham Pizzey, *Crosbie Morrison* (Melbourne: Victoria Press, 1992), 131–2.

[40] Sir Ian Clunies Ross, quoted at 186 in John L. Hopper, "Opportunities and Handicaps of Antipodean Scientists: A. J. Nicholson and V. A. Bailey on the Balance of Animal Populations," *Historical Records of Australian Science*, 7 (1989), 179–88.

[41] McIntosh, *Background to Ecology*, 188–9; Graeme Caughley, author's interview, 4 June 1992.

National Park, for example, were done in the 1950s.[42] Frederic Wood Jones's *The Mammals of South Australia* (three volumes, 1923–5), the major study of native Australian mammals during this period, is typical. It was a standard list of mammals aimed at a general audience. It too was produced with little institutional support. South Australian biologists did the work without pay, and the state printed it and made it available to schools and the public. The Preface lamented the lack of inexpensive but accurate books on the plants and animals of South Australia and the handicap that this was "to young Australia and so to the progress of Australian science."[43] H. H. Finlayson's expedition to the "Red Centre" in the early 1930s is another example of the persistence of an older form. It was sponsored by officials of the Northern Territory government interested in natural history, and its outstanding achievement was in this tradition: the rediscovery of a species described in the nineteenth century and never found again.[44] Finlayson even used a classic method of finding the animal. He hired Aborigines. He reported to the public as well as professionals, writing a series of newspaper articles, which were the basis for *The Red Centre: Man and Beast in the Heart of Australia,* and organized the book like a local natural history. Beginning with the land it went on to plants, animals, and people, combining interests that science had been separating for the last century – nature appreciation, biological collecting, and anthropology (see Chapter 7).

There were even fewer ecological studies in New Zealand, though not for lack of subjects. The island's plants and animals were part of the biota of the supercontinent Gondwana; many were found only on the islands; and they were daily being displaced by exotics. There were, though, fewer than a million Anglos in New Zealand, little money for research, and even less institutional support. The careers of Leonard Cockayne, the preeminent botanist in New Zealand in his generation, and G. M. Thomson, who wrote the first survey of exotics in New Zealand, show the possibilities and difficulties of the situation. Both were British immigrants and like Stephen Forbes had trained by apprenticeship and self-study. Both were important scientific figures in the islands. Each had a term as president of the New Zealand Institute. Thomson helped reform the Institute in the 1880s, was active in organizing the

---

[42] Information on the lack of surveys in the tropical north comes from the author's interview with John Calaby, research scientist with CSIRO, 4 June 1992.

[43] Frederic Wood Jones, *The Mammals of South Australia* (Adelaide: Government Printer, 1923–5), 1: Preface.

[44] H. H. Finlayson, *The Red Centre: Man and Beast in the Heart of Australia* (Sydney: Angus & Robertson, 1935), 96–7.

forest and bird society in 1914, and worked on the 1921–2 revision of the wildlife laws. Cockayne surveyed several islands as possible refuges for native species and advised the government on deer policy. Science was not Cockayne's or Thomson's profession, though. Cockayne was a horticulturist, Thomson science master of the Otago high schools.[45]

Though not trained in ecology, both worked in the field. Cockayne, like Forbes, studied ecological problems before the discipline was well defined; by 1900 he was publishing on one of its major topics, succession. During the next decade he gained field experience doing botanical surveys of native forests, primarily on islands. (That the government relied on his part-time labor and did not establish an office was one indication of limited resources.)[46] He was primarily interested in cultivated plants and was only "secondarily a field botanist," but he was one of the few in the colony, and in 1904 that brought him a distinction "unique among British and Empire botanists," an invitation to write the volume on New Zealand for the German botanical series *Die Vegetation der Erde*.[47] Delayed first by the need for field studies, then by World War I, it appeared in 1920. Conceptually, it was not new, but it was the only work that mapped the islands' plant communities in a modern context, and it crowned his career. The other milestone of the period was Thomson's *The Naturalization of Animals and Plants in New Zealand*, an encyclopedic book, never repeated and not yet fully supplanted. It described all the introductions above the level of microscopic life, from rainbow trout and deer to trees. It was very much in the tradition of natural history and had its conceptual foundations in that field. "[T]he biological question of the origin of species," Thomson said, "was the *raison d'être* of this work. . . . [T]he conviction early grew upon me that here in New Zealand was a field in which the accuracy of Darwin's views in certain directions could be put to the test."[48] Like Cockayne's, it was a notable effort, but also like his a solo project that did not grow out of a research

[45] C. A. Fleming, *Science, Settlers and Scholars*, Bulletin 25, Royal Society of New Zealand (Wellington: Royal Society of New Zealand, 1987), 31–3; Galbreath, "Colonisation, Science, and Conservation" (unpublished Ph.D. dissertation, University of Waikato, 1989), 12–14.

[46] Leonard Cockayne, "Report on a Botanical Survey of Kapiti Island" (1907); "Report on a Botanical Survey of the Waipoua Kauri Forest" (1908); "Report on a Botanical Survey of the Tongariro National Park" (1908); "Report on a Botanical Survey of Stewart Island" (1909); "Report on the Dune-Areas of New Zealand, Their Geology, Botany, and Reclamation" (1911). All were issued by the Department of Lands and Survey (Wellington: Government Printer).

[47] "Obituary," *Transactions of the New Zealand Institute*, 65 (1936), 457–67; Sheail, *Seventy-five Years in Ecology*, 109, 109–10.

[48] G. M. Thomson, *The Naturalization of Animals and Plants in New Zealand* (Cambridge University Press, 1922), 1: 503.

program or lead into one. It would be another generation before ecology, as a self-conscious field in a stable institutional setting, came to New Zealand.

Much of this is the usual story of a scientific speciality. An accumulation of information in an established field (natural history) led researchers to ask new questions, which they found new ways to answer using ideas and perspectives from the old area, other fields, and the culture at large. A new focus grew from their work, producing a new community of interest and then a new discipline. Ecologists, like everyone else, used the institutional apparatus available. That the field centered in the countries with the most resources and the best-developed scientific communities is hardly surprising, and the use of these ideas on the periphery equally familiar. The cost of the new fields, though, meant that there was an important shift in power relations at the turn of the century. The difference between center and periphery had been the possession of specialized knowledge, and it had seemed possible to develop that even in places like New Zealand. The great natural history museums had been concrete, or at least stone and timber, embodiments of that aspiration. Now the difference was the ability to train new workers and carry on research programs, and this was a different matter. Certainly there were in Canada, Australia, and New Zealand in the early decades of the century no counterparts of the local societies and cabinets of specimens that had marked the beginnings of natural history in the settler nations. These countries would find ways to acquire expertise that they could not afford to develop, but that would take time, and the agencies and their strategies would testify to the new conditions.

Not a Normal Science

So far the analysis makes ecology seem a normal part of modern science. Certainly the same analysis could be applied to most fields. Even before World War I, for example, Ernest Thomson had found it necessary to leave Nelson, South Island, and go to Britain to earn his Nobel Prize in physics, and by the 1960s the costs of research in this area were so high the nations of Western Europe had to pool their resources to buy their physicists the latest in atom smashers. Ecology, though, had its own peculiarities. For one thing, it did not comfortably fit into biology. For another, its explanations were not what the reductionist ethos of science called for. Ecologists also had a different relation to their subject matter, one they shared with other field biologists, which was at odds with science's ideal of detachment. The first of these was an open question, debated in the literature. The theoretically inclined discussed the second. No one talked about the third.

An early problem was ecology's place in the temple of science (a phrase of the time). Elton called it scientific natural history, others thought this an oxymoron. Some believed ecology to be a branch of general physiology; others considered it a fundamental division of biology; still others saw it as a new science that should incorporate sociology (as the science of human interactions).[49] Intellectually, it seemed to have no center, being divided in ways that suggested it was less a subject than an approach to one. Plant ecology and animal ecology had little in common, and there was a major methodological split between the study of a single species (autecology) and a system (synecology). It was not clear on what level – that of individuals, populations, or species – systems should be studied. Even defining research areas was difficult. (One reason for limnology's early rise was that streams and lakes had banks.) As late as 1952 an eminent ecologist, Lee Dice, said that the ecologist's need to rely on other areas of biology was great enough that it "has led certain biologists to the opinion that ecology is not a separate science, but mainly a point of view." He disagreed, but he had to fight the battle.[50] Ecologists so often met these problems by coining terms that there arose the derisive definition of ecology as that part of biology wholly given over to nomenclature. The terms, though, were also part of ecologists' struggles to put a veneer of conventional science on what seemed unsophisticated observation. In an age that equated science with quantitative measures and theories, the field was distressingly qualitative. Measurements might be precise but theories had little mathematical sophistication.

Then there were ecologists' explanations. Reductionism was the norm and the ideal in science. A phenomenon was explained when it was described in terms of the laws governing a more fundamental level of organization. The functioning of living organisms, for example, was to be stated in terms of the chemical reactions that supported digestion, movement, and so on. Chemical phenomena, in turn, were explained by appeal to the properties of atoms. At the philosophical extreme all was to be explained in terms of physics, which became the fundamental science (a position most popular among physicists). Ecological explanations ran the other way. Clements did not explain associations in terms of individual plants or their physiology but saw the parts (the plants) as forming a new whole (the community), which had to seen on its own terms. The properties of the system, in this view, emerge as the system

---

[49] McIntosh, *Background to Ecology*, 1–27.
[50] Lee R. Dice, *Natural Communities* (Ann Arbor: University of Michigan Press, 1952), 1. On the current state of this question see Ernst Mayr, *This Is Biology* (Cambridge, Mass.: Harvard University Press, 1996), 221–3.

becomes more complex, and they are a product of that complexity. They cannot be explained by reference to simpler levels of organization. "Emergence," the term for this view, was not quite respectable, but it was concealed in many positions and sometimes appeared in the pronouncements of established figures. The physiologist Claude Bernard, deeply committed to a materialist position and to reductionist explanations, implicitly appealed to it in saying that the living was not reducible to the nonliving.[51] In his text on game management Aldo Leopold remarked that "some of the properties of game species are *not* discernible in the individual bird or mammal, but become apparent only through the study of behavior of large aggregations of individuals, or game populations."[52] No one, though, developed this insight, if such it was, in ecology. "Holism," the popular term now associated with it, came not from an ecologist but a South African general, Jan Christian Smuts, who wrapped the idea in such a cloud of late Victorian idealism that few scientists paid any attention. Ecologists concentrated on making rigorous observations and formulating laws. They tried, in short, to fit in. They would not take emergence seriously until the 1970s.[53]

Ecologists' darkest secret was their passion for their subjects. Scientists were supposed to be objective. They were to assess all ideas and observations without regard to their own interests and have no emotional interest in the subjects of their research. As far as theories went ecologists had no more (or less) trouble with that stricture than anyone else, but when it came to the focus of their research it was a different matter. No one had a personal relationship or experience with atoms, compounds, rocks, or muscle tissue, but everyone had some connection to visible nature. Often it was a love of nature or some particular form of life that drew people to make their careers in field biology, and this passionate concern was common even in what seem the dull and dry areas of the study of visible nature. Consider plant taxonomists. They seem the extreme of dry, rigorous, imperial scientists – going to the field, finding new specimens, classifying them, and putting them on sheets of paper in filing cabinets. Edgar Anderson, a prominent member of the tribe, pointed out that far from being cold and detached "the average taxonomist is an incurable Romantic." He has an "appetite for virgin vegetation [that] rises directly from some deep emotional urge rather than from cool professional competence." In Central America, he said,

---

[51] Maurice H. Mandelbaum, *History, Man and Reason: A Study in Nineteenth-Century Thought* (Baltimore: Johns Hopkins University Press, 1971), 379–80.

[52] Aldo Leopold, *Game Management* (New York: Scribner's, 1933), 47.

[53] Jan Christian Smuts, *Holism and Evolution* (London: Macmillan, 1926). On emergence in biology see Ernst Mayr, *The Growth of Biological Thought* (Cambridge, Mass.: Harvard University Press, 1982), 63–4.

his colleagues would pass over any number of fields and cut-over forests offering important taxonomic problems to get to the cloud forest. And "[w]hen the taxonomist mentions the cloud forest, when he makes his plans to go there, when he climbs the lower slopes of the peak, his eye lights up and his voice has a richer vibrato; obviously he is somehow in love with the cloud forest." Not, Anderson admits, that it is not awe-inspiring. It is. But the taxonomist, rushing here and there, grabbing everything in sight, seems to wish to "make the cloud forest a part of his professional self, to get the forest into his grasp."[54]

Ecologists were much the same. Some in the Nebraska school came to believe that no one could understand the grasslands without having seen the great sea of grass as it was before it had been plowed up in the late nineteenth century – a Romantic notion if there ever was one.[55] They were not alone. Three years after founding the Ecological Society of America the members established the Committee on the Preservation of Natural Conditions for Ecological Study. The title suggested a rigorous professional outlook, but there was more to it than that. It was so active in ways that some members perceived as inconsistent with proper scientific detachment that after World War II the group was split off from the society as a private organization, the Nature Conservancy.[56] This strain was strongest in North America, but present elsewhere. British ecologists also discussed saving natural areas and set up a committee that, after World War II, became the British Nature Conservancy.[57] Finlayson regretted that the "old Australia is passing. The environment which moulded the most remarkable fauna in the world is beset on all sides by influences which are reducing it to a medley of semi-artificial environments, in which the original plan is lost and the final outcome of which no man may predict."[58] Leonard Cockayne wrote that "[f]ew incidents are more to be regretted in the settlement of new countries

---

[54] Edgar Anderson, *Plants, Man, and Life* (Berkeley: University of California, 1971), 45–6. The focus on undisturbed areas continues. See more recent comments by D. Adamson and R. Bucanan, "Exotic Plants in Urban Bushland in the Sydney Region," cited in R. L. Amor and C. M. Piggin, "Factors Influencing the Establishment and Success of Exotic Plants in Australia," in Derek Anderson (editor), *Exotic Species in Australia: Their Establishment and Success*, Proceedings of the Ecological Society of Australia (Adelaide: Ecological Society of Australia, 1977), 10: 15.

[55] See Tobey, *Saving the Prairies*. Victor Shelford's *Animal Communities of North America* (Chicago: Geographic Society of Chicago, 1913) and Shelford and Frederic Clements's *Bioecology* (New York: Wiley, 1939) recreate in the imagination the landscape of the continent before the whites arrived, as does Lee Dice's *Biotic Provinces of North America* (Ann Arbor: University of Michigan Press, 1943).

[56] Burgess, *The Ecological Society of America*.

[57] Sheail, *Seventy-five years of Ecology*.

[58] Finlayson, *The Red Centre*, 16. See also Baldwin Spencer, *Wanderings in Wild Australia* (London: Macmillan, 1928).

than the more or less complete destruction – unavoidable in many cases – of the fauna and flora. This is especially to be deplored when the members of these are of a rare or peculiar character. Such destruction has taken place in New Zealand to an extreme degree."[59] Three years later he said that his survey of Stewart Island was part of an effort "to put on record, before it is too late . . . the general aspect of the vegetation of typical or specially interesting districts." Speaking of the reserved areas, he said that it "now remains to see that these reserves are kept sacred." If the areas are "religiously guarded, our own people, and visitors from all parts of the world, will be able to see the wonderful plant-life of New Zealand and her unique birds exactly as nature planted the one and provided for the other."[60]

Sentiment was strongest among plant ecologists. One reason was that the destruction of plant communities was usually more visible than that of animals, but another was that plants seemed self-sufficient. A world of plants was plausible, one of animals was not, and plant ecologists thought of their subjects in just these terms. It was unnatural that their beloved plants should vanish down a cow's throat or in a fire. For Clements and the other grassland ecologists, prairie fires and buffalo were not part of the community, merely "disturbances" that "set back" succession. Leonard Cockayne, living in a land without native grazing animals, had the greatest opportunities, and he took them. He came to see primeval New Zealand as a plants' paradise and exotic ungulates, particularly red deer, as the snake in Eden. He argued that because New Zealand forests had developed "unexposed to the attacks of grazing and browsing animals, the moa (Dinornis) excepted" (and he tended to overlook the moas), they were uniquely vulnerable.[61] In the late 1920s he decided that since red deer were still protected, "long after their baneful influence in destroying great national wealth was plainly manifest, it is surely time to protest emphatically." "[T]hese priceless forests of ours are in imminent danger of being transformed into debris-fields and waste ground, and the water which they controlled become the master, pouring down the naked slopes after each rainstorm, bearing with it heavy loads of stone, gravel and clay to bury the fertile arable lands below and occasion floods in the rivers. Is the protection of deer

---

[59] Cockayne, "Botanical Survey of Kapiti Island," 2.

[60] Cockayne, "Report on Stewart Island," 2, 41, 42.

[61] Graeme Caughley, *The Deer Wars* (Auckland: Heinemann, 1983), chap. 6, provides an excellent analysis of Cockayne's ideas and their impact on New Zealanders' understanding of their biota. What follows is based on that work. Quote is from Leonard Cockayne, *Monograph on the New Zealand Beech Forests, II. The Forests from the Practical and Economic Standpoints* (Wellington: Government Printer, 1928), 2.

and the like to be permitted to lead to such disaster?"[62] Not if Cockayne could help it.

Even at the time there were objections to this scenario. Could we so easily dismiss a population of seven-foot-tall birds? More to the point, were the deer destroying the forest? That belief rested on an extrapolation from the fact that deer ate plants to the conclusion that this would mean the end of the plants. It seemed common sense, but was it true? Certainly grasslands had existed with many grazers, and Cockayne did not show that New Zealand plants lacked the ability to recover from grazing pressure. A few years before, G. M. Thomson had suggested that the simple models of replacement, superiority, and competition that underlay this idea were wrong. "Among naturalists of the last half century [including Darwin] there has been a strong belief that the introduced fauna and flora have been directly responsible for the diminution of many, and the disappearance of some indigenous animals and plants." Thomson had not found it so. On the basis of long experience he concluded, "Where man does not interfere with the vegetation, the indigenous species can hold their own against the imported forms." The amateur naturalist and grazier H. Guthrie-Smith came to the same conclusion a year before.[63] Cockayne himself had found that introduced plants took over open areas but were unable to penetrate the truly virgin vegetation but gave that no weight when considering deer.[64]

Finally, did vegetation prevent erosion? It seems obvious that it would, and that lack of vegetation would lead to erosion, but there were other elements to consider. The crustal collision that produced the Southern Alps is still pushing the mountains into the sky; there are frequent earthquakes; and the land is poorly consolidated. On the western slopes of the Southern Alps rainfall can be up to ten meters (roughly thirty feet) a year. Under these conditions, would any amount of plants hold the land together? Later reconstructions of the historical landscape, comparing settler descriptions, surveys, and drawings with the current state of the ground, suggested that it could not. Many cases of what seemed erosion from settler activity predated Anglo settlement.[65] No one talked about this at the time. It was not just that Cockayne was the expert on New Zealand vegetation or that his views seemed common sense. Ecology was only consolidating its intellectual base. Little could be

[62] Cockayne, *Beech Forests, II*, 10, 11.
[63] Thomson, *Naturalization of Plants and Animals*, 509; H. Guthrie-Smith, *Tutira: The Story of a New Zealand Sheep Station* (London: Blackwood, 1921).
[64] Cockayne, *New Zealand Beech Forests, II*, 3.
[65] Caughley, *Deer Wars*, 73–8.

brought to bear at the time on the question of how deer were related to forests and the mountains and valleys where they lived.

Conclusion

Ecology, using the tools of modern science and producing sophisticated theories, fundamentally changed the public's relationship to the culture's knowledge. But while authority shifted toward the experts, it was only a shift. For one thing, these theories seemed applicable and useful. Cockayne, speaking of deer and forests, addressed a problem that occupied not only scientists but forest managers, the new breed of game managers, hunters, and politicians in many areas of Anglo settlement. For another, the public assimilated the scientists' ideas, at least in part. Like Forbes we see the pond as a miniature world, and our view of the landscape has been shaped by Clements and Elton. We interpret what we see in terms of communities, trophic levels, and food chains. Like Finlayson and Thompson, we see exotics as agents of change. We also have their ambivalence toward knowledge and appreciation. We seek information and value the exact and rigorous study of nature, but refuse the detached stance required by science. If we do not, with Sir Thomas Browne, seek to "suck Divinity from the flowers of nature" we do look for our understanding to heighten our appreciation of nature's marvels and to ratify our vision of an interconnected world in which we cannot move a stone without the troubling of a star.[66]

We can pass over the philosophical arguments for and against this search. People will look for meaning and meanings in the world, and they will enlist their culture's knowledge to guide them. They have done so since antiquity. The Victorians used natural history to ratify their views, or averted their eyes from theories (such as Darwin's) that threatened their vision. Environmentalists appeal to ecology to support their positions. We should recognize that this is treacherous ground. We call ecology a "subversive science," and it is, particularly of simple ideas of human domination of nature, but it also undermines our own perspectives, using reductionist arguments in the service of a holistic vision. It speaks of what we see but, despite the apparently obvious subject matter and reliance on observation, does so in the context of a conscious and theory-laden construction. An understanding of ecology is not antithetical to an appreciation of nature, but ecologist and environmentalist are different social roles (though conservation biology is reshaping both science and social roles). This gets ahead of our story. The tension

---

[66] Thomas Browne, *Religio Medici* (1642; Oxford: Oxford University Press, 1964), 16.

between understanding and appreciation, the science of nature and the love of it, are an important part of both scientific and popular reactions to the world in the environmental era, but that is two generations and three chapters ahead.

SECTION THREE

FINDING FIRM GROUND

In the first half of the twentieth century, experts and the public traveled different roads in their search for an understanding of the land, and these often seemed to lead only over old ground. There was a renewed enthusiasm for the conquest of nature, while nature appreciation seemed to follow lines laid down in the 1890s. Scientists had new ideas but did not seem sure what they meant. Two of the three chapters in this section follow familiar trails; only in the last does a new one emerge.

The first chapter concerns the rise and fall of faith in science and technology that occurred as the settler societies turned to more intensive exploitation of the land. The high tide of optimism, in the 1920s, brought talk of irrigating the Australian inland with "inexhaustible" underground water, and there were several commercial reindeer farming ventures in the Arctic. These dwindled and died, and then drought in North America and Australia in the following decade not only dashed optimism but revived the debate about limits. In these events we can see how much had changed in the half-century since Powell and Goyder – and what had not. The next chapter, on the second wave of nature appreciation, shows new ideas appearing in the midst of old. The last one deals with a new departure. A small group of ecologists and game managers used ecology's focus on process and relationships in natural systems to develop a larger perspective on the world, and they began to apply it. One, Aldo Leopold, embarked on an even more radical search, for the ethical implications of this view of nature.

There was here the same interplay of common knowledge and ideas with unique conditions that we saw in earlier years, but circumstances were different. At the turn of the century the settlers' search for their place in the land had been closely tied to their developing national identities (particularly so in Australia and Canada). Now national identities were much better established and the years of conquest part of history.

Experience and science now spoke more strongly of restraint and learning to live with the land. With Leopold we can glimpse something beyond even that – the reintegration of knowledge and sentiment, not only for the individual but for the society. No one paid much attention at the time, for Leopold worked out his ideas during the Great Depression and World War II, and *A Sand County Almanac* appeared in the heady years of postwar expansion. The ideas, though, were there, and scientists and then the public would take them up in the continuing search for how to live with as well as on the land.

6

REACHING LIMITS, 1920–1940

Nature is not to be conquered save on her own terms.

Paul Sears, 1935[1]

The drought and depression of the 1890s shook the settlers' faith in their ability to shape nature to match their dreams, but rains and good prices a decade later restored it, and their plans became more expansive than those of the nineteenth century. In the United States and Canada there were plans to make reindeer and musk oxen North America's new beef, the hitherto useless Arctic tundra their pasture. Vilhjalmur Stefansson, explorer and publicist, even spoke of "abolishing the Arctic." In Australia a group known as "Australia Unlimited" spoke of tapping the "inexhaustible" reservoirs of underground water in the outback. Stefansson, who lectured around Australia on the "friendly Arctic," lent his name and fame to the project. All this sounds much like the events of fifty years before. The boomers of the 1920s, sketching great plans and hinting of riches, certainly resembled their predecessors, and the opposition was, as it had been, smaller, better grounded in facts and in theory, and less swayed by hope.

The visions, though, were new. Earlier ones had been of rural arcadias for virtuous farmers. Now the talk was of opportunities for investors. Development had meant people and work; it was coming to mean capital and science. This new knowledge, rather than the pioneer virtues of determination and hard work, would transform each country, producing new farms and new crops that would carry the settlers beyond the limits of conventional farming. The earlier generation of enthusiasts had appealed to experience and called on the wisdom of practical people to counter experts' judgment. Now everyone saw nature in science's terms,

---

[1] Paul Sears, *Deserts on the March* (Norman: University of Oklahoma Press, 1935), 3.

and pinned their hopes on it. The dissenters were a new breed as well, not self-taught experts but trained professionals, and they did not rely on statistics and observation but established theories and disciplines.

After the good years came the bad. The 1930s saw a savage drought in Australia and a legendary one on North America's Great Plains. Here ecology entered the public debate, contributing not only knowledge but a new perspective, seeing the world not as a collection of parts but as dynamic, interconnected systems, and looking not to the next year or generation but to the long-term health of the land. It had more authority than in the days of Goyder and Powell, but it did not sweep all before it. Expertise could reframe the public debate, but decisions on development were matters not of fact but of value and about what power people had or should have over the world around them. Schemes for farming the Australian inland or raising reindeer in the Arctic for North American dinner tables rested on a vision of human knowledge and power remaking the world. Calls for converting farms to grassland, abandoning marginal land, spoke of limits. While everyone would admit, if pressed, that humans were not as gods, few wished to say so, particularly when that admission would darken the great settler dream of opportunities for all and wealth for anyone willing to work for it. The frontier was gone. What hope if science could not make the deserts bloom and the Arctic yield its tribute? This chapter traces the arc from optimism to despair in North America and Australia. It begins with optimism, the plans and projects for Arctic development, a glittering dream that began in the 1890s and lasted almost to World War II, then Australian dreams of watering the outback – the extreme of optimism. It ends with the countercurrent of pessimism and the focus on limits that accompanied the droughts of the 1930s in the United States and Australia.

Optimism Renewed

"To Meet the Possibilities of the Country"

Plans for an Arctic reindeer industry and buffalo–yak–cattle crosses to produce a new animal for new pastures now seem the stuff of Sunday supplements, but in the early twentieth century they looked like a logical extrapolation from experience. The settler economies had from the first depended on transplanting and adapting crops and stock, and the process had continued through the great expansion. Governments had helped, introducing crops, establishing botanical gardens and then experiment stations. At the end of the nineteenth century the U.S. Department of Agriculture was sending collectors around the world to find new crops and new varieties of old ones, and setting up extension programs to bring the results of scientific studies to farmers. There were orange and grapefruit orchards in Florida and California, Indian cattle in southern Texas, and Russian apples in Minnesota.[2] Canadian agricultural scientists were producing wheat that would mature in the short growing seasons of central Alberta and Saskatchewan. Australia was changing its pastures with new grasses from around the world. Plans to raise reindeer, musk oxen, or crossbreeds as a new meat supply for Canada and the United States seemed only another step in the progress of civilization.

The Arctic enthusiasm began in Alaska and quickly spread to Canada. Between 1892 and 1901 Sheldon Jackson, first superintendent of education for the Territory, used government and private money to import slightly more than a thousand reindeer and Lapp herders to teach the indigenous peoples the skills of reindeer herding. His aim was not only to establish an industry but to provide the native peoples a source of income that would wean them from their "primitive" ways of life and integrate them into the modern economy and white civilization. The project began well, and by 1914 several native corporations had large herds.[3] Alaska was only an outpost of the United States, the North was Canada's destiny (or so Canadian politicians had been saying since Confederation). In 1907 Parliament appropriated $5,000 for a medical missionary, Wilfred Grenfell, to introduce domesticated reindeer into Labrador. His was the classic vision. We have, he said, "conquered" the North Atlantic and "forced its reluctant waters to pay us a handsome annual tribute" of fish, seals, and whales, but the land, "from which man generally first extracts payment, has as yet yielded nothing." If, he said, we cannot transform the land into pasture for our cattle, we "can reverse the process and transform the cattle into extraordinary ones to meet the possibilities of the country." Labrador, he thought, could export up to 250,000 reindeer a year, "a very large meat supply from a section of the country that seems to offer no other possible return for labour."[4] Others

---

[2] The annual *Report of the Secretary of Agriculture* and the subsequent *Yearbook of Agriculture* of the U.S. Department of Agriculture are rich sources of information on this movement. On the USDA's Office of Introductions, established 1889, see subsequent yearbooks and reports and David Fairchild, *The World Was My Garden* (New York: Macmillan, 1938).

[3] Margaret Lantis, "The Reindeer Industry in Alaska," *Arctic, 3* (April 1950), 27–44; Keith Murray, *Reindeer and Gold* (Bellingham, Wash.: Center for Pacific Northwest Studies, 1988).

[4] Wilfred Grenfell to George P. O'Halloran, Deputy Minister of Agriculture, 30 December 1909. See also W. T. Grenfell, "Dr. Wilfred T. Grenfell's Missions in Newfoundland and Labrador" (no publishing data), File 1888257, v. 1045, RG 17, Records of the Department of Agriculture, Public Archives Canada, Ottawa.

shared the vision, some with a degree of wishful thinking (and ignorance) excessive even for the time and place. In 1911, seeking official
support for a private venture in reindeer herding, Conrad Siem proposed to employ French-Canadians, "the happy and quaint creoles
of northern latitudes." They were "ill at ease today in the growing
cities . . . and in the turmoil of our usurping civilization. They are
longing for a closer communion with nature on the wide snow-covered
fields lit up by the bright moon . . . stars . . . and the scintillating northern lights." Having spent centuries under the "benign rule of the
Hudson Bay Company, they are accustomed to a feudal system resembling patriarchism. They will make good and faithful herdsmen."[5]

Nothing came of that, but there were several efforts to "transform the
cattle" and the crops. Grenfell's Labrador mission offered prizes for vegetable growing, a project that, a newspaper noted, "cannot but be of
interest to all loyal Canadians, who have at heart the improvement and
upbuilding of all parts of our vast Dominion." The government had
an officer in charge of northern agriculture looking into possibilities,
collecting articles on "vegetable growing in the Soviet Arctic."[6] The
Dominion Experiment Station crossbred yaks, bison, and cattle to get
an animal hardy enough for the Arctic, meaty and prolific enough to be
worth raising (yaks were added in an attempt to get around the sterility
common in cattle–bison crosses). The station's report proudly described
the last of its experiments, animals that were 50 percent yak, 25 percent
bison, and 25 percent domestic cattle, as "the only animals of their kind
in existence." This is almost certainly true. In 1923 the Canadians
shipped three yaks – two cows and a bull calf – to the USDA experiment
station at Sitka, Alaska, but the Americans had no better luck.[7]

Grenfell's enthusiasm outran his knowledge, and his reindeer project
collapsed, but others took up the cause. The most conspicuous was
Vilhjalmur Stefansson. A Canadian born of Icelandic immigrant parents,
he had been raised and educated in the United States and returned early
in the century to work as a museum collector. He learned from the Inuit
how to live off the country and travel in it, and was one of the last people
to add land to European maps – some small islands in the Arctic Ocean
– and speak to people who had never seen whites. It was a peculiarly
modern adventure. Exploration had always depended upon support,

[5] Conrad Siem, "Memorial on the Introduction of Domesticated Reindeer into Canada"
(title page says written at the request of R. B. Borden, Premier of the Dominion, no publication date), File 188257, v. 1045, RG 17, Public Archives Canada, Ottawa.

[6] London, Ontario, *Echo*, 9 June 1910, in Correspondence, 188257, v. 1045, RG 17, Public
Archives Canada, Ottawa.

[7] Box 55, BU 241, RG 84, Records of the Dominion Park Service, Public Archives Canada,
Ottawa.

and so explorers had always had to sell their projects, but now they had to sell them to the general public. Explorers now needed, in addition to technical expertise, physical courage, and endurance, a knowledge of public relations and public speaking. They lived between the wilderness and the media, using the discoveries of each expedition as material to raise money for the next, whose knowledge then had to be fed back into the publicity machine. Unless he died on an expedition, as the explorer aged he had to find new work or live on his fame.

By 1917 Stefansson was drifting in that second direction, abandoning the Arctic for the boardroom and lecture hall, making a new career as the "ambassador of the North."[8] In books and lectures he told the public that it was possible to live in the Arctic. Our picture of the region as a place of desolation and death came, he said, from the disasters of the nineteenth century, particularly the Franklin expedition. These had happened because the Victorian explorers had not known how to live in the North. We did. He spoke of the "friendly Arctic." Shifting the conventional rhetoric ninety degrees, he said that it was northward the course of empire made its way. With graphs, charts, and pages of argument he showed that civilization had been steadily marching northward since its first development in the Middle East. The Arctic would be the new Mediterranean, the ocean around which the civilization of the future would develop.[9]

He intended to be one of its business pioneers. In May 1919 he told a joint session of the Canadian Parliament about the bright prospects for reindeer and musk oxen, and later resigned his place on the Royal Commission on the Reindeer and Musk-Ox, appointed to look into the prospects of such an industry. He wanted to start his own herd. Backed by the Hudson's Bay Company, he persuaded the Canadian government to give him an exclusive lease on the southern part of Baffin Island. That scheme collapsed when the reindeer died or scattered and the Lapp herders quit. The Commission thought the problem was lack of suitable range, but overoptimism and lack of planning would probably be more accurate.[10] Stefansson formed the Stefansson Arctic Exploration and

[8] D. M. Bordais, *Stefansson: Ambassador of the North* (Montreal: Harvest House, 1963).

[9] Vilhjalmur Stefansson, *The Northward Course of Empire* (New York: Harcourt, Brace, 1922) and *The Friendly Arctic* (New York: Macmillan, 1921); Richard J. Diubaldo, *Stefansson and the Canadian Arctic* (Montreal: McGill-Queen's University press, 1978); Bordais, *Stefansson: Ambassador of the North*; William B. Hunt, *Stef: A Biography of Vilhjalmur Stefansson, Canadian Arctic Explorer* (Vancouver: University of British Columbia Press, 1986).

[10] A. E. Porsild, "Report on the Reindeer and the MacKenzie Delta Reindeer Grazing Reserve" (no publishing information, internal evidence suggests 1948), copy in file 111, v. 491, RG 109, Records of the Canadian Wildlife Service, Public Archives Canada, Ottawa.

Development Company, which had an even more ambitious agenda. It planned to land a group on Wrangel Island, off the Arctic coast of the Soviet Union, claim it for Great Britain, and establish a base for commercial exploitation of the area. It was a small but complete disaster. The four men who put ashore in the fall of 1921 for a two-year stay all died during the second winter, one on the island, the others in an unsuccessful attempt to travel over the ice to the mainland. Critics condemned the expedition as ill-planned and ill-managed, Stefansson's idea of a "friendly Arctic" as nonsense at best, murderous nonsense at worst.[11]

The Canadian government was undaunted. In 1922 the Royal Commission on the Reindeer and Musk-Oxen decided that earlier ventures might have failed because they were in unsuitable areas. It sent out A. E. Porsild, chief botanist of the National Museum of Canada, to find better ones. He spent seven months in Alaska, then traveled from Nome along the coast. The government bought 3,000 animals from Alaska and arranged to have them sent to an area Porsild had identified in the Mackenzie Delta. The botanist pressed on, looking for other pastures in the Keewatin District and on the west coast of Hudson's Bay. In 1935 he resigned to return to his usual studies, having, he said, spent "ten summers and seven winters in the Arctic getting Canada's first Government-owned reindeer off to a good start."[12] Disaster followed that "good start." The Alaskan industry collapsed in the late 1930s, even as the Canadians were launching theirs, and within a decade it was gone. Revisiting the Canadian project in the late 1940s, Porsild found it a shambles. He thought too little time had been spent convincing the native peoples that herding was "as profitable and sound economically as trapping for fur," but the next sentence of his report suggests other reasons. They "feel today that the life of a reindeer herder is more insecure and involves more hard work, isolation, and hardship than that of a hunter or trapper."[13] Others blamed weather, tundra fires that had destroyed the lichen the reindeer fed on, or wolves.

The underlying problem was something no one had counted on or even thought about – the ecology of the area. In 1953 A. Starker Leopold and F. Fraser Darling conducted an ecological survey of the

[11] Diubaldo, *Stefansson and the Canadian Arctic*, 136–86. A much more sympathetic account is Bordais, *Stefansson: Ambassador of the North*, 161–70. See also Hunt, *Stef*, 195–227, and Melody Webb, "Arctic Saga: Vilhjalmur Stefansson's Attempt to Colonize Wrangel Island," *Pacific Historical Review*, *61* (May 1992), 215–39.

[12] Porsild, "Report on the Reindeer," 3. Records of the government projects are in volumes 485–504, RG 109, Public Archives Canada, Ottawa.

[13] Porsild, "Report on the Reindeer," 4, 10–11. On the Alaskan herds the most accessible source is A. Starker Leopold and F. Fraser Darling, *Wildlife in Alaska* (New York: New York Zoological Society and the Conservation Foundation, 1953), 68–82.

industry and concluded that while the tundra fires of the 1930s had been a factor, the projects had been doomed from the start. The boosters had thought of the Arctic as pasture, assuming a given acreage would support a certain number of reindeer year after year. The Arctic, though, was not pasture, but a mosaic of food sources that the herds grazed through the year, finding different feed in different areas. The key constraint was the major source of winter feed, the branching lichen that was the understory of the spruce forests. It grew so slowly that the "pastures" either had to carry a very small number of animals or be left ungrazed for long periods. In Lapland, the scientists pointed out, an elaborate system of "herding and of rotating winter pastures – methods that were worked out empirically over centuries of time" – reduced grazing pressure to what the range would bear. Unfortunately, "neither Eskimo herders nor most of the whites (excepting the transported Lapps) were mobile enough to follow any such grazing schedule."[14]

The enthusiasm for remaking the Arctic and extending the "possibilities of the country" now seems so visionary that it is important to remember that it was grounded not only in the settlers' history of expansion but in their very recent "advances" in Arctic exploration. The record of Arctic expeditions in the nineteenth century had been, as Stefansson pointed out, a chronicle of misery and death. The Franklin expedition – two ships and 128 men lost without trace – was only the worst. Starvation, scurvy, and death were common. Then a few people had learned from the people who lived there, either the Inuit or, in the case of the Norwegians, the Lapps.[15] With these "new" methods Peary reached the North Pole (or some point well north of Greenland). Roald Amundsen sailed the Northwest Passage, then went to the South Pole. The only great disaster, Scott's south polar expedition, had clung to the old ways. Stefansson, like the pioneer farmers pushing on a few more miles into drier or colder country, extrapolated from this and his own experience to boost the "friendly Arctic," and it was only a short step from there to trying to extract profits from its "pastures."

"Australia Unlimited"

Australian plans to tap the "unlimited" underground reservoirs of the inland were as visionary as the Arctic ventures, and a better example of the technological optimism that led people to believe that humans could use everything on earth for their industrial economy. This was an old

---

[14] Leopold and Darling, *Wildlife in Alaska,* 77.
[15] An accessible introduction is Pierre Berton, *The Arctic Grail* (New York: Viking, 1988), 538–41.

dream, dating from the 1880s, when artesian wells drilled in the interior showed there was water. No one could say how much or where it
was from, but the very lack of information seems to have encouraged
speculation. In *The Dead Heart of Australia* (1906) J. W. Gregory suggested it had percolated up from the deeper layers of the earth. Others
thought it was stored rainfall moving through rock formations, though
where these led was unclear. A few thought it had come from the
Himalayas.[16]

In the years just after World War I tapping this resource and turning
the "dead heart of the continent" into pasture and field became an
enthusiasm. The definitive statement, which gave the movement its
name, was Edwin J. Brady's 1,083-page effusion, *Australia Unlimited*, published in 1918.[17] Brady was in the great line of boosters, the men who
had boomed the Prairie Provinces, the Great Plains, and the backblocks
of Australia and New Zealand. Australia's days of prosperity, he argued,
were in the future. The "wealth of today is but a beggar's moiety of the
unlimited wealth of the future which will be won by the application of
modern knowledge to local conditions." He scorned the "groundless
fear" that "haunts many minds, that those large unoccupied tracts of the
interior of Australia, which are yet imperfectly known, will not prove fit
for future settlement." The explorers, despite their failure and their
faults, have proved that the whole continent is good."[18] He attacked
head-on the popular conception of Australia as an arid land. That had
begun with the Burke and Wills expedition, which was misunderstood.
There was nothing inevitable about that disaster. People who knew the
country had predicted the result before the expedition had left. The
men had perished because they knew no bushcraft. The area where they
died, Brady triumphantly pointed out, was now pasture. Fiction writers
who needed excitement in their stories had perpetuated the idea of a
dangerous land, making "accidental happenings like dry seasons and
bush fires . . . appear the permanent conditions of the continent." In a
long section, "The Desert Myth," he explained in detail that Australia
had no actual desert. "Instead of a 'Dead Heart of Australia' there exists
in reality a Red Heart, destined one day to pulsate with life." There was,

---

[16] J. W. Gregory, *The Dead Heart of Australia* (London: John Murray, 1906), 273–341. An
example of the enthusiasm is Francis Fox, "Irrigation in Australia and What It May
Accomplish," in *Australia Today, 1911*, a special number of *Australasian Traveler*, 1
November 1910, 85–94. See also David Branagan and Elain Lim, "J. W. Gregory,
Traveller in the Dead Heart," *Historical Records of Australian Science, 6* (June 1986),
71–84; A. E. Faggion, "The Australian Groundwater Controversy, 1870–1910," *Historical Records of Australian Science, 10* (December 1995), 337–48.

[17] Edwin J. Brady, *Australia Unlimited* (Melbourne: George Robertson, 1918).

[18] Ibid., 53.

he admitted, little rainfall, but it was never low over the whole center at once. There were no rivers, but "Australian Nature, of her ancient Wisdom, has substituted permanent underground storage and flow for regular surface condensation and drainage."[19]

The great dissenter from this gospel was a geographer at the University of Sydney, Griffith Taylor. He had long held that Australia was arid and was perfectly willing to say so. In 1914, as a new member of the Commonwealth Weather Service, he wrote an article titled "The Physical and General Geography of Australia" for a handbook to be used by members of the British Association for the Advancement of Science, who were holding their first meeting in Australia. Under the circumstances he might have been expected to boost the country. Certainly another piece in the same volume, "The Climate of Australia," did. It rejected the idea that there was a desert anywhere on the continent and referred vaguely to unlimited artesian water. Taylor spoke of "desert," and included maps with much of the Centre labeled "tropical desert."[20] In *The Climatic Control of Australian Production* (1915), a bulletin of the Bureau of Meteorology, he was equally skeptical and in "Nature versus the Australian" (1920) called for a forthright recognition of the limits of irrigation in the inland. It would be possible to supply some small areas, but bores (in American English wells) would not change the face of the country.[21] Taylor later noted with a certain pride that Western Australians were so horrified by his ideas that they banned his textbook, *The Geography of Australia*, from the state's schools.[22]

His battle with the optimists has parallels with the one John Wesley Powell waged against Western boosters some fifty years before. In both cases believers in the promise of an undeveloped region faced a man appealing to expert knowledge, and there are similarities in the men's careers. Both had one kind of education and another kind of job, and each had been on a great adventure. Powell, though, was at the beginning of the era of specialized education and professional science – he had become a natural historian by apprenticeship – and at the center of Victorian exploration; his trip through the canyons of the Colorado was one of many that mapped the American West as settlers poured into the country. Taylor was of the first generation of trained professionals.

[19] Ibid., 39, 630, 631.
[20] Griffith Taylor, "The Physical and General Geography of Australia," 86–121, and H. A. Hunt, "Climate of Australia," 122–62, in G. H. Knibbs (editor), *Federal Handbook* (Melbourne: Government Printer, 1914).
[21] Griffith Taylor, *The Climatic Control of Australian Production*, Bulletin 11, Commonwealth Bureau of Meteorology (Melbourne: Victorian Government Printer, 1915); idem, "Nature versus the Australian," *Science and Industry*, 2 (August 1920), 459–72.
[22] Griffith Taylor, *Journeyman Taylor* (London: Robert Hale, 1958), 139.

He had two degrees from the University of Sydney – one in mining engineering, another in geology and physics – and he used an 1851 Exhibition scholarship (given to bright colonials to allow them to continue their education at "home") to earn another at Cambridge. The great age of exploration was almost over. He was a member of the 1910–12 Scott Antarctic expedition, seeking the last prize of heroic geography, the South Pole. (Fortunately for his career, he was not included in the Pole party.)

The two men, however, undertook and fought their battles in very different spirits. Powell seems to have disliked controversy; it was Taylor's element. At a time when Australia excluded Asians and newspapers warned of the "Yellow Peril," he said that a study of cephalic indexes showed "the Mongol has a more compact and higher type of brain than the white man, and has evolved further." He recommended some intermarriage with the Chinese to improve the Australian "race." Higher education was still geared to the classics, and he said that Latin was not necessary for culture. Natural historians in Australia, not to say the general public, had not championed evolution, and Darwin was in eclipse. Taylor said that evolution was one of the things every schoolchild should learn.[23] Powell had gone quietly to defeat; Taylor waged a vigorous campaign and gave as good as he got.

In 1921 he became a professor of geography at the University of Sydney, the first such chair in the country, and it was from this vantage point, as "Professor Taylor" (as the newspapers commonly called him), that he fought "Australia Unlimited." The high point was Stefansson's visit in 1924. It was a grand tour. The explorer gave eight lectures in Sydney, sixteen in Melbourne, two in Geelong, four in Ballarat, and three in Adelaide, on three subjects, "Abolishing the Arctic," "The Friendly Arctic," and "My life with the Eskimos." Advertisements promised that each presentation would be different and urged people to buy tickets for several performances.[24] After this he traveled overland from Adelaide to Alice Springs with a party of boosters. It was, he told reporters, part of a new project. He had spent many years in the Canadian Arctic, studying areas that were thought uninhabitable because they were too cold. Now he proposed to shift his attention to those considered too dry. His reports, sent back to the newspapers

---

[23] Box with items 284–400, Series 4, Griffith Taylor Papers, National Library, Canberra; Marie Sanderson, *Griffith Taylor: Antarctic Scientist and Pioneer Geographer* (Ottawa: Carleton University Press, 1988), 110. The best description of the climate controversy is J. M. Powell, *Griffith Taylor and 'Australia Unlimited'* (Brisbane: University of Queensland Press, 1988). My thanks to Professor Powell for a copy.

[24] *Adelaide Register*, 26 June 1924; Pamphlet File, Mitchell Library, Sydney.

during the journey, were relentlessly upbeat. The salt pan that formed the dry bed of Lake Eyre was daunting, but Stefansson invited readers to take the train from Los Angeles to Salt Lake City. There they would see some real desert. He thought the exhilarating climate would be a great attraction; the health of one member of the party had improved since leaving the end of the rail line at Oodnadatta. Three hundred and fifty miles north of Oodnadatta he said that "if this is desert we have been travelling through an oasis the entire day." He compared it to Utah as the Mormons found it. (The ambiguity of this endorsement may have been lost on Australian readers.) Three days later the newspapers reported that Mr. Nelson, representing the Northern Territory, has risen in Parliament to ask Prime Minister Bruce about these reports. The explorer, he said, had "not yet been successful in locating the 'alleged desert' in Central Australia," and he asked that this information be sent to Professor Taylor with a request that he "desist from his perpetual slander on Central Australia." The prime minister said he was heartened by evidence that the country was better than had been thought, but he declined the responsibility of giving advice. Stefansson's return and his report to Parliament were a triumph for the optimists.[25]

Taylor faced his critics with facts, venom, and sarcasm. These were appropriate, for this was less a scientific battle than one over visions of things not seen yet believed in. The architects of Australia Unlimited drew on geology for their hopes of unlimited underground water, ignoring Gregory's calculations, made twenty years before, that there was too little water to irrigate more than a small part of the country.[26] His figures looked much like Powell's for the Great Basin. A second defense was faith in the future. Stefansson often appealed to this argument. Before leaving for Australia he had told a reporter that he was confident that the Centre could be developed. When Taylor wrote suggesting that he make no judgments until he had seen the land, Stefansson brushed him off. The pessimists, he said, were wrong when they assumed that if they saw no solution there was none. Those who followed us might be more ingenious.[27] Others were even more adamant. A newspaper dismissed one of Taylor's articles on the Australian desert by saying that one does

---

[25] Stefansson sold reports to the *Melbourne Argus* and the *Sydney Morning Herald*. Reports appear in the *Argus* on 7, 17, 21, 23, 24, 26, 28, 30, 31 July and 2, 5, 6, 9, 12 August, with follow-ups on 22 September and 6 October. The Parliamentary exchange is in the *Sydney Morning Herald*, 31 July 1924. Stefansson's report is "Central Australia; Report by Dr. Vilhjalmur Stefansson" (Melbourne: Victorian Government Printer, 1924; for Commonwealth Government). Powell, *Griffith Taylor*, has a full account.

[26] Gregory, *Dead Heart*, 276–8.

[27] Stefansson to Taylor, 20 May 1924, box with items 359–410, Series 4, Taylor Collection, National Library, Canberra.

not argue with a scientist, "one merely knows that this is wrong." Charles Barrett, one of the Melbourne nature writers, held out for agriculture in the Centre "despite all that Science may say concerning it."[28] Taylor, like Powell, lost his battle and in 1928 left Australia for a position at the University of Chicago. He went on to Toronto, not returning home until he retired. He lived, though, to enjoy his victory. In retirement he was the grand old man of Australian geography, his heresies the orthodoxy of the new generation.[29]

The debates, though, were different than they had been in Powell's and Goyder's day. Then there had been a common commitment to the social vision of a country of independent farmers. Indeed, the radical changes Powell proposed in the land laws had been made to allow small farms and individual ranches in the arid region.[30] This was a durable dream – even after World War II Australia offered returning veterans a place on the land – but it was fading even in the 1920s. The very name "Australia Unlimited" was redolent of corporate enterprise, and the group's plans were for a heavily capitalized industrial agriculture that would enrich the country, not farms for the people. The confident assertions that human endurance alone would conquer nature, so apparent in the 1880s, were missing in the 1920s. No one claimed that plowing or tree planting would make it rain or moderate the climate. Then people had appealed, easily and often, to the experience of "practical men," and even Powell had relied on what Utah farmers had grown as a guide to the climate. Now people accepted the scientists' vision. Even Brady's rhetoric about the "ancient Wisdom" of Australian Nature, which had "substituted permanent underground storage and flow for regular surface condensation and drainage," was based on geology, which did not show the surface, or anything anyone could see. It was an interpretation of evidence filtered through theory.

Not everything, though, had changed. The central "science" in these debates was not geology but climate, and it was a rich mix of atmospheric physics, statistics, and ideas from classical civilization. At the turn of the century Queensland planters had fought the "white Australia" policy on the grounds that white people could not labor in the tropics. The new laws would harm the sugar industry. In 1906 J. W. Gregory found it a heartening sign that whites were working and living in the heat of

---

[28] Taylor, *Journeyman Taylor*, 202; *Sydney Morning Herald*, 17 January 1927.

[29] See "Obituary" in *Australian Academy of Science, Yearbook, 1964*, 46 (no publishing data, 1964); Taylor, *Journeyman Taylor*, 332–3.

[30] John Wesley Powell, *Report on the Lands of the Arid Region*, U.S. Congress, House of Representatives, 45th Congress, 2d Session, H. R. Exec. Doc. 73 (Washington, D.C.: U.S. Government Printing Office, 1878), vii–viii.

Hergott Springs near Lake Eyre.[31] Even Griffith Taylor (in what seems to have been his only approach to conventional wisdom) thought the tropics unhealthy for white families, and as late as 1939 Ian Clunies Ross, a member of the national Committee on Scientific and Industrial Research, worried that large parts of tropical Australia, with their high heat and humidity, were "most unsuitable for white settlers and particularly white women."[32] In the 1890s, though, equations describing air movement and heat transfer began to replace sketchy notions of heated air and condensation. Weather began to be studied in terms of movement in the atmosphere of a rotating earth heated by the sun. By the time Taylor joined the Commonwealth Weather Service it was showing fronts, those unseen realities made by coordinating barometric readings, and the public was learning to read weather maps.[33] As was the case before (and since), science had the greatest influence when it buoyed people's hopes, the least when it troubled their dreams, and they appealed to it selectively. That the land set limits on what people could do was no more popular in twentieth-century Australia than it had been in nineteenth-century America. Ideas, though, were not all. In the next decade drought over large areas of the United States and Australia threatened not plans for development but vast areas of established agriculture, and it renewed the debate about limits and living.

Optimism Dashed

Dust Bowl

Mile-high, those gloomy curtains of dust are the proper backdrops for the tragedy that is on the boards. The lustful march of the white race across the virgin continent, strewn with ruined forests, polluted

---

[31] Gregory, *Dead Heart*, 32–4. On climatic ideas see Clarence Glacken, *Traces on the Rhodian Shore* (Berkeley: University of California Press, 1967), 80–115.

[32] Griffith Taylor, *The Control of Settlement by Temperature and Humidity*, Bulletin 14, Commonwealth Meteorological Service (Melbourne: Victorian State Printer, 1916). See also idem, *The Australian Environment, Especially as Controlled by Rainfall*, Advisory Council of Science and Industry, Memoir No. 1 (Melbourne: Government Printer, 1918); I. Clunies Ross, "Blanks on the Map," in J. C. G. Kevin (editor), *Some Australians Take Stock* (London: Longmans, 1939), 83. This sentiment had largely faded by this time, even for Australia. See Warwick Anderson, "Geography, Race and Nation: Remapping 'Tropical Australia,' 1890–1930," *Historical Records of Australian Science, 11*, 4 (1997), 457–68.

[33] Taylor relied on Gregory's surveys, which were the most recent. Gregory, *Dead Heart*, 332–3. On the development of weather theory see Robert Marc Friedman, *Appropriating the Weather* (Ithaca, N.Y.: Cornell University Press, 1989).

streams, gullied fields, stained by the breaking of treaties and titanic greed, can no longer be disguised behind the camouflage which we call civilization.

Paul Sears, 1935[34]

Dust storms and drought were nothing new on the Great Plains, but Sears was right about this particular crisis. Those "gloomy curtains of dust," the storms that began in the summer of 1932, were a start of a catastrophe of almost biblical proportions. The dark clouds blotted out the noonday sun and sifted grit into every corner and onto every plate. A few darkened the skies over Washington, D.C., and covered ships in the Atlantic with a film of soil from the stricken region. They peaked in the late 1930s – sixty-eight in 1936, seventy-two in 1937 – but there were still seventeen in 1941. Photos created an iconography of the Depression, searing the agony of plains farmers into the national memory: clouds, drifts of dirt, ruined fields, and the faces of "Okies" moving west, seeking the golden land of California, living in tents and off charity.[35] Even today the "Dust Bowl" is a potent symbol, invoked to frame arguments about science and knowledge, nature and humans.[36]

Since the early nineteenth century, when the treeless area was called the "Great American Desert," people had argued about farming the Great Plains. The Dust Bowl renewed the debate and gave it point, but now ecology began to guide the discussion. There was no single figure like Powell, and no document like his report, but the New Deal's commissions, investigations, reports, and studies raised the question of what the plains would, in the long run, support. The discussion began in disciplines like agricultural economics, which sought to rationalize land use and found in the stricken areas their greatest challenge. Economists found in ecology "a like-minded emphasis on accepting limits to economic expansion, although in ecology's case the limits were those inherent in nature – the capacity of the earth and its network of life to sustain man."[37] They also found ecologists. Frederic Clements, the great theorist of plant succession, his student John Weaver, and Paul Sears, who had taken his Ph.D. at Chicago and taught for a decade at Nebraska,

---

[34] Sears, *Deserts on the March*, 168.

[35] Storm statistics from Donald Worster, *Dust Bowl* (New York: Oxford University Press, 1979), 15. Worster provides a good introduction to the whole episode.

[36] William Cronon, "A Place for Stories: Nature, History, and Narrative," *Journal of American History*, *78* (March 1992), 1347–76, outlines the debate and offers a good starting point for any discussion.

[37] Worster, *Dust Bowl*, 198.

became involved.[38] Clements called for a national plan of "desettlement, land classification, and resettlement" that would be guided by "a detailed survey of every farm and ranch [to] determine the best use of its various parts. This demands not only the fitting of the three types of plant communities, crop, grass and forests, into a unified system, but also placing the emphasis where climate and soil indicate it belongs."[39] Here was Goyder's Line on a grand scale.

As state botanist, Sears helped write Oklahoma's land use laws during the drought, but he is better remembered for *Deserts on the March* (1935), a book conventionally described as a neo-Malthusian text, but whose intellectual ancestor is more clearly George Perkins Marsh's *Man and Nature*. Like Marsh, Sears was concerned with the earth's limits and humans' destruction of the land that supported them. He argued for a long view, generations rather than years, saw soil as the basis of civilization, and appealed to the example of other cultures. He had, though, a different science and grounded his analysis in an analogy drawn from it. The grasslands, he pointed out, were complex mixtures of species. As such, they favored stability over production, endurance over accumulation, survival over riches. Farms were monocultures that sought to maximize immediate returns of one thing, and in so doing sacrificed stability and resilience. This had implications for public policy, but Sears refused to draw radical conclusions, appealing instead to conservation. He charged that society had recklessly given away resources on terms no business would have considered, and called for policies that had something better than the "perspective of the newborn" and a "sense of continuity and proportion."[40] He did not challenge the practices of settlement. That would have required him to analyze the ideas of private ownership and question the great dream of continued growth.

The fate of a USDA documentary film, *The Plow That Broke the Plains*, suggests that his caution was justified. Produced in 1937 for the Resettlement Administration, a New Deal agency, it presented a dramatic picture of the destruction of the soil. As a farmer crumbled a clod in his hands and the wind scattered it, a dramatic voice-over warned that the western plains were a "natural grassland," a land of "high wind and sun." Plowing, the movie said, destroyed the grass that held the soil in place; the frantic expansion during World War I and into the 1920s, and

---

[38] Ibid., chap. 13; Ronald Tobey, *Saving the Prairies* (Berkeley: University of California Press, 1981), chap. 7.

[39] Tobey, *Saving the Prairies*, quoting Clements, 207. Frederic E. Clements and Ralph W. Chaney, *Environment and Life in the Great Plains* (Washington, D.C.: Carnegie Institution, 1937), 51.

[40] Sears, *Deserts on the March*, 218.

the dry years, brought the inevitable disaster. It did not condemn farmers or farming, but concluded with an appeal for relief. Congressmen, local officials, and ordinary citizens denounced the movie as a fraud and a libel. The drought, they said, was a natural disaster, and only temporary. The rains would come again and the land yield its bounty.[41] Americans were willing to admit that this was a crisis, but not that it was a permanent problem or that they had had a hand in producing it. Even, perhaps especially, in the stricken areas there was enormous opposition to any suggestion that nature placed limits on people's actions. They clung to optimism and faith in human knowledge and will, and resisted the contrary arguments from science, social science, and the experience of dry years. They almost had to, for they were defending not a dream of settlement but the lives they had built on the plains. The USDA withdrew the movie from circulation.

The concept behind all this – climate – remained elusive. In 1941 the USDA titled its yearbook *Climate and Man* but did not resolve the problems or even state them clearly. The thick volume had hundreds of tables, charts, and graphs, but it was data in its simplest form – records of rainfall, wind speeds, and hours of sunshine. There was little discussion of what this meant for farming or settlement. Of the Great Plains it said that "precipitation trends come in long swings of above and below normal," but quickly added that this was not absolutely certain, for "the period covered by this summary is a comparatively short one (only about five decades)."[42] On permanent settlement in the arid regions it was silent. There was little else it could do, for it could not resolve the fundamental difference in perspective that was at the base of these arguments. Farmers focused on the short term, on what would grow this year and the next few years. They had to. Mortgages, taxes, and bills came due far more regularly than rain. Good land was land that yielded enough, regularly enough, to keep them going. This, however, was to some extent a matter of judgment, and the farmers' was colored by an almost unconscious assumption that good years were the norm, bad ones the exception. Once a generation, a catastrophe might wipe out the region's economy, but that was not something they could, or did, acknowledge. Ecologists and administrators, on the other hand, took a long-term view and looked at the land as well as the crops. The land was to provide not just for this year or generation but into the distant future.

These differences color even scholarly debate on the issue. Consider two books published in 1979, Donald Worster's *Dust Bowl* and Paul

---

[41] Worster, *Dust Bowl*, 31–4, deals with this, but the literature is very extensive.
[42] U.S. Department of Agriculture, *Climate and Man*, Yearbook of Agriculture, 1941 (Washington, D.C.: U.S. Government Printing Office, 1941), 693.

Bonnifield's *The Dust Bowl.*[43] About all they agree on is that the culture and its way of relating to the land emerged from the drought unchanged. Worster described the "dirty thirties" as a chronicle of human greed and ignorance that damaged the land. Bonnifield saw the dust storms as a natural phenomenon that the settlers bravely endured, more hindered than helped by their government, until the returning rains wrote a happy ending to their story.[44] The difference is perspective. Worster saw the Dust Bowl in the context of European settlement, the world economy, and the long-term health of the land – which was preserved or destroyed by society's technology and care. Bonnifield looked at people's experiences during the dry years. The land, even the society, was simply the backdrop for individual action.

The new perspective of farming and grazing as the interaction of social system and ecosystem was not peculiar to Americans and the Dust Bowl. The same view appears in the work a British scientist, Francis Ratcliffe, did on the Australian drought of these years. He reached many of the same conclusions as the grasslands ecologists and, at least as clearly as the Americans, pointed to the disjunction between human economic and social systems on the one hand and the rhythms and requirements of the land on the other. That he did so without formal training in ecology says something about the movement of that field's perspective through the scientific community and its use in public policy.

To Define the Problem

If you take on the job, all you will be expected to do is to define the problem. That is often half the battle. When we know exactly what is happening, and why it is happening, we should be able to decide whether the solution lies with botanists or soil experts, whether it must depend on the control of the rabbit, or is fundamentally a question of stock management.

A. E. V. Richardson, 1935[45]

Such was the description of the job Australia's Committee on Scientific and Industrial Research wanted Ratcliffe to do. It was, like the American investigation of the Dust Bowl, to be the study of people and

---

[43] Worster, *Dust Bowl*; Paul Bonnifield, *The Dust Bowl* (Albuquerque: University of New Mexico Press, 1979).

[44] Cronon, "A Place for Stories," analyzes these books and their relation to environmental history.

[45] Quoted in Francis Ratcliffe, *Flying Fox and Drifting Sand* (1938; reprinted Sydney: Angus & Robertson, 1947), 188.

the land, but while the Americans could assume that weather and its patterns were at the heart of the matter, the Australians could not. There were also institutional differences. The American government deployed commissions, academic ecologists, and teams of specialists. Ratcliffe, who thought of himself as a general biologist, worked alone. Nor was the CSIR like an American agency. It had developed in conjunction with the academic specialities it depended upon and carried on research, usually in conjunction with the land-grant schools. The Australian bureau had developed out of the mobilization of science for war during World War I and continued (becoming the CSIR in 1926) as a way to apply it to economic development. Its job was to identify crucial problems and find people who could address them, which often meant bringing them in from abroad.[46]

Ratcliffe was one of those experts, trained abroad, brought in for a specific task. He had been one of Julian Huxley's students just after World War I (Charles Elton had been another), and had come to Australia in 1929 to study, for the CSIR, the damage flying foxes (large fruit-eating bats) did to orchards on the east coast.[47] It is worth going back to that project, for Ratcliffe's approach and solution show a vision of humans and nature very different from the common view. On the surface it seemed a straightforward matter of pest control. Certainly the growers felt that way. They saw the problem, Ratcliffe said, as the existence of "a pestilential animal which required extermination, and the solution a practical method of wholesale destruction."[48] He thought that although there was "a considerable number of odd facts about the natural history of the Australian species . . . anything like an accurate picture of their population as a whole and what might be called their economy was conspicuously lacking. It was my task to provide such a picture."[49]

His methods did not seem particularly innovative. He traveled around Queensland, tracing migrations, collecting specimens, examining food habits, and estimating numbers, range, and natural enemies, much like any natural historian. He carried a copy of Leach's *An Australian Bird Book*, checking off the species he saw, and the account he wrote of this and his drought work, *Flying Fox and Drifting Sand*, was in the tradition of the British traveler in the colonies – the metropolitan in the provinces

---

[46] On these years see Boris Schedvin, "Environment, Economy, and Australian Biology, 1890–1939," in Nathan Reingold and Marc Rothenberg (editors), *Scientific Colonialism: A Cross-Cultural Comparison* (Washington, D.C.: Smithsonian Institution Press, 1987), 101–26. On the CSIR's focus see Schedvin, *Shaping Science and Industry* (Sydney: Allen & Unwin, 1987), 75–90, and chaps. 5–7.

[47] Huxley's introduction to Ratcliffe, *Flying Fox*, vii.

[48] Ratcliffe, *Flying Fox*, 5.   [49] Ibid., 5.

and what he found. His conclusions, though, were anything but conventional. Extermination, he decided, was not practical and not needed. The four species of bats ranged up and down 2,000 miles of coast, following the blooming of the eucalyptus. Though they occasionally fed in orchards, fruit was not their favored food. Shooting or frightening off flocks when they caused damage was all that needed to be done and all that could be done.[50]

Ratcliffe framed the situation in terms like those that game managers in the United States were beginning to use (Chapter 8). His view was very different from the farmers'. For them the species was the pest and the solution to get rid of it. This was the conventional attitude, which had often led to spending enormous sums over long periods. It assumed that fields could be made perfectly safe (and that they should be). Ratcliffe saw the problem not as the species but as the behavior of a part of the population. He rejected, if only implicitly, the idea of complete control, weighing various goods and asking if the cost to society was worth the farmers' higher profits. He admitted that his solution was "tinged with defeatism," and presented it as a practical measure, but it had a radical core. It asked that people live with nature.[51] The word "ecology" does not appear in Ratcliffe's writing, but this was an ecological perspective, concerned not with entities but relationships, not species but populations, not immediate and human interests but the long-term health of the system.

Ratcliffe had fallen in love with Australia while studying flying foxes, and he happily returned to work on the drought. Here too he used natural history terms but followed a new line of thought to a radical end. The difference in perspective was even more apparent. He had, he said, spent a year mulling over the problems of pasturing sheep in the inland when he finished his work, and he had found a solution, but "it was not the answer I wanted or hoped to find."[52] The key was the wide variations in rain from year to year. They made conventional grazing impractical. When drought came, the graziers could not retreat to the coast, for that would require them to reduce and rebuild their flocks at unpredictable intervals for an unknown time. Their bills did not stop with the rains, and if they sold their flocks they would lose their breeding programs and with them the quality and consistency the market required. Besides, if everyone sold stock when drought came and bought again when it rained, they would first drive prices down, ensuring that they got little, then drive them up, making the cost of rebuilding prohibitive. They could maintain their flocks if they could drive them to new pastures, but

[50] Ibid., 124, 5–11.   [51] Ibid., 5.   [52] Ibid., 17.

their property was defined by lines on the ground, which they could not change. The current strategy amounted to keeping enough sheep that they destroyed not only the annuals but the perennials. This destroyed for a generation the land's ability to carry stock in any but the best of seasons.

The settlers thought of droughts as events, breaks in the run of "normal" years. "The lesson of experience, verifiable by records now covering something over half a century, can be summarized in the following words (and to give them the emphasis their importance demands they should be printed in block type fully an inch high): THE AUSTRALIAN INLAND MUST EXPECT A SMASHING DROUGHT EVERY DECADE, AND LESSER DROUGHTS MORE OFTEN." That included much that people considered pasture, and in the "centre of the continent, the 'Dead Heart of Australia', it would be more accurate to say that good seasons are the exception, drought the rule."[53] The upshot (and Ratcliffe printed this in italics) was that "*the fodder reserve of the semi-desert country is nowhere sufficient to stand up indefinitely to the strain that must be placed upon it*. . . . The essential features of white pastoral settlement – a stable home, a circumscribed area of land, and a flock or herd maintained on this land year-in and year-out – are a heritage of life in the reliable kindly climate of Europe. In the drought-risky semi-desert Australian inland they tend to make settlement self-destructive."[54] The bores that allowed stock to graze the whole of their paddocks meant they could destroy the perennial vegetation. "The depletion of the bush, closely followed by erosion and drift, has since proceeded hand in hand with the 'improvement' of the country."[55]

The answer was not more of the same, nor was it science as people conceived of it. "Those who regard land policies as sacrosanct (being, as they are in Australia, products of human idealism) can only suggest that the forces of science should be mobilized to grapple with the problem of the inland." It could not be done. Botanists can find plants and can suggest how to encourage fodder species; "they have not the magic to refashion the vegetation of a vast and varied region to withstand the demands of an arbitrary and over-exacting system of exploitation." The only thing that would preserve the country was "[c]onsciously to plan a decrease in the density of the pastoral population of the inland." This would seem laughable to many, but by 1960 it "may be an urgent and obvious necessity." The authorities in the United States, he said, had been forced to realize that settlement must be reorganized on a less intensive basis, and Australia had to follow.[56] Ratcliffe, it should be

---

[53] Ibid., 322–3, 319–20.   [54] Ibid., 322, 323.   [55] Ibid., 325.
[56] Ibid., 323, 325, 330–1, 331.

said, was not alone. In *Australia's Dying Heart: Soil Erosion in the Inland,*
written in 1940, a South Australian grazier indicted land tenure, private
ownership, financial responsibility, and public welfare as the root causes
of agricultural depression in the inland pastures.[57] Ratcliffe was unusual
in laying out so clearly the disjunction between an economic system
adapted to short-range goals and a natural system that insisted on the
long run. He was prescient in stressing the key role of unpredictable but
inevitable events in determining land use and far ahead of most of his
contemporaries in seeing agriculture and settlement in these terms.

He was too optimistic, though, in pointing to the United States as an
example, or in thinking that people would change. The authorities may
have realized that settlement had to be reorganized on a less intensive
basis, but they did not do it. The New Deal helped farmers and farm
communities, and it resettled some people on better land (or at least
different land), but it did not reshape the land system or change land
use. When the rains came back to the Great Plains in the early 1940s,
so did optimism and plows. After World War II farmers tapped the
Ogallala Aquifer, lying just to the east of the Rockies, and hailed it as an
inexhaustible reservoir. No one spoke of "Nature's wisdom," but otherwise the rhetoric might have come from *Australia Unlimited.*[58] In Australia
the government offered farms to veterans, and in the postwar boom millions of acres, never before farmed, were "broken" for wheat.[59] The professional debate changed, but not entirely. *Land Utilization in Australia*
(1950), for instance, warned of simple measures and, citing Taylor's
work, pointed out that rainfall in Australia varied more than in other
parts of the world, but it did not move beyond Ratcliffe's analysis.[60]
Indeed, it walked softly around the questions he had raised. In the areas
where rainfall was marginal, there were "innumerable instances of
financial disaster," but the authors could only say that "[s]ome day an
attempt may be made to ascertain whether, over a long period, the
expenditures made on areas of this type have ever been subsequently
recouped." They only suggested that "[p]ossibly a nomadic system of
occupation and management over vast areas is required."[61]

[57] Jock H. Pick, *Australia's Dying Heart: Soil Erosion in the Inland* (Melbourne: Melbourne University Press, 1942), 9–10.
[58] John Opie, *Ogallala: Water for a Dry Land* (Lincoln: University of Nebraska Press, 1993), gives a full account of this episode.
[59] William Lines, *Taming the Great South Land* (Sydney: Allen & Unwin, 1991), 197–224; Geoffrey Bolton, *Spoils and Spoilers* (Sydney: Allen & Unwin, 1981), 147–57.
[60] S. M. Wadham and G. L. Wood, *Land Utilization in Australia* (2d ed., Melbourne: Melbourne University Press, 1950).
[61] Ibid., 100. In the late 1970s, studying the pastures of western New South Wales for the state and the CSIRO (the CSIR's successor), Graeme Caughley declared that rainfall in

Conclusion

The dreams and disasters of the interwar years speak of change, persistence, and confusion. The great expansion was over, but old habits and habits of thought died hard. Some believed we had to take more conscious account of nature's limits, but they were not sure quite how to do it and even less certain of the impact on society. This is hardly surprising. From the sixteenth century, Europeans had been building economies by extracting more resources from more of the earth. From the early nineteenth century, the settler societies had built their economies and identities on the promise of continually growing wealth available to all.[62] The plans of this period were an attempt to continue the old quest by new means. Corporations, marshaling capital and deploying science, would substitute for individual pioneers and their grit and hard work. Expansive plans to move beyond the Temperate Zone to the desert and the Arctic were evidence of arrogance but also of desperation. Familiar land had yielded what it could. Now unfamiliar land must yield new crops if expansion was to continue.

Reactions to the these dry years show the changing perspective and changing conditions. Goyder and Powell, condemned as pessimists and obstacles to progress, had shared with their critics a faith in small farms and individual farmers. Both the New Deal experts and Ratcliffe had questioned the viability of European farming over vast areas where it had been established for two generations. Even the optimists now spoke of hanging on; no one planned further plowing or more pastures. The earlier debate had been about the present and the immediate future; in the 1930s the long term was a presence or a specter that boosters had to dismiss. Science had been the skeptics' weapon; now it was the common ground for understanding nature. Australia Unlimited's faith in vast underground pools depended on accepting geologists' ideas

the area varied so erratically from month to month and season to season that there was no such thing as "normal." The system was, in the modern scientific sense of the term, chaotic – locally unpredictable but globally stable. Author's interview with Graeme Caughley, June 1990. See also Graeme Caughley, "Ecological Relationships," in Graeme Caughley, Neil Shepherd, and Jeff Short (editors), *Kangaroos: Their Ecology and Management in the Sheep Rangelands of Australia* (Melbourne: Cambridge University Press, 1987), 159–87. On this project see also Parliament of Australia, Senate Select Committee on Animal Welfare, *Kangaroos* (Canberra: Australian Government Publishing Service, 1988). On chaos theory see James Gleick, *Chaos: Making a New Science* (New York: Penguin, 1987), 48.

[62] The widespread and continuing appeal of that vision is apparent in modern policy debates about worldwide development. For an example from another part of the neo-Europes see Warren Dean, *With Broadax and Firebrand* (Berkeley: University of California Press, 1995), chaps. 7–9.

about the earth – though not their conclusions. Arctic herding drew on the example of animal breeding to justify its ventures. The new view was not dominant then and it is not universal now. We are still, with somewhat more realism, engaged in that quest for new frontiers and, like our grandparents, find it easier to believe in human ingenuity than to change our ways.

Great plans and great disasters were very obvious, but there was a less visible current, a renewed interest in the significance and noneconomic value of the land for the individual and the society. Like the programs for development in these years, this was a reprise of themes popular a half-century before, and much of the activity followed old lines. As people ceased to work in nature they sought a connection with it through play or amateur study. There were, though, new ideas. They did not change policy, but they are important. In these years a few people established a new direction that has come, in the past generation, to set the terms of the continuing debate about how the settlers are to live in and with their lands.

7

NATIONAL NATURE, 1920–1940

Issues of climate played out in the interwar years as vast social dramas. The search for national nature – the meaning of the land for the settlers – was a much quieter process. There were no expeditions to the deserts or Arctic islands, no new technologies, nothing like the Dust Bowl or the Australian drought. Simple activities like watching birds and hiking were important here, and the greatest disturbance was a debate among scientists and administrators about the control of predatory animals. Shifts in sentiment, however, were no more conclusive than debates over climate. Indeed, they were less so, for they involved not events or theories but attitudes and values. The issues, though, were at least as important as those concerning about "the possibilities of the country." They were at the heart of that generation's search for a place in the land and the land's place in their culture, and the decisions that emerged would shape policy over vast areas and serve as the foundation for environmentalism, the next part of the settlers' response.

This chapter is a survey of old ideas and the search for new ones in these years. In the nineteenth century national nature had been on a large scale. There were maps and scientific reports and national symbols that ranged from emblematic animals and plants to landscape paintings. The emphasis was now shifting to a smaller scale and to individual experience. "Nature" had been vistas, large herbivores, and enormous trees. It was coming to mean personal – even if intermittent – contact with nature. There was as well a greater recognition of the impact of settler society on the land. Parks became rallying points for nature protection, and almost all birds and animals received some legal protection. A few, mainly scientists, even argued for the preservation of "varmints" as part of the life of the land.

These things happened in all our countries, but in different ways, and we will start with Australia, go to New Zealand, and end with North America. In the first two we see established ideas developed; in the last new ones are visible, if not yet important. The theme, though, is not progress but the continuing influence of the land on the people and people's use of the culture's knowledge in their search for meaning. This shows more clearly a divergence that appeared at the end of the nineteenth century. To say that the northern settlers saw nature and society as separate worlds and yearned to cross the chasm between, while their cousins to the south saw all as part of one world that they had constructed, oversimplifies and distorts, but it is a useful exaggeration. Particularly in the United States, nature was the "other," a different realm, and wilderness was becoming not only a place of spiritual and psychological significance but an object of pilgrimage. In Australia and New Zealand people did not seek wilderness, they went to the bush, where they found recreation and mateship rather than transcendence. The split appears in literature, arguments over predator poisoning, wildlife policies in national parks, and even organizations – the Victorian Town Planning and National Parks Association, for instance, fused elements that were separate or separating in North America.[1] Environmentalism would make this more apparent. For the present we can see its beginnings and the continued development of distinct national ideas.

Australia: New Meanings for the Bush

Australians' interest in nature was distinct. Its focus was the bush and pioneer settlement, and hiking – the local variant was called bushwalking – assumed an importance it had nowhere else in the Anglo world. Bushwalkers and their clubs were the backbone of national parks and nature preservation. Wildlife was less important and wildlife policy a less visible focus of national sentiment than it was in North America. The Americans were even then developing a new academic, but applied profession to preserve wildlife for the hunt – game management – and the government was taking unprecedented steps to save favored species (Chapter 8). To the Australians, however, action did not appear necessary. There was much open land, and visible species, with the exception of the koala, did not seem to be in danger. States enlarged their lists of protected animals, and some included native plants but did little to see that these were respected.[2] There was also the belief, dwindling but still

---

[1] This is part of that different division of human and natural worlds, and remains true today. See, e.g., P. Ali Menon and Harvey C. Perkins (editors), *Environmental Planning in New Zealand* (Palmerston North: Dunmore, 1993), and Australian Bureau of Statistics, *Australia's Environment* (Canberra: Government Printer, 1993).

[2] Graham Pizzey, *Crosbie Morrison: Voice of Nature* (Melbourne: Victoria Press, 1992), 122–68. For New South Wales see Birds and Animals Protection Act, 1918, No. 21, and Amendment Act, 1930, No. 12; Queensland Animal and Bird Protection Act, 1921, and

potent, that the native species were doomed to vanish before the better-adapted European introductions.[3] It was not, certainly, that Australians were not interested. Newspaper columns and amateur natural history societies continued, and the Gould Leagues flourished. In 1938 Crosbie Morrison convinced a publisher to launch a nature magazine, *Wild Life*, and when he began a series of radio talks to boost it he became a minor celebrity.

The one exception, the campaign to save the koala, shows the conventional core of wildlife sentiment. The fur trade had been hard on the koala, and by the early part of the twentieth century the species was almost gone in large parts of the Southeast. Public and official reaction, though, was sketchy at best. In 1903 New South Wales declared a two-year closed season on a number of native species, including the koala. (The others were red kangaroos, wombats, echidnas, sugar squirrels, and flying squirrels.) Queensland proclaimed protection for the koala from 1902 to 1910, and in 1912 South Australia put it on the list of animals wholly protected.[4] The decline continued. By the 1920s there were only remnants in Victoria, and Frederic Wood Jones thought it "rather doubtful" that any remained in South Australia.[5] Sentiment, though, was changing, and in 1924 a larger than normal kill in New South Wales triggered a public campaign. Three years later the Queensland government brought things to a head by putting the unemployed to work killing koalas. Local activists rose in protest, and the *Brisbane Courier* joined the cause, printing the kill in black-bordered columns. It reached 534,738 before the government acted, and only in 1937 did the state extend permanent protection.[6] By then South Australia and Victoria were moving beyond protection to restoration. Victoria had created an island refuge at the turn of the century, and in 1935 it began

---

Heber L. Longman, "Protection of Fauna in Queensland," in James Barrett (editor), *Save Australia: A Plea for the Right Use of our Flora and Fauna* (Melbourne: Macmillan, 1925), 191–9; South Australia Animal Protection Act, 1912, No. 1106, and amendments; Victoria, The Game Act, 1928, No. 3689; and F. I. Norman and C. A. D. Young, "Short-Sighted and Doubly Short-Sighted Are They," *Journal of Australian Studies*, 7 (1980), 2–24. The lack of interest among the public, though, may be judged from Geoffrey Bolton's comment, in *Spoils and Spoilers* (Sydney: Allen & Unwin, 1981), 130, that it was "a mark of courageous eccentricity when in 1927 a Gosford nurseryman began to deal in Australian natives." He finds no general interest until after World War II.

[3] For a discussion of these views see Pizzey, *Crosbie Morrison*, 258–66.

[4] Native Animal Act, 1903, New South Wales Statutes; Animals Protection Act, 1912, No. 1106, South Australia Statutes. On Queensland see AGS/N350, Corr. re native Animals, 1902–1910, Queensland State Archives, South Brisbane.

[5] Frederic Wood Jones, *The Mammals of South Australia* (Adelaide: Government Printer, 1923), 1: 187.

[6] Fauna Protection Act, 1937, No. 22, Queensland Statutes.

to transfer koalas from areas where they were still abundant to ones where they had been wiped out.[7] Interest was high enough that koala parks were established in Sydney and Brisbane. The Sydney park organized a club and in the late 1930s published, irregularly, *Koala* magazine.[8] Koalas were becoming national darlings, a status they continue to enjoy.[9] That the animals were dwindling and distinctively Australian helped save them, but other Australian animals, just as unique and even more threatened, never received a tenth of the public concern and interest. Koalas had the inestimable advantage of being cute. Soft, furry, and gently rounded, they are about the size of a large teddy bear and lethargic enough to play the part. (Modern koala parks do a brisk business in selling tourists souvenir photos taken while they hold a koala in their arms.) It was humane sentiment and sentimentality that mobilized the Australian public, not the animal's place in Australian nature.[10]

"The Cult of the Outdoors"

Bushwalking, which had begun in the 1880s as part of the common Anglo enthusiasm for outdoor recreation, was by the interwar period a force for nature preservation as well as a way to nature. Clubs campaigned for more and better national parks, and the plans they developed shaped proposals and policy into the postwar era.[11] This was unique. New Zealand had bushwalkers, but there was also skiing and mountaineering. In Canada hiking was only one of a suite of outdoor recreations, and while it was regionally important in the United States there were other national groups working for wildlife or defending the

[7] See, e.g., J. McNally, "Koala Management in Victoria," Wildlife Circular No. 4, Fisheries and Game Department, Victoria, 1957. Copy courtesy of Libby Robin. See also Jones, *Mammals of South Australia*, 1: 187.

[8] File on Koalas, Box MLK 3387, Dunphy Papers, Mitchell Library, Sydney. The library's card catalog indicates an irregular publication of *Koala* from 1937 to 1940.

[9] Comments on koalas appear in *Conference of Authorities on Australian Fauna and Flora* (Hobart: Government Printer, 1949), 11, 24, 34. On more recent developments see tourist brochures, toy stores, and children's books.

[10] On humane treatment see Harriet Ritvo, *The Animal Estate* (Cambridge, Mass.: Harvard University Press, 1987), 125–66, and James Turner, *Reckoning with the Beast* (Baltimore: Johns Hopkins University Press, 1980), 39–59.

[11] Myles Dunphy, "Some Reminiscences," 73, File 16, Box MLK 3304, Myles Dunphy Papers, Mitchell Library, Sydney. This is a later paper; more contemporary evidence is Dunphy to Noel Griffith, 29 September 1931, Correspondence, 1931–1932, 1943, Box 10, Myles Dunphy Papers, Add on 1823, Mitchell Library, Sydney. A more accessible source of Dunphy's ideas is Patrick Thompson (editor), *Myles Dunphy: Selected Writings* (Sydney: Ballagirin, 1986). On Victoria see Tom Griffiths, *Secrets of the Forest* (Sydney: Allen & Unwin, 1992), 76–84.

parks.[12] Myles Dunphy, a Sydney teacher and architect, was bushwalking's great champion and publicist. Born in Melbourne in 1891 and
raised near Sydney, he began walking in the Blue Mountains in his teens,
and as an adult he made the sport his cause. He wrote hiking guides for
the railroads and the New South Wales Tourist Department, advised
novices, and developed equipment. In 1914 he and two friends organized the Mountain Trails Club, and in 1927 he helped found the
Sydney Bushwalkers, the year – he later said – in which he and his friends
coined the word "bushwalking."[13] In 1932 he laid out plans for the
Greater Blue Mountains National Park that included "primitive areas,"
a term he borrowed from the U.S. Forest Service. The next year he
helped form the National Parks and Primitive Areas Council (acting as
its secretary until it disbanded in 1965).[14] In 1934 he and other New
South Wales bushwalkers got the Land Department to establish the
Tallowa Primitive Reserve in the Southern Highlands, "the first Australian reserve officially to recognize wilderness as a form of land
tenure."[15]

He was not alone, and interest extended beyond New South Wales.
In Victoria James Barrett, for many years president of the Victorian Federation of Walking Clubs, pressed plans much like Dunphy's. (He had
toured the United States in 1929–30 and, like Dunphy, used and
adapted American ideas.) The Victorian Town Planning and National
Parks Association proposed parks in the Victorian Alps that complemented Dunphy's for the range in New South Wales and were even more
ambitious. In the late 1930s the group suggested that all land in the
eastern highlands more than 4,000 feet above sea level be reserved for
recreation. The government rejected the plan, and World War II put it
on hold, but it reappeared. Queensland activists organized a state

---

[12] Bushwalking is still uniquely important. The Australian Conservation Foundation's
Green Pages, 1991–1992 (Melbourne: Australian Conservation Foundation, 1992) lists
thirty-two bushwalking clubs (some with many branches; the Federation of Victorian
Walking Clubs, for instance, has forty-nine clubs). There are, by contrast, thirty-nine
field naturalist societies, twenty-three devoted to native forests or rain forests, twenty
organized around birds, and fifteen interested primarily in wildlife. On New Zealand
see Records of Environmental and Conservation Groups of New Zealand, Inc., Record
92-071, Turnbull Library, Wellington. See also Roderick Nash, *Wilderness and the
American Mind* (3d ed., New Haven, Conn.: Yale University Press, 1982), 168–9, 190.

[13] Dunphy, "Some Reminiscences," reprinted in Thompson, *Myles Dunphy*, 15–16.

[14] John Yeaman, compiler, "Historical Notes Relative to Blue Mountains National Park
Trust," mimeographed, undated document (internal evidence suggests early 1970s),
Box MLK 3324, Dunphy Papers, Mitchell Library, Sydney.

[15] Land Conservation Council, *Wilderness: Special Investigation, Descriptive Report* (Melbourne: Land Conservation Council, 1990), 14. Dunphy's writings on the parks are in
Thompson, *Myles Dunphy*, 165–94.

National Parks Association in Brisbane in 1930. It was led by Romeo Lahey, who – ironically – worked for a timber company. He had been early in the field, fighting for Lamington National Park, established in 1915, and was president of the state association from 1930 until 1961, the year he died.[16]

Bushwalking's Anglo roots went deeper than administrative categories borrowed from the United States. Dunphy's exhortations would have been familiar to any North American who had read John Muir and Theodore Roosevelt. "Listen! You overworked clerks with lacklustre eyes and twitching nerves. You are short-winded, effete, stale, and partly dead." Head for the bush! "To be out with a swag is good medicine. . . . Without realizing it you will learn initiative, honesty, industry, cleanliness, trustworthiness, order, cheerfulness and how to remain healthy. These make character." Bushwalking was "the Cult of the Outdoors – craft, plus Appreciation plus Homage to the Cause of Life and Nature: a sort of sensible religion, in the practice of which the bushman acknowledges his Maker and justifies his existence" by using his talents for self, his comrades, the cult, and his country.[17] In 1931 Dunphy described the Mountain Trails Club as an association of those "who instinctively reject roads and beaten tourist routes in favour of the canyons, ranges, and wildest parts of this country. It is for those who love the forests and the broad, open life of the Bush, and who prefer to make their own trails, who have a definite regard for the welfare of wild life and the preservation of the natural beauties of the country." The group sought, therefore, "to combat the destruction of all things naturally Australian which casually goes on everywhere and for which posterity will hold this generation to blame."[18]

The accent was distinctly Australian. Bushwalkers hymned the glories of the bush, but they did not find in it the kind of sacred space that North Americans did in places like the Yosemite Valley. There was no Australian counterpart of Ansel Adams, who was making Half Dome and El Capitan into icons. Their passion was for the bush itself, and they valued it as a place of human activity – mateship. Bushwalking also pre-

---

[16] On Victorian parks see Sarah Bardwell, "National Parks in Victoria, 1866–1956" (unpublished Ph.D. dissertation, Monash University, 1974). A summary of early action is the introductory chapter of Dick Johnson, *The Alps at the Crossroads* (Melbourne: Victorian National Parks Association, 1974).

[17] Myles Dunphy, "Campcraft and Trailing," three-volume unpublished manuscript, dated 1930, Box MLK 3335, Dunphy Papers, Mitchell Library, Sydney. Quotes from chapter "On Walking" and at end of Section One. In "On Walking" Dunphy uses American outdoor cult as an example for Australians.

[18] Dunphy to E. B. Hawkes, 20 October 1931, File Correspondence, 1931–2, 1943, Box 10, Dunphy Papers, Add on 1823, Mitchell Library, Sydney.

served to a greater degree the amateur natural history tradition (which reflected its greater hold in Australia). Its most distinctive feature, though, was its appeal to pioneer values and the pastoral frontier, which in North America were more closely associated with hunting. The bush-walkers' ideal was the swagman, the nineteenth-century bush worker who tramped from station to station. In a 1930 manuscript on campcraft Dunphy considered, and rejected, various packs and frames. The best way to carry a load was the old Australian way – with a kit rolled in a canvas cover slung over the shoulder ("the swag"), food in a "tucker bag," and in the free hand a "billy can" for "brewing up" (a triumph of sentiment over sense). The swag, he said, was the symbol of Australian brotherhood, and "the swagman is welcome wherever human hearts are right and simple."[19] Melbourne bushwalkers, historian Tom Griffiths said, "were attracted to the romantic notion of the swagman as a free Australian spirit wandering the bush tracks and were delighted when mistaken for one."[20] It is unlikely that Sierra Club hikers in the 1930s would have had the same reaction if a farmer had offered them a job picking fruit – but then there were no farms in the Sierra high country, where the club organized its outings. Dunphy and his fellows also had a sympathy for illegal squatting that was rare in the United States, where nature advocates saw the wilderness as shrinking and casual settlement a danger to it.[21] "The urban escapees," Griffiths said, "romanticised the male bush worker and generally welcomed his presence in the forests. There had not yet developed that enmity that we know today [1992] between bushwalking conservationists and timber workers."[22]

"A Story of Men and Cattle"

The evolving Australian sensibility is even more apparent in a popular nature story of the period, Frank Dalby Davison's *Manshy*. Privately printed and sold door-to-door after publishers rejected the manuscript, it won the Australian Literature Society's Gold Medal as the best book of 1931, was picked up by the prestigious firm of Angus and Robertson, and went on to become a schoolroom classic.[23] In form it was perfectly normal. It was a "realistic" animal story of the type produced a genera-tion before by Seton and Roberts. (Davison had lived in the United

---

[19] Dunphy, "Campcraft and Trailing," section titled "Humping the Swag."

[20] Griffiths, *Secrets of the Forest*, 80.

[21] Dunphy to Hawkes, 20 October 1931, File Correspondence, 1931–2, 1943, Box 10, Dunphy Papers, Mitchell Library, Sydney. See also Thompson, *Myles Dunphy*, 177–8.

[22] Griffiths, *Secrets of the Forest*, 80.

[23] A. A. Phillips, Introduction to the Angus and Robertson edition, 1975. See also Hume Dow, *Frank Dalby Davison* (Melbourne: Oxford University Press, 1971), 4–7.

States from 1908 to 1914 and may have been familiar with their work.) These had not, however, caught on in Australia, and the book stands without obvious predecessors or successors in the national literature.[24] It was unlike the essays of Donald Macdonald or Charles Barrett, and even less like the children's fairy stories May Gibb wrote.

Manshy is the story of a red heifer, born in the backblocks of Queensland, who escapes the roundup to make a life in the bush. It is told from the animal's point of view, and Davison gives her mental processes that are not human but go beyond instinct. There is no obvious appeal, as there was with Seton and Roberts, to animal psychology. Here Davison relies on stockmen's knowledge – he had worked with cattle as a soldier-settler in Queensland after World War I. Like Roberts, Davison does not name the central character of his story and, like Seton, makes its desire for freedom the mainspring of the plot. The heifer and her relation to the land and people, though, are distinctly Australian. In North American stories feral animals and stock that stray into the wild always come to a bad end. Nature does not even tolerate wild ones touched by the human world. To one of her "wild children," which humans have "snatched away," she "turns a face of stone."[25] Not only does Davison make a feral animal the central figure, he presents no dangers in the wild. The heifer's only enemy is the system of industrial ranching, with its relentless drive for "beef-on-the-hook." This, though, is as implacable as the wild doom in any of Roberts's tales. Ranchers are filling up the country, fencing it in, and hiring shooters to kill off the "scrubs." At book's end the heifer is alone, waiting at a waterhole for companions who will never return.

Their authors saw nature stories as a way to help people bridge the gap between humans and nature. Roberts saw them as a way to return "to the old kinship of earth."[26] When Seton spoke of the system of industrial ranching, as he did in his first major literary success, "Lobo, King of Currumpaw," he made the same point. "Lobo" recounts Seton's work as a "wolfer" on a southwestern cattle ranch, killing first the pack Lobo leads, then the wolf's mate, finally the "old outlaw" himself. He finds in

[24] Dow, *Frank Dalby Davison*, 3–4. Davison's own *Dusty* (Sydney: Angus & Robertson, 1946) is perhaps closest. Another example, not quite in the genre, is Henry Lamond, *White Ears the Outlaw* (Sydney: Angus & Robertson, 1949), originally published in serial form in the *Australian* in 1934.

[25] Charles G. D. Roberts, "The Return to the Trails," in *Watchers of the Trails* (Boston: Page 1904), 58–9. A general analysis can be found in Thomas R. Dunlap, "'The Old Kinship of Earth: Science, Man, and Nature in the Animal Stories of Charles G. D. Roberts," *Journal of Canadian Studies*, 22 (Spring 1987), 104–20. Note the contrast with Davison, who deals with a bit of wild Australia in *Dusty*, the story of a half-dingo, half-kelpie.

[26] Charles Roberts, *The Kindred of the Wild* (Boston: Page, 1902), 29.

the old wolf's love for his mate, the passion that betrays him into Seton's traps, an opening into the universe. He (though this is not described) hangs up his traps and rifle, abandons the poison bottle, and becomes a friend of the animal world.[27] Where the North Americans saw two worlds – wild nature and human society – Davison saw only one. His sub-title for *Manshy* was *A Story of Men and Cattle*. All were enmeshed in the system – humans, cattle, the land itself.[28]

"Great Red Landscapes, Wide Open to the Splendour of the Sun"

H. H. Finlayson's *The Red Centre* (1935) was another marker of a new Australian sensibility.[29] Like *Manshy* it had an established Anglo form – the naturalist's report of his expedition – and its dramatic moment was one common in natural history, the discovery of a rare specimen. What marks it off is its engagement with the desert, a landscape outside Anglo ideas of beauty and a part of the continent the Australians had shunned. Anglo Australia was a country on its own periphery. In the nineteenth century, in an evocative usage, "inside" had referred to the settled areas on the coast. Art and literature did not speak of the Centre. Paterson and Lawson had written of, and the Heidelberg painters depicted, the inland pastures that ran in an arc around it. In 1916 J. W. Gregory said that "Australia still suffers from the misleading idea that the continent consists of a narrow ring of fertile land around a vast internal desert" (an interesting observation, for a decade before he had given the Centre – the term first appeared in 1899 – the name that would describe it for two generations: "the dead heart of the continent").[30] In the early 1930s Finlayson had collected natural history specimens in the land of the Luritja people, in European terms the southwest corner of the Northern Territory, overlapping into Western Australia and South Australia. The explorers had found hardship there; Gregory had considered it useless land; *Australia Unlimited* had seen it only as raw material to be

---

[27] Ernest Thomson Seton, *Wild Animals I Have Known* (New York: Scribner's, 1898), 17–54.
[28] Frank Dalby Davison, *Manshy: A Story of Men and Cattle* (Melbourne: privately printed, 1931).
[29] H. H. Finlayson, *The Red Centre* (Sydney: Angus & Robertson, 1935), 38. "Red Centre" is now the term used in advertising tourist excursions to Ayers Rock and other attractions in the interior.
[30] J. W. Gregory, *Australia* (Cambridge University Press, 1916), 16; William Ransom, "Wasteland to Wilderness," in O. J. Mulvaney (editor), *The Humanities and the Australian Environment* (Canberra: Australian Academy of the Humanities, 1991), 10–11; J. W. Gregory, *The Dead Heart of Australia* (London: John Murray, 1906). On the expedition and the term see David Branagan and Elain Lim, "J. W. Gregory, Traveller in the Dead Heart," *Historical Records of Australian Science*, 6, 1 (1989), 71–84.

reshaped for human use. Finlayson found beauty. It was, he said, a com-
pelling land, "often attractive, and sometimes highly picturesque."[31] He
admitted that it offered no ease, no recreation, certainly no fruitful
fields. The early explorers had called it a desert, with reason. A "daily
threat of death would spoil the charms of a Paradise, and it is not pre-
tended that any part of the Centre is that."[32] Vegetation was sparse and
low, animals few, small, and nocturnal, and it was hot and dry. Still, "[i]n
retrospect, it is not the passing discomfort of the temperature that comes
to mind, but rather a vision of great red landscapes, wide open to the
splendour of the sun."[33]

Though he might not have put it in quite this fashion, Finlayson was
doing what the Heidelberg painters had done when they adapted the
techniques of French Impressionism to celebrate Australia's harsh and
glaring light. He was extending the nature aesthetic. His project,
though, was even more daring than theirs. They had had the vegetation
of the bush and the ennobling hardships of pioneering. There was no
white settlement in the Centre, and the "scenery" was beyond anything
that Europeans celebrated. This was the bare bones of the continent,
stripped, dry, open to a sun even harsher than Streeton's "purple noon's
transparent might," and hidden from settlement or even appreciation
by its own distances and the lack of water. The values Finlayson found
were equally conventional in form, equally radical in their implications.
He appealed to the primitive and the picturesque and to the power of
nostalgia. Even here, far beyond fields and flocks, the settlers were
changing the land, destroying unaware this harsh Eden. "[M]uch evidence of the past history of the life of the country slips suddenly into
obscurity. The old Australia is passing. The environment which moulded
the most remarkable fauna in the world is beset on all sides by influences
which are reducing it to a medley of semi-artificial environments, in
which the original plan is lost and the final outcome of which no man
may predict." When he assessed that loss, though, he left convention. If
"the devastation which is worked to the flora and fauna could be assessed
in terms of the value which future generations will put upon them, it
might be found that our wool-clips, and beef and timber trades have
been dearly bought." In a country that had, in the common phrase,
"ridden to prosperity on a sheep's back," this was heresy.[34]

---

[31] Finlayson, *Red Centre*, 22.
[32] Ibid., 28.
[33] Ibid., 38. Whether Finlayson was echoing Paterson's line from "Clancy of the Overflow,"
     "the vision splendid of the sunlight plains extended," may be of interest to literary
     analysts.
[34] Ibid., 16.

Finlayson's lament sounds like those John Muir had made about the destruction of California, and his appreciation of the desert resembles that of other Anglos, but there are differences as well. One is the Australian's much deeper immersion in natural history. It is not simply that he was a scientist. The form of the book was straight from natural history. It began with the geography and topography and went on to the landscape, animals, and Aboriginal humans. A chapter on the rediscovery, and securing the first live specimens, of the Plains Rat Kangaroo, *Caloprymnus campestris*, interrupted the flow, but this was because these events were the highlight of the expedition. The species had been described by John Gould in 1843 on the basis of "three specimens sent home from an unknown locality in South Australia" by the governor, Captain George Gray, and never found again.[35] The narrative then treated white civilization on the same terms as before, with chapters on the camel, the car, and the Europeans. This is the same unity that appeared in Gilbert White's *Natural History of Selbourne* and Davison's *Manshy*. Everything is part of the life of the land and all is grist for the naturalist's mill. Finlayson thought the settlers would regret their actions, but he did not see them contaminating the land. Nor, despite his critique of Anglo-Australian actions, did he see nature as a place from which to criticize society or find the desert valuable because civilization found it useless.[36]

The Red Centre is an exploration of ideas as well as of territory. Finlayson appreciated nature as much as any North American, but he worked within a less developed intellectual tradition. North American nature writing was part of a literary and social mainstream. Walden Pond was a place of pilgrimage, Muir and Burroughs literary and public figures. Nature as a national symbol was well worked into the culture, and U.S. and Canadian writers had by now developed distinct variants within the genre of nature essay and nature story. Nature's place in Anglo Australia was much less clear and its literature much thinner. Finlayson used a nineteenth-century literary form from Britain, the local natural history. Another difference, perhaps the crucial one, was that Finlayson was grappling with the desert. The Anglo settlers had spent the nineteenth century wishing the arid lands away and the early part of the twentieth looking for technology to abolish it, but while the desert was only one American landscape, it was the overriding reality of Australian settlement. Like the North in Canada it was a presence that affected everything. The Australians had ignored it. Finlayson was a

[35] Ibid., 96.
[36] Patricia Limerick, *Desert Passages* (Albuquerque: University of New Mexico Press, 1985), provides an introduction to American writing.

pioneer in seeking to come to terms with it. Like his collections and survey, it was incomplete. Much was found, not all was understood. The "old Australia" was precious, but Finlayson could not yet say exactly why.

New Zealand

"The Overthrow of the Old World and the Slow Re-establishment of a New Equilibrium"

In New Zealand as well, people were finding value in native nature. At the turn of the century, Parliament set aside a few offshore islands as refuges for vanishing native birds, seeking to save pieces of what they were coming to see as a unique land. In 1914 prominent citizens, several of them members of Parliament, organized the New Zealand Forest and Bird Protection Society. It fell apart, but in 1923 E. V. Sanderson formed the Native Bird Protection Society (which eventually took over the old name and then was allowed to add "Royal" to the title). It was supported by a rising sentiment for protection. A letter in the *Wellington Post* in 1926, signed simply "New Zealander," caught the new tone of biological nationalism. It condemned the government's allowing collectors of the Whitney expedition to take 846 specimens of native birds on the grounds that the interests of "native born New Zealanders" were being disregarded in order that "gaps in foreign museums" might be filled. These birds were "our property and the heritage" of the next generation. Three years later the Native Bird Protection Society raised the alarm when a party arrived from "that most anti-British of all American cities, Chicago," with "the avowed object of collecting specimens of our native birds." It applauded the decision to refuse permission for this work. (It did not say whether the patriotism behind this decision had to do with empire or nation – or if there was yet a difference.)[37] It also took more positive measures. It cooperated with the New Zealand Forestry League and the acclimatization societies to control exotics, prevent erosion, and rationalize the wildlife laws. Following the lead of the

---

[37] In 1934, after the founder of the original Forest and Bird Protection Society died, the new society took the old name, later adding "Royal" to its title. Quote from clipping, *Wellington Post,* 14 May 1926, in Department of Internal Affairs, Wildlife Files, 46/29/18, New Zealand Archives, Wellington. Progress report in File 56, Royal Forest and Bird Protection Society Papers, Turnbull Library, Wellington. See also File 425, Ms. papers 444, Royal Forest and Bird Protection Society. On the persistence of British sentiment, including nature sentiment, see Ross Galbreath, "Colonisation, Science, and Conservation" (unpublished Ph.D. dissertation, University of Waikato, 1989), 125–7, and idem, *Working for Wildlife* (Wellington: Bridget Williams, 1993), 13–14.

Gould Leagues and the Audubon Society, it produced materials for nature education in the schools.[38]

These visions of a land transformed and a heritage in danger rested on what could be seen and on common sense. It was conventional wisdom that all was vanishing before the triumphant Europeans. In the early 1920s two books appeared challenging that view and presenting a more complex story. They were H. Guthrie-Smith's *Tutira: The Story of a New Zealand Sheep Station*, the record of the author's forty years of nature observation and sheep raising, and G. M. Thomson's *The Naturalization of Plants and Animals in New Zealand*.[39] Thomson's was a conventional catalog, notable here mainly for recognizing that the clash of natives and exotics was a biological event almost without parallel in the world. *Tutira*, though conventional in form (it is a late descendant of Gilbert White's *Natural History of Selbourne*), is more radical in its appreciation of this theme. It is classic amateur natural history, a small area closely and lovingly observed over a long period of time, all the world in a small landscape. For forty years Guthrie-Smith, a Hawkes Bay grazier, had observed nature on his sheep station. He had, he admitted, a passion for small happenings, and he had traced the comings and goings of exotics and natives, the sudden efflorescence of some under the new regime of sheep farming, the decline of others, the unexpected resurgence of a few. Like Finlayson he told his story from the ground up. The early chapters dealt with the rocks, lakes, and soils of the station he bought in the 1880s. Next he discussed the vanished forests and the Maori who lived and fought there and placed so many names on the landscape and wrote of the white men who came to "break the land" and the successive waves of European plants and animals that accompanied them. His grand theme was one foreign to White but common to the settler lands – the mixing of suites of very different plants and animals. It was an enthralling spectacle. "So vast and so rapid have been the alterations which have occurred in New Zealand during the past forty years, that even those, like myself, who have noted them day by day, find it difficult to connect past and present – the pleasant past so completely obliterated, the changeful present so full of possibility."[40] Each ride over the station "has been for forty years a fresh page in the story

[38] See files of the Forest and Bird Protection Society for this period in Turnbull Library, Wellington, and Department of Internal Affairs files on birds, native wildlife, and sanctuaries in Records of the Department of Internal Affairs, New Zealand National Archives, Wellington.

[39] G. M. Thomson, *The Naturalization of Plants and Animals in New Zealand* (London: Blackwood, 1921).

[40] Guthrie-Smith, *Tutira*, vii.

– to be continued in our next – of the overthrow of the old world, and the slow re-establishment of a new equilibrium."[41]

One thing his observations should have overthrown was the common wisdom. New Zealanders, like Australians, had believed that the native species, unable to compete in the "battle for life" with the more aggressive and "better adapted" Europeans, were doomed. Hutton and Drummond's comments from 1905 were typical: "[W]e now lament the loss of many rare and remarkable birds, and would bring them back or at any rate, would stop the flight toward extinction." Alas, "the ancient Fauna, as a whole, and as it existed in its original state, is fast departing. . . . Try as we can, we can never bring back the birds that have refused to abide with us. . . . Their habitation has been laid waste, and their glory has departed."[42] Guthrie-Smith found that some native plants and animals did vanish, and many more dwindled in numbers, but there was no inexorable retreat before better-adapted European species. Instead there was an ebb and flow. He had, as one might expect, kept careful track of pasture plants. Some of the aliens succeeded, others failed, but "[e]ach time a plant has overrun central Tutira it has been a native." Each part of the station, in fact, had its own story, and he described in some detail how one area had "grassed itself in its own way, selecting and rejecting, and clothing itself finally with the fodder-plants suited to its particular requirements."[43]

The aliens, he found, did not spread ahead of humans but depended on them. Weeds appeared mainly on that small part of the run that could be freed from bracken, and "90 per cent of aliens have appeared about the homestead, the gardens, orchards, garden-paths, and roads."[44] As for the native birds, he believed that most would survive, though in diminished numbers, and those that vanished would not do so for "the reasons so often assigned; they are not less vigorous than their acclimatised rivals, they will neither be ousted by imported species or annihilated by imported vermin. The fact is that our imported birds are not bred to stand from thirty to seventy hours of tropical downpour. . . . In these storms species whose forebears have not been accustomed to face seven, fourteen, and twenty inches in three consecutive ceaseless days' rainfall, perish in great numbers."[45] Thomson made the same point. Naturalists had commonly believed "that the introduced fauna and flora have been directly responsible for the diminution of many, and the disappearance of some, indigenous animals and plants." However, "where

[41] Ibid., 384.

[42] F. W. Hutton and James Drummond, *The Animals of New Zealand* (Christchurch: Whitcombe & Tombs, 1905), 24.

[43] Guthrie-Smith, *Tutira*, 178, 179.   [44] Ibid., 244.   [45] Ibid., 204.

man does not interfere with the vegetation, the indigenous species can hold their own against the imported forms." The abundant plants that impress visitors are mainly *"weeds of cultivation."*[46]

That these observations were radical is apparent in the record of deer policy.[47] By 1920 concern about the growing herds was turning to alarm, but attempts to meet the situation were all common sense and old ideas applied to anecdotal observation. People assumed that European species were superior and that native plants, having evolved free of grazing pressure, were uniquely vulnerable. They extrapolated from the fact that deer lived on vegetation to the conclusion that they were destroying the forests. The remedy was equally conventional. The government removed protection in district after district, counting on farmers and sportsmen to thin the animals out. This became policy following a two-day conference in Christchurch in 1930. It was a gathering at which representatives of the Native Bird Protection Society, the acclimatization societies, the Farmers' Union, the Alpine Club, the Canterbury Philosophical Institute, the Forestry League, the Sheep Owners' and Farmers' Federation, and the New Zealand Institute (the national scientific body) conferred with officers from several government departments. They heard from Cockayne and accepted his predictions of disaster. The government removed all protection from deer and mounted an extermination campaign. Major Yerex, in charge of the operation (the rank was from World War I) planned it like a military campaign. His "deer cullers," as they were officially known, were to sweep through the country, district by district. When they were finished, the deer would be gone.

This was the established logic of pest control. It failed here as clearly as it had when applied to rabbits in Victoria or to coyotes in Montana, and the New Zealanders were no more successful in coming to grips with the situation. The problem was that they lacked concepts even to study what was going on. This was natural history's limitation, as science and as a way to understand nature. It could deal with nature's pieces, but not with the processes that bound them together on the land. Thomson and Guthrie-Smith could see that the native species were not simply vanishing and everyone could see that deer were increasing, but lacking understanding of the dynamics of population growth and the environ-

---

[46] Thomson, *Naturalization*, 514, 509. A more recent study of conventional myths and the lack of scientific support is Carolyn King, *Immigrant Killers* (Auckland: Oxford University Press, 1984), 9–13. I thank Dr. King for a copy.

[47] The best account is Graeme Caughley, *The Deer Wars* (Auckland: Heinemann, 1983), 7–33. See also Ross Galbreath, *Working for Wildlife* (Wellington: Bridget Williams, 1993), 64–80.

mental factors that governed it they could little more than wonder – or endorse the extermination campaign. A few, as we have seen, were groping toward this vision. Like Ratcliffe, seeing the contradiction between the economic and biological systems, Davison, Finlayson, Guthrie-Smith, and Thomson could see the layers of relationships that bound things together, even if they could not describe them in any detail. Others, though, with access to new concepts, were starting to use new methods to work toward a more exact knowledge.

North America: New Meaning for the Wild

> I set out later as a teenager with Roger Tory Peterson's *Field Guide to the Birds* and binoculars in hand, as all true naturalists in America must at one time or other.
>
> Edward O. Wilson, 1994[48]

In these years a few North Americans were moving most quickly toward a new vision of nature, but before turning to this radical work we need to consider its conventional context, the continuing development of nature appreciation. The most striking evidence of that was a new generation of field guides to birds. They were, in one sense, the successors of Chapman's and Leach's work, but they were so much better – or people so much more interested – that they can fairly be said to have opened nature appreciation to a much larger audience. Certainly they paved the way for the nature boom that came after World War II. The exemplar is Roger Tory Peterson's *A Field Guide to the Birds* (1934), arguably the single most significant publication in American nature literature. It made birding, as enthusiasts came to call it, the hobby of millions, and it became the flagship of an entire library that now covers everything from clouds and seashells to animal tracks.[49] At first glance it looked like any bird book, but it was distinguished by Peterson's single-minded and skillful use of text and illustrations (he was an artist as well as a birder) to one end: putting a name on what was seen in the field. It included only information that would focus the reader's attention on "impressions, patterns, and distinctive marks," and the paintings were done with an eye to what could be seen at a distance.[50] Ducks, often seen

---

[48] Edward O. Wilson, *Naturalist* (Washington, D.C.: Island Press, 1994), 14.

[49] Roger Tory Peterson, *A Field Guide to the Birds* (Boston: Houghton Mifflin, 1934; original reprinted in 1996, following Peterson's death). Any large bookstore in the United States will have two dozen of this series, not to speak of competitors, in its nature section. Peterson's impact is discussed in Felton Gibbons and Deborah Strom, *Neighbors to the Birds* (New York: Norton, 1988), 299–300.

[50] Peterson, *Field Guide*, xix.

far off, were shown as outlines filled in with patches of black and white, and there were plates showing them overhead, as duck hunters saw them. Species usually seen in flight were shown that way, and similar species were grouped together in the same pose. The pictures, Peterson said, would often tell the story, and where they did not the reader "should select the picture that most resembles the bird he saw, and then consult the text."[51] That was full of phrases like "No other . . . combines the characters of *yellow wing-patch* and *black throat*" (Golden-Winged Warbler) and "the only *bluish* Warbler with a *yellow* throat and breast" (Parula warbler). In the first edition Peterson passed over the robin as "[t]he one bird that everybody knows," but in the second edition (1947) he described it in the same fashion: ". . . recognized by its gray back and *brick-red* breast. . . . Robins walk on lawns with erect stance."[52]

This is a mixture of folkbiology, science, and earlier bird books. The central concept, pattern recognition, is the way people have always sorted out birds – first by shape and size to pin them down to a group (gull, tern, or hawk, for example), then by characteristics like color patterns or behavior that mark out particular species. It was not, as Peterson admitted, a new idea even for bird books. The "overhead flight patterns of hawks had been worked out years ago, notably by Ernest Thompson Seton," and "Dr. John B. May, in Forbush's *Birds of Massachusetts*, was the first to revive and popularize the idea."[53] The birds were arranged and named by modern taxonomic practice. They appear in the texts, if not always in the plates, arranged by families and genuses. The scientific names were included, and the popular ones were drawn from the American Ornithological Union's English-language list. Successive editions reflect an increasing formality in allocating popular names. The 1934 edition, for example, identified *Falco peregrinus* as the "duck hawk." The 1947 revision had "duck hawk or peregrine falcon"; the 1980 edition identifies the bird as the "peregrine" and relegates "duck hawk" to smaller print and parentheses.[54]

The same pattern also appeared in Neville Cayley's *What Bird Is That?*, published in Australia three years before.[55] It supplanted Leach's hand-

---

[51] Ibid., vi, 92, 21.

[52] Quote on p. 171 of the first edition of Peterson's *Field Guide* (1947, same publisher). Australian readers, if they have a copy of Graham Pizzey's *Field Guide to the Birds of Australia* (Sydney: Collins, 1980), can compare the description of a comparably common bird, the Willie Wagtail: "Black and fan-tailed, with white underparts sharply cut off below breast; note *white eyebrows and slim white whiskermark*" (264).

[53] Peterson, *Field Guide*, v–vi.

[54] On changes see also R. Tod Highsmith, "What's in a Name?" *Living Bird, 14* (Spring 1995), 30–5.

[55] Neville Cayley, *What Bird Is That?* (Sydney: Angus & Robertson, 1931).

book, which had never been completely satisfactory. Besides its long sections on things like avian anatomy, which were not useful for field identification, it had treated only Victorian birds. That left out a quarter of the species in Western Australia and lesser percentages in the other states.[56] Like Peterson, Cayley provided a complete list for a large area – in this case all of Australia – and concentrated on field identification. He did not have a full system of field marks and patterns, and his illustrations were less helpful than Peterson's, but in one way he was more attuned to the amateur. Rather than listing the birds in taxonomic order, he grouped them in an intuitive fashion. There were sections on forest birds, nocturnal birds, ground feeders, birds of the tree trunks and branches, and so on. He paid less attention to scientific names, or more to popular ones. (This is still true for Australia. Graham Pizzey's *Field Guide to the Birds of Australia* [1980] lists far more popular names for each species than do modern American guides.)[57]

In both hemispheres people used these books to put names on what they saw afield, but they did it in different contexts. Bird-watching in the United States was almost wholly disassociated from amateur natural history. Identification was becoming an end in itself. The "big day" of the spring migration was becoming an event and the "life list" central to the sport. This was the equivalent of hiking as many miles or over as many peaks as quickly as possible – nature as a stage for competition with others or oneself. Australian bird-watching retained (and retains) a closer connection to nature study. More than their Audubon equivalents, Gould League publications emphasized systematic observation, field notes, and projects, and even recent Australian field guides do not have the checklists or "life lists" of American publications. There is one in Peter Slater's *The Birdwatcher's Notebook* (1988), but it is part of a data set, with places to record multiple sightings, information on relative abundance, and observations on nests (start, finish, number of eggs, hatching, and fledging). Slater dismisses listing as "twitching," at the "solipsist" end of the spectrum of activities, and passes on with the phrase "At a more serious level . . ."[58]

---

[56] J. A. Leach, *An Australian Bird Book* (Melbourne: Whitcombe & Tombs, 1911). For the state of New Zealand amateur studies see Perrine Moncrieff, *New Zealand Birds and How to Identify Them* (Auckland: Whitcombe & Tombs, 1925). It is much less advanced.

[57] Pizzey, *Field Guide.*

[58] Peter Slater, *The Birdwatcher's Notebook* (Sydney: Weldon, 1988), 51, 70. An English work, contemporary with Cayley's, shows the continued Australian debt to Britain: E. M. Nicholson, *The Art of Bird-watching* (New York: Scribner's, 1932). Despite the publication data, this was printed in Britain and was not intended for the United States. It does not even deal with species identification, leaving that for other books. Australian newspaper columns and Vincent Serventy's correspondence with amateurs throughout Australia in the postwar period (Vincent Serventy Collection, National Library, Canberra) show this strain.

"Conserve Our Native Species!"

Since the late nineteenth century, when the Audubon Societies were organized around the issues of bird protection, saving nature had been part of nature appreciation. Concern increased as people came more and more to see birds as part of the larger whole, native nature. This issue was not as important as it was in New Zealand, but it was becoming visible among at least a small core of biologists, ecologists, and game managers. The success of the ring-necked pheasant, brought to Oregon in 1889, had set off a low-level but continuing enthusiasm for new species to replace "shot-out" native game birds or add novelty to the bag, and when, in 1928, John Phillips reviewed the record of introductions, he found two schools of thought, "widely at variance. One of these, the conservative, represented by such eminent naturalists as Joseph Grinnell of California and many others, believes in preserving at all costs the present or rather the original status of native birds and harmless mammals, and points out the great dangers incurred in the importation of new species in other parts of the world, and especially the danger of the spreading new diseases." The other would try anything without considering the dangers or the suitability of the species to the country.[59] The conservatives did speak of the disasters of acclimatization, but they also were concerned to save what they saw as a still-living native nature. Of the game birds of California, Grinnell wrote: "Conserve our native species! There are none whose qualities are superior: they are part of the natural heritage of our land, and have been serviceable in the past; we are responsible for their preservation."[60]

Grinnell, a central figure in the defense of native nature in this generation, was one of those who had made the transition from natural history to ecology. Born in 1877 in the Indian Territory, where his father was a doctor at an agency post, he had grown up on reservations and in Pasadena, California – somewhat more rural then than now – and had retained into adulthood his childhood interest in nature. When he

[59] John Phillips, "Wild Birds Introduced or Transplanted in North America," U.S. Department of Agriculture, Technical Bulletin 61 (Washington, D.C.: U.S. Government printing Office, 1928), 5. This carried over into Canada. See Acts and Legislation, Box 1, RG 109; Transplantation of Exotic Wildlife, Box 43, RG 109; memo of Hoyes Lloyd, 7 September 1923, Box 35, RG 84 – all in Public Archives Canada, Ottawa. See also W. F. Lothian, *A History of Canada's National Parks* (Ottawa: Environment Canada and predecessor agencies, 1977–87), 4: 18–19.
[60] Joseph Grinnell, Harold C. Bryant, and Tracy I. Storer, *The Game Birds of California* (Berkeley: University of California Press, 1918), 44. On the larger dimensions of this concern see Lee Clark Mitchell, *Witnesses to a Vanishing America: The Nineteenth Century Response* (Princeton, N.J.: Princeton University Press, 1981).

joined the Klondike gold rush in 1898 he spent more time collecting birds than panning for ore.[61] Back in California, an introduction to Miss Annie Alexander provided a way to turn his interest into a profession. Alexander, heir to a Hawaiian sugar fortune, was passionately interested in natural history. She had used her wealth to get around the social prejudices against women in the field, and she proposed to do more. When she met Grinnell she was engaged in endowing what became the Museum of Vertebrate Zoology at the University of California, Berkeley. She decided Grinnell should be its director. He began in 1908, when the museum was established, and died in office in 1939.[62]

He was a consummate museum builder, meticulous about the smallest details. He and Miss Alexander corresponded about specimen labels and ink, for he was determined to have permanent records. As director he inspected the facilities regularly and wrote tart notes when things were not up to snuff. He kept the museum away from anything that did not directly serve research, refusing to have public exhibits and giving away ones that were pushed on him. He trained his students to take complete field notes, and the volumes are still shelved, in alphabetical order, in the museum.[63] He made the MVZ one of the finest research institutions in the country and trained a generation of mammalogists and ecologists.[64] He was also a passionate advocate for nature. He was concerned about native birds, and argued for leaving dead trees on Berkeley campus because they provided nest holes and food for birds. He was concerned about the impact of sheep on high mountain pastures – one of Muir's subjects as well. He wanted to save predatory mammals as part of the country's wildlife and was behind the first organized protest of predator poisoning (discussed later).

Like Frank Chapman, shooting the birds he loved, or Edgar Anderson's plant taxonomists, putting the cloud forests into herbaria, his

---

[61] Elizabeth Grinnell (editor), Joseph Grinnell, *Gold Hunting in Alaska* (Chicago: David Cook, 1901).

[62] Hilda Wood Grinnell, "Joseph Grinnell: 1887–1939," *Condor, 42* (January–February 1940), 3–34. Michael Smith, *Pacific Visions* (New Haven, Conn.: Yale University Press, 1987), discusses California science.

[63] See, e.g., Grinnell to C. Hart Merriam, 27 November 1907, in Rescued Correspondence, 1907, Box 78, W. L. McAtee Papers, Library of Congress, and Joseph Grinnell, "The Methods and Uses of a Research Museum," *Popular Science Monthly, 77* (August 1910), 163–9, reprinted in *Joseph Grinnell's Philosophy of Nature* (1943; reprinted Freeport, N.Y.: Books for Libraries, 1968). The museum files, collections of notes, and ordered correspondence testify to his methods. Files of museum workers testify to his inspections. Files of the Museum cited with the permission of Dr. David Wake, Director.

[64] On Grinnell's influence see Tracy I. Storer, "Mammalogy and the American Society of Mammalogists," *Journal of Mammalogy, 50* (November 1969), 785–93.

passions for nature and for understanding it served each other, and he carried this mixture into the museum. He began faunal studies in Yosemite National Park soon after the museum was established. One of his students, Harold Bryant, went to work for the National Park Service, where he helped found and was first head of the nature interpretation program. Another, George Wright, was even more influential. Wright, who minored in vertebrate zoology under Grinnell while earning a degree in forestry, became a park ranger. In 1928 he suggested a survey of wildlife conditions in the parks. (That he was independently wealthy and offered to pay for it may account for his superiors' quick approval.) His campaigned for wildlife, and in 1930 Horace Albright, the agency's chief, called for the preservation of native flora and fauna and urged superintendents to eliminate or curb non-native forms and guard against new introductions in terms that echo his.[65] Two years later, on the basis of that survey, the Service established a new wildlife division (Bryant was on the committee approving it) with Wright as chief.

The record of that work, which became the division's first publication, called for a new and broader vision.[66] The Service, it said, should not concentrate on species tourists wanted to see but work to save all, for each "is the embodied story of natural forces which have been operative for millions of years and is therefore a priceless creation, a living embodiment of the past."[67] Instead of presenting wildlife as a spectacle or a sideshow, it should give visitors an authentic experience of the wild and teach them to look for it and appreciate it. Wright went on to develop plans to protect the trumpeter swan in Yellowstone and, with the Bureau of Biological Survey, to save the birds' breeding grounds outside the park. His death in 1935 in a car accident removed a strong voice for wildlife, but the office and ideas lived on. The Service took more interest in the Yellowstone grizzlies, and gradually (and with lapses) phased out garbage dumps for bears.[68] In 1939 Park Service officers told a Senate special committee on wildlife that the agency was also seeking to eliminate or hold to a minimum introduced plants and

---

[65] Horace Albright to Superintendent and Concessionaires, 11 November 1930, File 720, Entry 7, RG 79, Records of the National Park Service, National Archives, Washington, D.C.

[66] George M. Wright, Joseph S. Dixon, and Ben H. Thompson, *Fauna of the National Parks of the United States*, Fauna Series No. 1, U.S. Department of the Interior (Washington, D.C.: U.S. Government Printing Office, 1933).

[67] Ibid., 54.

[68] U.S. Department of the Interior, *Fading Trails: The Story of Endangered American Wildlife* (New York: Macmillan, 1943). Alston Chase, *Playing God in Yellowstone* (Boston: Atlantic Monthly Press, 1986), touched off the most recent round of debate over park wildlife policy. See also Frank C. Craighead, *Track of the Grizzly* (San Francisco: Sierra Club, 1979).

animals. Four years later it sponsored a book to alert the public to the problem of endangered species.[69]

"Concern for the Coyote Itself"

Outside the parks the new ideas were most evident in the campaign Grinnell and his colleagues mounted in the 1920s against the Bureau of Biological Survey's predator poisoning. To suggest that predators might have a place in the life of the land was a radical idea. Exterminating them was common practice and official policy, approved by even wildlife's defenders, who thought that "vermin" should be killed to save the "good" animals. The refuges run by the Audubon Society excluded, at the point of a shotgun, bird-eating hawks. In 1915 Congress had established an office of predator and rodent control in the Biological Survey to help ranchers eliminate "varmints," and by the mid-1920s the only wolves were a few loners, and populations of bears and mountain lions were dwindling. As attention shifted to the ubiquitous coyote, Grinnell became more concerned.[70] He was reluctant to take up the issue, for his beloved museum was a state institution and so subject to the wrath of rural legislators, but he worked behind the scenes and at the 1924 annual meeting of the American Society of Mammalogists, four speakers – his friends and associates – charged that the Survey was using poison widely and carelessly, taking a needless toll of nontarget wildlife, and hindering scientific study by destroying entire populations. Two Survey biologists replied.[71]

For almost a decade the argument went on, moving from coyotes to the general use of poison. The ASM formed a committee to investigate the original allegations, and, when that group split, it appointed a second, which also deadlocked. By this time it was 1929, and A. Brazier Howell, protesting that the society was treating the agency with kid

[69] U.S. Congress, Senate, Special Committee on the Conservation of Wildlife Resources, *The Status of Wildlife in the United States*, Report No. 1203, Senate, 76th Congress, 3d session (Washington, D.C.: U.S. Government Printing Office, 1940), 362, 352.

[70] Thomas R. Dunlap, *Saving America's Wildlife* (Princeton, N.J.: Princeton University Press, 1988), chap. 4, has a full account of the program and protests. This discussion relies on that. For a different view of the episode's significance see Donald Worster, *Nature's Economy* (San Francisco: Sierra Club, 1977; reprinted Cambridge University Press, 1985), chap. 13.

[71] The critics were Joseph Dixon, who had worked at the MVZ since 1915; Lee Dice, a Grinnell student; and Harold Anthony and C. C. Adams, two of his friends. Papers from this session appear in the February 1925 issue of the *Journal of Mammalogy*, 6: Lee R. Dice, "Scientific Value of Predatory Mammals"; E. A. Goldman, "The Predatory Mammal Problem and the Balance of Nature"; Joseph Dixon, "Food Predilections of Predatory Mammals"; and C. C. Adams, "The Conservation of Predatory Mammals."

gloves, circulated a petition calling for an end to the current program. He also wrote two articles for *Outdoor Life*: "The Borgias of 1930" and "The Poison Brigade of the Biological Survey." W. C. Henderson replied for the Bureau in "The Other Side of the Poison Case." E. Raymond Hall, one of Grinnell's students, weighed in with a two-part article, "The Poisoner Again." Harold Anthony, with the American Museum of Natural History, blasted the Survey in *Science*, Howell reinforced the indictment, and E. A. Goldman replied. There was another session on the subject at the ASM's 1930 meeting, and the two groups agreed on a joint inspection of the agency's field operations. That broke down in a welter of mutual suspicions. In 1931 *Condor*, a West Coast ornithological journal published at the MVZ, detailed the Survey's use of thallium-poisoned grain to kill birds.[72] In the end the policy remained. The public was not ready to accept the "varmints" as part of the life of the land, and the stockmen, the program's political supporters, were opposed to any change. They wanted to wipe out the last coyote and prairie dog, even if it took tax money from around the country to do it. The Depression gave the work a boost. There was strong pressure to help farmers and ranchers; poison was the cheapest method of control, and hiring people to spread poisoned grain to get rid of prairie dogs and ground squirrels provided work for the rural unemployed. The issue gradually faded, and there was a reconciliation of sorts. Ira Gabrielson, who had resigned from the ASM in 1931 in protest against the accusations leveled at his agency, re-joined a few years later – though he did not, as one of the protestors privately noted, offer to pay back dues.[73]

Gabrielson's action and his opponents' muted protest were characteristic of the whole episode. Despite all the harsh words and bold titles in *Outdoor Life*, all the participants pulled their punches. They had to. This was a family feud. The ASM was full of the Survey's critics, but the agency had organized it, in 1919, and elected its old chief, C. Hart Merriam, its first president. The Survey, besides killing coyotes, did a great deal of scientific work, and virtually all the protestors had worked on one or another of the Survey's projects. The debate was also inconclusive because there was not much evidence.[74] The protestors had no studies on the effect of poisoning on coyote populations or of coyotes on rodent populations, and discussions of coyotes and sheep were

---

[72] Dunlap, *Saving America's Wildlife*, chap. 4.

[73] Gabrielson File, MVZ; Harold Anthony to A. Brazier Howell, 7 May 1936, Howell File, Department of Mammalogy, American Museum of Natural History, New York.

[74] E. Raymond Hall would have disputed this. In an interview with the author in the summer of 1980 he stated clearly his conviction that the protestors had clear evidence of the damage Survey operations were doing to Western wildlife.

backed by little but anecdotal evidence. The Survey spoke of millions
lost to marauding predators, but in private Gabrielson told his colleagues that estimates of stock lost to predators were guesswork.[75] Debate
over the Kern County Mouse Epidemic of 1927 shows the state of
scientific understanding. In January of that year mice living in the fallow
fields of a dry lake bed in Kern County, California, swarmed out over
the countryside. They came in incredible numbers. People killed them
by slamming two by fours on the ground; there was a layer of crushed
animals on the roads; and at one warehouse there were two tons of
poisoned rodents. Grinnell sent Hall to investigate, and he returned
blaming the Bureau. Its recent poisoning in the area had killed off
the predators that kept the rodents in check. Investigators from the
California Department of Agriculture put the blame on good weather
and the fallow fields. They had provided food and cover that allowed
the irruption. Lacking quantitative evidence and some understanding
of the population dynamics of mice (particularly whether predators or
food or something else normally held the population in check and how
much the vegetation shielded the mice from predators) it was impossible to say who was right or if anyone was.

In addition, everyone had mixed feelings. The protestors accepted
the proposition that "stock-killers" could not be tolerated on ranch land.
They did not, either as a practical or philosophical matter, object to
predator control – only to the blanket use of poison. Survey officers, in
private, admitted there was much to the other side's complaints, and
they often admired the wildlife they were killing. In "The War on the
Wolf" (1942) Stanley Paul Young said that "[w]here not in conflict with
human interests, wolves may well be left alone. They form one of the
most interesting groups of all mammals, and should be permitted to
have a place in North American fauna." Vernon Bailey, who spent many
years in Survey predator and rodent control operations, also invented
humane traps, and his private papers include photos of trapped wolves.
The caption on one is "A Big Gray Wolf, in the snow. Caught and Held
in two no. 2 Steel Traps, Feet Frozen but no less painful." Under another
is typed "Yes, he killed Cattle, to eat, but, Did he Deserve This?" (punc-
tuation and capitalization in original).[76]

The most eloquent voice of protest within the Bureau was Olaus
Murie's (on his later work see Chapter 8). In 1931, while a supervisor

[75] See report of the conference of field agents in Ogden, Utah, 23 April 1928, in "Report
of Conferences, 1928–1941," in General Files, Division of Wildlife Services, Records of
the U.S. Fish and Wildlife Service, RG 22, U.S. National Archives, Washington, D.C.
[76] Stanley Paul Young, "The War on the Wolf," Part II, *American Forests, 48* (December
1942), 574; Vernon Bailey Papers, Folder 12, Box 5, Smithsonian Institution Archives,
Washington, D.C.

of predator and rodent control work in Jackson, Wyoming, he complained to a superior that on a joint inspection tour people had tried to convince him that the Survey was right instead of giving him the information he needed. Some had even offered him bribes, "the chance to hunt ducks without a license, etc." He had reviewed the mammalogists' complaints and "to tell the truth I could not see that the queries of the Society had been adequately met." Five months later, writing to A. Brazier Howell, he said that he was "very fond of native mammals, amounting almost to a passion . . . [and] would make considerable sacrifice for the joy of animal companionship and to insure that other generations might have the same enjoyment and the same opportunity to study life through the medium of the lower animals." He wanted even the "so called injurious rodents around." There had to be control, he said, but it needed to be put on a sound basis. We are "passing around an appalling amount of misinformation about the effects of predators on game. I have been awakening to this fact only in the last few years."[77] In the next few his doubts would lead him out of the Survey. He quit in 1945 to work for the Wilderness Society.

In 1952, looking back at the controversy, he suggested that the "scientists who became so concerned at that time did not, I believe, understand their own motivation. . . . The big issue put forth was that 'innocent animals' were being killed incidental to poisoning operations. Deep in their hearts, if they had thought it out fully in the formative years of the opposition, was concern for the coyote itself."[78] It almost certainly was, but to think it out fully required concepts and evidence neither protestors nor poisoners had. People spoke of the "balance of nature" and said that coyotes controlled rodent populations, or they claimed that humans had destroyed the balance and now must control pests, but ecologists were just beginning to formulate concepts and conduct field research that would enable them to address the issue. Only when they did would it become possible to defend predators as part of the life of the land and see the value of "the coyote itself."

The debate spilled over to the national parks, where predators were not "in conflict with human interests." Indeed, their presence there could be defended as part of the "nature" the parks were to preserve. A year after the mammalogists' began the debate, the annual conference of American park superintendents reduced the money that rangers received from selling the pelts of predators trapped in the parks and cut

[77] Olaus Murie to W. C. Henderson, 9 January 1931, Box 265, Predatory Animals, Report on Poisoning File, Murie Papers, Conservation Center, Denver Public Library, Denver, Colo.; Murie to A. Brazier Howell, 7 May 1931, Department of Mammalogy, American Museum of Natural History, New York.
[78] Olaus Murie to C. C. Presnall, 7 December 1952, in Miscellaneous P File, Murie Papers, Denver Public Library, Denver, Colo.

the list of outlawed animals to three – wolves, coyotes, and cougars. Three years later, after Joseph Grinnell spoke to them, they passed a resolution against trapping in the parks. In May 1931 Park Service Director Horace Albright pledged "total protection to all animal life," saying that predatory animals "have a real place in nature" and would be killed only where they were causing "actual damage" or threatening a species that needed special protection. The parks, he said, should preserve "examples of the various interesting North American mammals under natural conditions for the pleasure and education of the visitors and for the purpose of scientific study."[79] By the mid-1930s the agency had virtually ended predator control in the parks, except in Alaska, and was defending predators as part of the life of the land.

The commissioner of Dominion Parks, J. B. Harkin, believed his country led the way. In 1931 he said that "[t]he policy respecting predatory animals in the Canadian National Parks which has been in effect since 1928 has apparently since been adopted for the United States National Parks. . . . This is apparently a case of two administrations having independently reached similar conclusions." Perhaps, but it seems more likely to have been a response to the mammalogists' argument.[80] Only four months after the *Journal of Mammalogy* printed the papers from the 1924 meeting, Harkin argued for predator preservation in terms that echoed the Bureau's critics:

> Predatory animals are of great scientific, educational, recreational and economic value to society. It is now generally recognized that they should be preserved. . . . [I]t is considered most desirable to maintain public sanctuaries or parks for their preservation in favourable isolated areas. . . . Scientists of note state it is worth years of effort to secure just such areas where all controls were absolutely prohibited. . . . Other scientists are pointing out that much of the antipathy [toward these animals] is not backed up by evidence obtained from a study of food habits.[81]

He later sent park superintendents a paper by C. C. Adams, one of the protestors, calling for the preservation of wildlife, particularly native wildlife, in the parks. He circulated Grinnell's address to the conference

[79] C. C. Presnall, "Condensed Chronology," File 719, Entry 7, RG 79, U.S. National Archives, Washington, D.C.; Horace M. Albright, "The National Park Service's Policy on Predatory Animals," *Journal of Mammalogy, 12* (May 1931), 185–6.

[80] Harkin to Wize, Box 37, U300, RG 84, Public Archives Canada; Albright, "The National Park Service's Policy on Predatory Mammals," 185–6. On ties between Canada and the United States see Harrison F. Lewis, "Lively: A History of the Canadian Wildlife Service," typescript, no date, Records of the Canadian Wildlife Service, RG 109, Public Archives Canada, and 1939 memo by Lewis and C. H. D. Clarke on proposed Biological Survey of Canada, Box 38, U300, RG 84, Public Archives Canada.

[81] Harkin to Cory, 20 May 1925, Box 35, RG 84, Public Archives Canada.

of American superintendents, put an end to the practice of letting wardens keep the pelts of animals killed in the park, and began requiring written justification for the killing of all but wolves, coyotes, cougars, and wolverines.[82] He wanted "to have wild life conditions in the National Parks remain as nearly as possible in a natural condition," for it was this that gave them "their chief charm and attraction." There was in the parks "a special need for a policy of preserving the balance of nature, exercising only the control necessary to maintain that balance against such human interference as may unavoidably occur."[83]

Wilderness

Wilderness was another focus for new ideas. The concept already had multiple meanings and considerable mystique in American culture, but the interwar years saw a deeper and more focused discussion that developed around the new need to protect wilderness. The parks, established to preserve scenery and developed as playgrounds, were not serving this purpose, and it was in the U.S. Forest Service that debate began and there that the first administrative actions were taken. The agency was better known for cutting down forests than preserving them in their natural state, but the growing popularity of outdoor recreation suggested it look at this use as well. Just after World War I two of its agents, Aldo Leopold and Arthur Carhart, found themselves looking at areas in the southwestern national forests suitable for summer camps. The Bureau was interested in development, but they recommended that some areas be left alone. Carhart, who had been transferred to the upper Midwest, applied his ideas to the Superior National Forest. There were some like-minded people already there, and by the end of the decade a group led by Sigurd Olson was working to preserve an area on the Minnesota–Ontario border for primitive travel. It became the Boundary Waters Canoe Area. In the Southwest Leopold saw his superiors set aside the Gila Wilderness area about the time he was transferred to Madison, Wisconsin, in 1924.[84] Others took up the cause, and in the 1930s Robert

---

[82] Box 37, U300, RG 84, Public Archives Canada.

[83] Harkin to Superintendent of Wood Buffalo National Park, 12 January 1925, Box 35, U300; Announcement of 3 October 1928, Harkin to Superintendents, Box 36, U300, Records of the Dominion Park Service, RG 84, Public Archives Canada.

[84] Kurt Meine, *Aldo Leopold* (Madison: University of Wisconsin Press, 1988), 177–8, 194–7, 224–6, 244. Leopold's own thoughts are most accessible in the essays of this period collected in Susan L. Flader and J. Baird Callicott (editors), *The River of the Mother of God* (Madison: University of Wisconsin Press, 1991), and David E. Brown and Neil B. Carmony (editors), *Aldo Leopold's Southwest* (1990; reprinted Albuquerque: University of New Mexico Press, 1995).

Marshall, in charge of the New Deal's Indian Forest Service, emerged as the leader and inspiration of a small group seeking to save wilderness. In 1935 he, Leopold, Olson, and a few others founded a group, the Wilderness Society, as a forum for discussion, a way to advance public awareness and education, and a lobby for legal protection. Marshall died in 1939, but the society survived, and he became the postwar movement's patron saint.[85]

Like the biologists arguing for the coyote, wilderness advocates had a cause they were not quite able to articulate, and for the same reason. They were beginning an exploration of ideas and values beside which Finlayson's extension of aesthetics to the Red Centre or Guthrie-Smith's celebration of the great change in New Zealand's biota were conventional exercises. That nature should have value in and of itself, and that it should be appreciated and visited rather than occupied or used, ran against the dominant trend of the society and raised questions even more fundamental than the movement that had created the parks. The advocates' arguments, to be sure, drew on deep philosophical roots in Western thought and were grounded in science, but they moved in new directions and suggested odd destinations. Indeed, it could be argued that the implications of these ideas have been the main subject of the modern defense of nature and that we have not yet fully explored them even now.

Conclusion

Compared with debates on climate, the exploration of nature's meanings seems tentative, undefined, and inconclusive. It was. Stripped of its associations with physiology and temperament, climate had a comfortable solidity. It seemed part of everyone's life, and even the scientific instruments used to measure it – wind and rain gauges and thermometers – were scaled to humans and reported understandable quantities. Consequences were equally evident. Crops grew or they did not, farms flourished or went under the auctioneer's hammer. Reasons for saving native wildlife or protecting large areas of land from development were more abstract. The beauty of the Red Centre or the drama of a new world forming on a New Zealand sheep station was even harder to see. These discussions, though, were just as important as those about development or disaster, for they were part of the settlers' continuing quest for a home in the land.

This generation had the same resources as its predecessors, the cul-

---

[85] On Marshall and his career see James M. Glover, *A Wilderness Original* (Seattle: The Mountaineers, 1986).

tural forms they had inherited from Europe and shaped or developed to their needs, but they were deepening their search. National nature had been an obvious thing, a few very visible or distinctive plants and animals or striking features of topography or climate. Now it was coming to encompass a wider range of creatures and to emphasize immediate experience. It had been a matter of using the land, at first without regard for the future, then for the economic future. Now it was looking at the very long term – at least in the American ecologists' reaction to the Dust Bowl and Ratcliffe's to the South Australian drought, in Wright's call for the preservation of all species as the heritage of evolution, and in the defense of the coyote and the exploration of the meanings of wilderness. This ultimately hinged on quantitative studies, scientific theories, and data. In grounding this long-term vision, work on wildlife, particularly game, was important. It is to this, the intellectual foundations of the environmentalist defense of visible nature in science, that we now turn.

8

AN ECOLOGICAL PERSPECTIVE,
1920–1950

One of the requisites for an ecological comprehension of land
is an understanding of ecology, and this is by no means co-
extensive with "education"; in fact, much higher education
seems deliberately to avoid ecological concepts. An understanding of ecology does not necessarily originate in courses
bearing ecological labels; it is quite as likely to be labeled
geography, botany, agronomy, history, or economics. This is
as it should be, but whatever the label, ecological training is
scarce.

Aldo Leopold, 1949[1]

Ecological training was scarce, and scattered among disciplines, but it
was there. Ecology as a body of knowledge may have been inadequate
to guide policy or convince the boosters, but it was sufficiently developed to raise troubling questions – witness Ratcliffe's analysis and Sears's
discussion. It could now provide a coherent, if not detailed, vision of the
settlers and their lands and suggest that living with, rather than just on,
the land might be necessary for the settlers' own survival. The spread of
this perspective, its development within science (which gave it authority
in the culture), and its critique of settler culture were the most important changes of the interwar period.

Ecology spread from a small base. It began as a community of interest rather than a scientific discipline, and grew less because society supported its research than because it supported individuals who were
thinking about these problems on their own. Some of the people were
field researchers without much (or any) scientific education, often
working on the practical problems of producing game, who found in

---

[1] Aldo Leopold, *A Sand County Almanac* (1949; reprinted New York: Ballantine, 1966),
262.

conditions on the land the keys to population growth and decline. Others were ecologists by training or inclination, people who had moved into the field. Still others were administrators trying to apply what their subordinates were finding out. By the time World War II diverted their attention and dried up their funds they had developed a quantitative understanding of nature's processes that allowed them to discuss human interactions with nature on a deeper level than rules of thumb or folkwisdom.

This work was done largely in the United States, simply because that country had a combination of problems and the people to address them. Wildlife and natural areas were of more concern than in other settler lands, and they were more deeply – or at least obviously – threatened. The American scientific community was the largest, best organized, and most sympathetic to the kinds of investigation that would lead to an eco-logical comprehension of the land. In the postwar years Americans would play prominent roles in spreading this perspective and the research methods associated with it to their colleagues in the other settler countries, and popular education would take it to the public. The bulk of this chapter deals with game populations in farm country and wild nature in the national parks. Studies on game led to a new profes-sion, game management, that made significant contributions to our understanding of nature's processes. Research done by the National Park Service's Wildlife Division, which included pioneer studies of mam-malian predators and their prey, gave scientific backing to the move to save them as part of America's wild nature. The last part of the chapter deals with Aldo Leopold's career and his changing ideas, which devel-oped within this professional matrix. In the course of his professional life, from 1909 to his death in 1948, Leopold went from conservation to an ecological appreciation of the land, helped found game manage-ment, and provided a critique of our treatment of the land, based in science, which became central to the environmental movement in the United States.[2] His ideas and example are now cited in all the settler countries.

New Ideas about Game

The game crisis of the 1920s was a product of a generation of action based on what seemed self-evident facts. Early in the century the Ameri-

---

[2] Leopold provided a short account of some of his development in the 1947 Foreword intended for *A Sand County Almanac*. It finally appeared in J. Baird Callicott (editor), *Companion to "A Sand County Almanac"* (Madison: University of Wisconsin Press, 1987), 281–8. The best study of Leopold's intellectual journey is Susan Flader, *Thinking Like a Mountain* (Columbia: University of Missouri Press, 1975).

can states had adopted the sportsmen's program of game restoration. That had been based on the idea that killing animals reduced their populations and that game, therefore, could be restored by reducing those deaths. To that end the states outlawed market hunting, cut bag and possession limits, shortened seasons, and hired wardens to reduce the human toll. To save game from other menaces they put bounties on predators. The federal government had helped by making it illegal to transport across state lines game that had not been killed in accordance with state law, and its predator control work in the West, though done primarily to protect stock, also served the sportsmen's interests. By the 1920s it was clear that this was not the answer. Ducks, geese, and quail continued to decline, while deer populations from Pennsylvania to Arizona went through inexplicable booms and equally puzzling plunges.

The most ambitious program – and the one that most clearly shows the state of knowledge and understanding – was the Migratory Bird Treaty between the United States and Canada, which committed both countries to sustainable hunting and continuing management of waterfowl. It grew out of the inability of states or provinces to save these species on their own. Because the birds migrated, any local jurisdiction that cut the bag limit only left more for people in less scrupulous areas. National action was difficult, for both the U.S. Constitution and the British North America Act of 1867 left authority over wildlife to the states or provinces, but in 1913 Congress tried to find a way around that. It set a federal hunting season on waterfowl, arguing that their migration across state lines involved them in interstate commerce, which was a federal responsibility. North of the border J. H. Fleming, an ornithologist, wrote to J. B. Harkin, commissioner of Dominion Parks, saying that the new act made possible a "useful Federal law along the same lines in Canada, in fact concurrent legislation is necessary if either country is to benefit, and something approaching uniformity has long been desirable."[3] That would require a treaty, but since opponents of the new law were already seeking a court case to test its novel perspective on wildlife and trade, that seemed the only sensible remedy. Besides allowing a common policy, it would provide solid legal ground. Treaties, under both countries' fundamental laws, superseded local regulation. A draft was ready by 1916, and by 1918 both countries had ratified it and passed enabling legislation. Courts on both sides of the border quickly turned back legal challenges.[4]

---

[3] Fleming to Harkin, 8 June 1913, Box 1, Acts and Legislation, RG 109, Records of the Canadian Wildlife Service, Public Archives Canada, Ottawa.

[4] Legal notes in Boxes 1–3, Acts and Legislation, RG 109, Public Archives Canada, Ottawa, include material from both sides of the border and show the legal challenges and decisions.

The treaty was based on commonsense ideas of wildlife and nature – the connection between individual deaths and population – and the vague notion that northern Canada was a duck factory, producing new flocks each year. It was not that legislators ignored science; there was little that applied. Elton's work, recall, was still a decade off. Only one of the negotiating team, Gordon Hewitt, had a modern scientific education – he had earned a Ph.D. in zoology in 1909 from the University of Manchester – and when he came to Canada as Dominion entomologist he was one of the first professionally trained biological scientists in the country.[5] One other, Maxwell Graham, had significant legal responsibilities for wildlife regulation. He was one-third of the Animal Division of the Dominion Parks Branch of the Department of the Interior. The American negotiators, T. S. Palmer, Henry Henshaw, and E. W. Nelson, were all self-trained naturalists employed by the Bureau of Biological Survey, which at that time had no experience or authority in wildlife law enforcement.

At the start this lack of theory was not a handicap. It might not be sufficient to cut the number of ducks shot each year, but it was a necessary first step, and further action would require gathering information, for which the Biological Survey and the Dominion Parks Branch were well suited. As government agencies they commanded resources unavailable to academic departments. This was apparent in their first major project, tracing migration routes. Everyone knew birds flew north in the spring and south in the fall, but no one knew the process in any detail. To find out, the two agencies organized the first large-scale bird banding study in the world, a scientific project that depended on organization and administration to gather many bits of information. The agencies worked out a numbering system and had bands made, in different sizes for different kinds of birds. They recruited and trained volunteers across the continent to band the birds, and they kept the records of that work. To get the bands back they mounted a public education campaign to inform hunters of the project and ask them to mail bands from birds they shot to the Survey in Washington. (Because the Americans had more resources and hunters they kept the records.)[6]

[5] Leland O. Howard, *A History of Applied Entomology* (Washington, D.C.: Smithsonian Institution Press, 1930), 186–7; Janet Foster, *Working for Wildlife* (Toronto: University of Toronto Press, 1978), 127–47.

[6] Frederic C. Lincoln, "Bird Banding in America," *Annual Report of the Smithsonian Institution, 1927* (Washington, D.C.: U.S. Government Printing Office, 1928), 331–54; idem, "A Decade of Bird Banding in America: A Review," *Annual Report of the Smithsonian Institution, 1932* (Washington, D.C.: U.S. Government Printing Office, 1933), 327–51. On the Canadian program see Harrison Lewis, "Lively: A History of the Canadian Wildlife

By the late 1920s returned bands were showing a natural phenomenon no one had seen or could see, a set of highways in the sky, used each spring and fall since the retreat of the Pleistocene ice. As birds flew north each spring from South and Central America, Mexico, and the Gulf coast – each species from its own area and at its own time – their needs and the land funneled them along a few broad routes, which the scientists called flyways. One ran along the Pacific coast, another on the east slope of the Rockies, a third up the Mississippi valley, and a fourth on the Atlantic coast. As they reached what is now the northern United States, they branched out again. Some species dropped off to nest in prairie potholes, others on the prairies themselves, still others in the northern forests, and the final voyagers came to rest on the shores of the Arctic Ocean. In the fall adults and the young of the year flowed south along the same routes, coming together in the flyways, spreading out again on the wintering grounds.[7]

While biologists sketched in details, the flocks continued to dwindle. In the early 1930s the situation appeared critical. The response was more of the same. The U.S. and Canadian governments shortened seasons and lowered bag limits even more, and the Americans began an ambitious program of making wetlands into wildlife refuges. Late in the decade they organized a continental waterfowl census (with support from corporations and a sportsmen's group, Ducks Unlimited). By then there was something new in the air. Graham Cooch of the Canadian Wildlife Service recalled working on that project, spending a summer "driving between potholes and wading through brush and marsh," meeting all the "leaders of wildlife management" at a time when "almost everything that was being done was brand new." He was, he said, not "the only recipient of this biological technology transfer." It continued into the early 1950s, and "[m]any of the senior scientists of the Canadian Wildlife Service owe an enduring debt to our mentors from [this period]."[8]

The work they were learning about came from studies of more sedentary wildlife populations, species that a single researcher could look at. One crucial species was deer. They were the continent's common big game, and restoring them had been a major element of the sportsmen's

Service," unpublished manuscript, RG 109, Records of the Canadian Wildlife Service, Public Archives Canada, Ottawa, 67.

[7] Lincoln, "Bird Banding in America" and "A Decade of Bird Banding in America" provide early material. On later developments see U.S. Department of the Interior, *Flyways* (Washington, D.C.: U.S. Government Printing Office, 1984).

[8] F. Graham Cooch, "Canadian Connections," in Department of the Interior, *Flyways*, 305, 307.

program. The results caused consternation and forced a reexamination of the assumptions behind game restoration. As seasons were shortened and predators killed, deer recovered, but the "balance of nature" seemed not to be operating. On ranges across the country their populations exploded, reaching unprecedented numbers. They destroyed the browze and then starved.[9] The classic horror story, repeated in conservation and hunting journals for the next thirty years, involved the Kaibab deer herd. Early in the century the federal government had set aside a large area on the North Rim of Arizona's Grand Canyon as the Kaibab National Forest and made it a game preserve. The Biological Survey's hunters covered the area, killing off the coyotes, wolves, mountain lions, and bobcats. The deer herd grew. In 1919 the U.S. Forest Service rangers in charge of the area began warning that there were too many deer and recommended that more be killed. State officials balked, fearing reaction from Arizona sportsmen. Forty years of preaching had convinced them that short seasons and low bag limits meant good sport. The National Park Service joined in. The Kaibab deer spent part of the year in Grand Canyon National Park, and agency officials wanted a large herd to attract tourists.

In the winter of 1924–5, while this argument went on, the animals began to die. In the spring their carcasses littered the forest, and that summer E. Raymond Hall found animals too weak to avoid a man on foot, so thin their ribs showed through the coat. Seedlings had been eaten to the ground and browze nibbled as high as a deer standing on its hind legs could reach. The forest looked, in Aldo Leopold's later description, "as if someone had given God a new pruning shears, and forbidden Him all other exercise."[10] For the next decade deer continued to die, the forest to suffer, and authorities to argue about what should be done.[11] What gave the Kaibab its reputation was that other areas had the same experience. Pennsylvania had begun applying the sportsmen's remedies early in the century. Soon it began paying for farmers' fencing, then letting them shoot animals in their fields. Vernon Bailey, surveying the situation for the Biological Survey in 1929, found starving deer, stripped forests, ruined crops, and little hope, for he thought the woods could not support even the animals that remained. The deer irruptions attracted attention beyond the border. Dominion

---

[9] A catalog of irruptions can be found in Aldo Leopold, Lyle K. Sowls, and David L. Spencer, "A Survey of Over-Populated Deer Ranges in the United States," *Journal of Wildlife Management*, *11* (April 1947), 162–77.

[10] Leopold, *Sand County*, 140.

[11] Thomas R. Dunlap, "'That Kaibab Myth,'" *Journal of Forest History*, *32* (April 1988), 60–8. A recent, more extensive discussion is Louis S. Warren, *The Hunter's Game* (New Haven, Conn.: Yale University Press, 1997).

Park Service files contain many memos on the U.S. experience, and on the copy of Bailey's report from Pennsylvania is clipped a note for the director: "Mr. Harkin – very important." The records of New Zealand's Royal Forest and Bird Protection Society have a collection of these documents as well.[12]

The other focus of concern was upland game birds, particularly the bobwhite quail, which was declining over much of its range. The quail's decline was less dramatic than deer population explosions and crashes, but research in this area was even more important in shaping ideas about management and species' relationship with the land. It was here that theories about population dynamics and the relation of species to their environment were worked out in a way that could be applied to visible nature, the animals people could see and among whom they lived.[13] Quail were a focus of early research in part because they were a popular and widespread game species – that attracted research funds – and because their life history made it possible for a single researcher to work on a defined population. Quail lived on farmland, which made it easy to get around the study area. Their ranges were small – a bird might live and die within a quarter-mile of where it was hatched – and each fall they formed coveys that persisted until the spring. Winter censuses were thus fairly simple. Finally, they lived across much of the eastern United States, and so could be studied under a variety of conditions.

Quail research ratified Lepold's observation that an "understanding of ecology does not necessarily originate in courses bearing ecological labels." The work he hailed as the first successful attempt to use "science creatively as a tool to produce wild game crops in America," a "rounded-out system of control of all actionable factors, based on a preceding scientific life-history investigation," began as a very practical project, directed by practical people.[14] In 1923 a group of wealthy sportsmen approached E. W. Nelson, chief of the Bureau of Biological Survey, seeking the agency's help in restoring quail on their Georgia shooting plantations. They offered to pay for a four-year study and provide a plantation as a test site, if in return they could get some practical advice. Nelson agreed and appointed W. L. McAtee, head of the Survey's Food

---

[12] Boxes 36–9, U300; memo in Box 36, RG 84, Records of the Dominion Park Service, Public Archives Canada, Ottawa. These also record Canadian interest in the concurrent irruption of New Zealand deer and the eradication program that began in the 1930s. In New Zealand, see Files 221 and 230, Ms. papers 444, Royal Forest and Bird Protection Society Papers, Turnbull Library, Wellington.

[13] On the importance of this work see David Lack, *The Natural Regulation of Animal Numbers* (London: Oxford University Press, 1954), 154–60, 170–8.

[14] Aldo Leopold, *Game Management* (New York: Scribner's, 1933), 16, 20.

Habits Laboratory, to supervise the project. He, in turn, chose Herbert Stoddard to carry out the fieldwork.[15] Nothing in these arrangements suggested a new approach or science on the cutting edge, nor did the people involved. Nelson was an old-line naturalist who had begun work for the Survey in 1890. McAtee, hired in 1904, specialized in analyzing the stomach contents of birds and mammals, a technique that had been new in the 1870s. Stoddard had left school at fifteen to become a farmhand and turned his interest in nature into a profession by collecting and mounting specimens for the Milwaukee Museum.

His study, though, was rigorous and quantitative. For five years (the fieldwork ran a year past its original schedule) he counted eggs and coveys and calculated mortality at every stage from egg to adult. He banded birds to check mobility and range, then mapped covey territories against types and quality of food in each season, shelter from predators and weather, and necessities like gravel for grit and dirt for dust baths. He burned areas to see how fire, then regrowth, affected the birds' use of the area.[16] It was very much what Elton was recommending in *Animal Ecology:* field studies that would "revolve around censuses, the structure of the population by age and sex, the birth-rates and death rates, movements, as well as the influence on these of outside changes and interrelations."[17] Stoddard, though, had his study under way when *Animal Ecology* appeared, and he did not read the book until 1931, when Leopold lent him a copy. He sent it back six months later with a noncommittal comment, "I must make a point of getting his books; his comes nearest of being the sort of ecology I can appreciate."[18] If the defining character of self-conscious ecology is the "application of experimental and mathematical methods to the analysis of organism–environment relations, community structure and succession, and population dynamics," this was ecology.[19] Its perspective certainly was. It focused on processes rather than pieces, populations rather than species, and it included the land as an active agent. Though Stoddard did not speak in these terms, he was, like the Nebraska ecologists, constructing nature from measurements, not analyzing by common sense what he saw on the land.

[15] Herbert L. Stoddard, *Memoirs of a Naturalist* (Norman: University of Oklahoma Press, 1969), 70–110, and Memos, Box 80, Waldo L. McAtee Papers, Library of Congress, Washington, D.C.

[16] Herbert L. Stoddard, *The Bobwhite Quail* (New York: Scribner's, 1931).

[17] Charles Elton, *Animal Ecology* (London: Sidgwick & Jackson, 1927; reprinted London: Methuen, 1966), xi.

[18] Stoddard to Leopold, 5 December 1931, Correspondence – Stoddard, Aldo Leopold Papers, University of Wisconsin Archives, Madison.

[19] Sharon E. Kingsland, "Defining Ecology as a Science," in Leslie A. Real and James H. Brown (editors), *Foundations of Ecology* (Chicago: University of Chicago Press, 1991), 1.

Rather than making his understanding abstract, this approach allowed him to see past obvious facts to underlying processes. Predation is a case in point. People had believed that since hawks ate quail, they reduced the quail population and that, therefore, if we killed the hawks there would be more quail. By the same logic poison or weasels should have erased the "rabbit menace." Looking at the life history of the quail rather than events in its life, Stoddard found that cotton rats and ants, which ate eggs and young, had more effect than hawks on the number of adult quail and that the key control was not predators but the land. Quail had thrived in the late nineteenth century because the farms of that period had met all their needs. Mixed crops and weeds provided food year-round, snake fences and woodlots cover and nesting sites. The roads provided dirt for baths (which kept down parasites) and gravel for their crops. Modern, straight fences, close plowing, paved roads, and clean fields caused their numbers to fall.[20]

That this pioneer study developed outside the discipline and, in fact, outside academic research is evidence of how new ideas moved through the larger scientific community. Disciplines allowed focus by cutting down on communications with others, but they did not have airtight borders. Projects and individual curiosity created networks of informal learning that crossed these boundaries. People working in museums, departments, and government agencies had contact, in person or through professional journals, with others working on the same problems. At times they deliberately fostered a new community of interest around a particular topic, which might, if it was important or interesting enough, become the nucleus of a new discipline. This was happening even as Stoddard finished his fieldwork. In 1928 the Sporting Arms and Ammunition Manufacturers' Institute hired Leopold (who had just resigned from the Forest Service) to study game conditions in the upper Midwest, and it decided to underwrite studies of the Hungarian partridge, northern bobwhite, ruffed grouse, and Gambel's quail. The Survey was to organize and oversee the work, and Leopold, for the Institute, and Stoddard, for the Biological Survey, visited universities in the Midwest to decide which ones would do the fieldwork.[21] This fusion of private interests and government, academic research and federal agencies had been developing in the land-grant system since the late nineteenth century, and it had been commonplace for a generation.

What was not commonplace was the results. They showed not only that "common sense" was a poor guide, but that the things people had thought important (predation being the most obvious example) were

---

[20] Stoddard, *Bobwhite Quail, 4.*   [21] Stoddard, *Memoirs of a Naturalist,* 217–25.

only secondary. Work on the northern bobwhite, studies to replicate Stoddard's work at the other end of the bird's range, was part of Leopold's education. He was just forming what would become the country's first department of game management, at the University of Wisconsin, and he put his first graduate student, Paul Errington, on the project. Each winter from 1929 through 1933 he counted every covey in his study area each week, looked for evidence of predators, and tried to identify and quantify causes of death. He found that neither the type nor the number of native predators seemed to affect winter survival rates. What did was cover. Some areas consistently carried a high population through the winter; others did not.[22] By the time Leopold was writing *Game Management* (1933), he was using this and other studies to argue for a new stance. Killing predators might increase the game, but the case had to be made in each instance. It could not be assumed. What was certain, he warned, was that it was useless to forge ahead in the belief that "the issue is merely one of courage to protect one's own interests, and that all doubters and protestants are merely chicken-hearted."[23]

Game Management: Crusade into Profession

Game management must become a profession if it is to become a fact. ... [I]t has long been an empirical art in Europe, but the attempt to adapt that art to biological principles and to American conditions and traditions is new.

Aldo Leopold, 1933[24]

It had to become a profession, but Leopold did not envision, or develop, an academic one like ecology. He looked instead to the applied agricultural sciences that had grown up in land-grant colleges and the U.S. Department of Agriculture. Rather than independent units working on theory, linked by journals and professional associations, departments

---

[22] Paul Errington, "Bobwhite Winter Survival in an Area Heavily Populated with Grey Foxes," *Iowa State College Journal of Science, 8* (1933–4), 127–30; idem, "Predation and Vertebrate Populations," *Quarterly Review of Biology, 21* (June 1946), 221–45. *Of Predation and Life* (Ames: Iowa State University Press, 1962) summarizes Errington's mature thought.

[23] Leopold, *Game Management*, 252, 230–52. These results seem to have been noted in New Zealand as early as 1935; see E. V. Sanderson to T. Gilbert Pearson, 12 March 1935, File 221, Ms. papers 444, Royal Forest and Bird Protection Society Papers, Turnbull Library, Wellington. The persistence of older attitudes may be seen in the forceful declarations against them in Carolyn King, *Immigrant Killers* (Auckland: Oxford University Press, 1984), 130–2.

[24] Leopold, *Game Management*, 413, viii.

in these fields applied their professional understanding to local problems. They were tied to a national research community through some bureau in the USDA and to the farmers by the extension service (which was part of the state land-grant college). Ideally, research was local and its results could be applied in the area by people who were not scientists. The information it yielded was universal and, through the USDA, available to all. Game managers knew the form well, for many of them had degrees in these fields and had worked in conservation agencies and land-grant schools.[25] Leopold, in fact, saw the new field as a part of that system. Game managers "must install a similar machinery for research and education," he said, "or else use the agricultural machinery already set up. Since game is largely an agricultural by-product, the latter course seems by far the best."[26] His own department was in the university's College of Agriculture, and, besides research, the plans he made for his new position included demonstrations, farmers' short courses, lectures, and portable exhibits for county fairs.[27]

The Biological Survey was the obvious federal agency, and in these years it played that role. It supervised and organized research projects from Stoddard's quail study on, and some of its employees joined the move for a new discipline. This was professional practice but also agency politics. In the 1920s the rapid growth of the agency's predator and rodent control responsibilities alarmed some old-line employees, who feared the Survey was losing its place as a scientific agency. They supported game management as a counterweight. One of the most active was W. L. McAtee. He began collecting publications on game research in the early 1930s, then convinced his superiors to support a continuing bibliography of research publications and research. Survey employees were prominent in the discussions that led to the formation of the Wildlife Society in 1937, were at least 15 percent of the initial membership, and had a representative on every committee. McAtee became one of the two trustees and editor of the *Journal of Wildlife Management*. Ira Gabrielson even sent a letter to all employees urging them to join. This, he said, is your society.[28]

[25] A. Hunter Dupree, *Science and the Federal Government* (Cambridge, Mass.: Harvard University Press, 1957), chap. 8.

[26] Leopold, *Game Management*, 406.

[27] Leopold Papers, Box 1, University of Wisconsin Archives, Madison.

[28] Many letters and memos on the Survey involvement, including Gabrielson's letter, are in Box 80, Waldo Lee McAtee Papers, Library of Congress, Washington, D.C. About half the initial membership worked for the federal government. The largest groups were in the Forest Service, the National Park Service, and the Biological Survey, but others were scattered in the New Deal conservation agencies. Rudolf Bennitt, "Nature and Scope of Training Graduates in Wildlife Conservation," in American Wildlife Institute,

The early work of the new Wildlife Society's membership committee shows how new it was. Its task, it said, was to define "the field . . . setting . . . objectives, and the intelligent direction of activities toward their attainment." Guidance had to be provided in "a field as devoid of directional tradition, as heterogeneous in make up and as inadequately charted as is Wildlife Management."[29] "Heterogeneous" was the word. The first group of officers and committee members held degrees in ornithology, fisheries biology, forestry, plant physiology, economic zoology, plant ecology, zoology, biology, and botany. Two – Herbert Stoddard and E. V. Komerak – had no higher education at all. To address this situation the committee proposed a "'semi-closed' professional type of society organization. . . . two classes of membership, 'Active and Associate,' and membership requirements [that] imply the necessity for 'substantial' training and/or experience requirements as a basis for Active membership." It provided a detailed guide for translating field and job experience into academic qualifications.[30]

Defining the field included saying what its relationship was to the "pure" science of ecology. Applied sciences were supposed to be using theory to produce practical results, but the world was more complicated than that. Game managers published in the ecological literature, and the introductory chapter of the postwar text on animal ecology, W. C. Allee, Orlando Park, Alfred E. Emerson, Thomas Park, and Karl P. Schmidt's *Principles of Animal Ecology* (a production of the Chicago school), presents a history of ecology's development that makes clear the great debt theorists owed to practical work.[31] Leopold certainly con-

Transactions of the Seventh North American Wildlife Conference, 1942 (Washington, D.C.: American Wildlife Institute, 1942), 495–505. A membership list in *Journal of Wildlife Management*, 2 (April 1938), Supplement, gives names and addresses. Not all people can be identified and the figures on employment are therefore minimums. The New Deal's conservation efforts were at least as important. The Duck Stamp Act of 1934 and the expansion of the refuge system created a demand for more professional managers, as did the wildlife conservation work that the government built into the Civilian Conservation Corps, the Resettlement Administration, the Soil Conservation Service, and many other agencies.

29   "Abstract of the Report of the Membership Committee," *Journal of Wildlife Management*, 2 (April 1938), 70.

30   Ibid., 70–2; "Summary of the Report of the Second Annual Meeting," *Journal of Wildlife Management*, 2 (April 1938), 67; General Files, Department of Wildlife Ecology, 1933–48, 9/25/3, Box 1, University of Wisconsin Archives, Madison.

31   W. C. Allee, Orlando Park, Alfred E. Emerson, Thomas Park, and Karl P. Schmidt, *Principles of Animal Ecology* (Philadelphia: Saunders, 1949). Other examples include Paul Errington and H. L. Stoddard, "Modifications in Predation Theory Suggested by Ecological Studies of the Bobwhite Quail," in American Wildlife Institute, *Transactions of the Third North American Wildlife Conference* (Washington, D.C.: American Wildlife Institute, 1938), 736–40, and Paul Errington, "Some Contributions of a Fifteen-Year Local Study

sidered those in his group as something more than apprentices and practitioners. On population cycles, a subject of practical and theoretical interest, he said that scientists had been studying these

> in the hand-made glass-bottle environments of the laboratory. This is proper – they will some day extend their controlled experiments to the hills and fields. But the game manager faces it here and now. . . . It is unlikely that the game manager will find the explanation of cycles, but his field observations are the main reliance of the scientists who will.[32]

He did not, though, see game managers as scientists. In *Game Management* he said that the field was "applied ecology" but that the pure science was not useful to the manager's job of "making land produce sustained annual crops of wild game for recreational use."[33] Later he told his colleagues, "We deal with Science, but we have no prospect of inventing new tools or powers."[34] This seems to have been the common attitude. A 1942 survey of directors of graduate programs in game management showed that "even in our own opinion the scientific reputation of wildlife work still seems to be somewhat in doubt in over half the schools covered in this inquiry." The directors produced this, for they were concerned that their students be as adept at "selling" ideas as developing them.[35]

New Ideas and the National Parks

Game management was one center of ecological thinking about visible nature; research in the Park Service's Wildlife Division was another. This was more closely tied to academic zoology and ecology. The bureau's first chief, George Wright, had studied under Joseph Grinnell. He had chosen Grinnell's assistant, Joseph Dixon, to work on the initial survey

---

of the Northern Bobwhite to a Knowledge of Population Phenomena," *Ecological Monographs*, *15* (January 1945), 3–34; Charles Elton, "On the Nature of Cover," *Journal of Wildlife Management*, *3* (October 1939), 332–8.

[32] Leopold, *Game Management*, 71.

[33] Ibid., 3, 39.

[34] Aldo Leopold, "The State of the Profession," *Journal of Wildlife Management*, *4* (July 1940), 343–6.

[35] Bennitt, "Nature and Scope of Training." A similar survey in 1978 found that "[a] truism at least a quarter of a century old is that wildlife management is largely people management." It went on to discuss, in terms Leopold might have used, the manager's need to know both science and people. John A. Kadlec, "Wildlife Training and Research," in Howard P. Brokaw (editor), *Wildlife in America* (Washington, D.C.: U.S. Government Printing Office, 1978), 485–97. See also William Van Dersal, "The Viewpoint of Employers in the Field of Wildlife Conservation," in American Wildlife Institute, *Transactions of the Seventh North American Wildlife Conference*, 506–17.

of wildlife conditions in the parks, and three of the early staff members had worked or been trained at the Museum of Vertebrate Zoology.[36] Carl P. Russell, who became chief after Wright's death in 1935, had studied under Lee Dice, a Grinnell student, and his assistant, Victor Cahalane, had worked with Dice at the Cranbrook Institute of Science. The Division's interest, though, was not entirely scientific. Its work was intended to guide the Park Service in the task of restoring nature in the parks, re-creating the landscape as it had been before whites arrived. It was also to offer authoritative arguments against the considerable, and quite vocal, opposition to that vision.[37] Many people wanted nature but not animals that ate other animals. Ranchers worried about "sanctuaries" for "stock-killers," and hunters protested that the "vermin" would take "their" trophies. The new approach baffled even people like retired director Horace Albright, who had supported the Division's initial moves. By the late 1930s he felt that things were going too far. "I find that the impression is quite widespread that the National Association of Audubon Societies and perhaps other organizations are more interested today in saving the predatory species of birds and mammals than giving reasonable consideration to the species that are regarded as very important by the general public." He hoped that the Park Service would not

---

[36] They were E. Lowell Sumner, Adrey E. Borell, and Theodore H. Eton, Jr. All were listed on the first page of the memo the Division sent to the Bureau of Biological Survey in 1935, asking that they be put on the mailing list for the Survey's new bibliographical guide to wildlife management research. File 720, Entry 7, RG 79, Records of the National Park Service, National Archives, Washington, D.C. Grinnell's commitment to ecology and the interests of his students may best be judged by the Correspondence of the Museum of Vertebrate Zoology, Berkeley; used by permission of Dr. David Wake, Director. Alfred Runte, personal communication, has suggested that while Grinnell's interest in wildlife was constant, his enthusiasm for ecological concepts developed by his students may have varied directly with the credit they gave him for his inspiration.

[37] George M. Wright, Joseph S. Dixon, and Ben H. Thompson, *Fauna of the National Parks of the United States*, Fauna Series No. 1, Department of the Interior (Washington, D.C.: U.S. Government Printing Office, 1933), articulated that vision. On policy see "Report of the U.S. National Park Service," in U.S. Congress, Senate, Special Committee on the Conservation of Wildlife Resources, *The Status of Wildlife in the United States*, Senate Report 1203, 76th Congress, 3d Session (Washington, D.C.: U.S. Government Printing Office, 1940), 350–2; Victor H. Cahalane, "The Evolution of Predator Control Policy in the National Parks," *Journal of Wildlife Management*, 4 (July 1939), 229–37; and File 720, Entry 7, RG 79, National Archives, Washington, D.C. Predator control in the parks ended by 1935, with the partial exception of Mt. McKinley, where some poisoning was done into the early 1950s. On this see Samuel J. Harbo, Jr., and Frederick C. Dean, "Historical and Current Perspectives on Wolf Management in Alaska," in Ludwig N. Carbyn (editor), *Wolves in Canada and Alaska*, Canadian Wildlife Service Report Series No. 45 (Ottawa: Canadian Wildlife Service, 1983), 51–64; Cahalane, "Evolution of Predator Control Policy."

take this attitude.[38] It did, and it hired Adolph Murie to investigate what was happening and suggest what might be done.

With Murie we cross a divide. When his older brother, Olaus, had graduated from college in 1912, ecological training was in its infancy. Olaus collected specimens and sampled areas for museums, then joined the Biological Survey. Adolph (B. A., 1923) did these jobs too – including a faunal survey in Alaska with his brother – but there were programs in animal ecology, and he earned a Ph.D. under Lee Dice at the University of Michigan. He apparently urged his brother to join him, for Olaus received an M.S. from Michigan in 1927, but this does not seem to have been a crucial experience. Olaus remarked later that he had learned ecology studying elk in Wyoming with Adolph.[39] In 1930 Olaus was working as a predator and rodent control supervisor in Wyoming, guiding the joint inspection team looking into the mammalogists' allegations raised in 1924 and writing letters about his dissatisfaction with predator poisoning. Adolph was in graduate school, and he signed A. Brazier Howell's petition condemning the Survey's poisoning program.

They wound up, though, doing almost parallel studies of predators and their prey. Between 1927 and 1932 Olaus worked on a project titled "Food Habits of the Coyote in Jackson Hole, Wyo.," under the direction of W. L. McAtee. Adolph wrote *Ecology of the Coyote in the Yellowstone*, based on research he conducted from 1935 to 1937.[40] The populations overlapped; both projects began in response to complaints that predators were killing off the more desirable species, and both aimed to see what effect coyotes had on elk populations. The difference in research methods shows the change that was coming with ecology. Olaus analyzed stomach contents and scats, which allowed him to say what the area's coyotes ate but not what effect they had on the elk population. For that he had to rely on deductions from his field observations. Adolph used stomach contents and scat analysis, but he also looked at coyotes' interactions with everything from bison and moose to porcupines and squirrels, and examined cover and weather throughout the year. He

---

[38] Albright to A. E. Demaray, Acting Director, 24 November 1937, File 720, Entry 7, RG 79, National Archives, Washington, D.C.

[39] Olaus Murie to Marcus Ward Lyon, 30 March 1935, Murie file, Department of Mammalogy, American Museum of Natural History, New York.

[40] Olaus J. Murie, "Food Habits of the Coyote in Jackson Hole, Wyo.," USDA Circular 362 (Washington, D.C.: U.S. Government Printing Office, 1935); Adolph Murie, *Ecology of the Coyote in the Yellowstone*, Fauna Series No. 4, National Park Service, U.S. Department of the Interior (Washington, D.C.: U.S. Government Printing Office, 1940). The following discussion is based on Thomas R. Dunlap, *Saving America's Wildlife* (Princeton, N.J.: Princeton University Press, 1988), chap. 4.

reconstructed historical patterns of wildlife abundance through park records. He examined current conditions by counting animals and calculating calf survival. He refined his methods in the late 1930s, working on wolves and Dall sheep in Mt. McKinley National Park (now Denali). There, by collecting more than 800 sheep skulls and comparing the living and dead populations in terms of age, sex, known disease, and injuries, he tested the argument that predators selected old, weak, ill, or otherwise defective individuals. He watched wolves in their dens over two summers, providing the first extensive observations on their behavior in the wild.[41]

Olaus and Adolph came to the same conclusion: individual elk were vulnerable, the population was not.[42] This is puzzling if we use conventional models, which see science in terms of facts and theories and in which conclusions flow with the inevitability of a theorem from objective data. Conclusions, though, can depend on perspective, as we saw with Ratcliffe. It is also possible for one person to be convinced of something on the basis of evidence that other observers would not accept as "proving" the case, as witness E. Raymond Hall's position on the Survey's poisoning work. Olaus Murie and Hall were both honest and not self-deluded; their conviction was the result of years in the field, experience that did not appear on resumés and could not be connected in a formal way to theories. Adolph's work carried more weight in the world because he had methods that allowed him to count – a decisive modern argument.

The defense of predators through scientific study was a defense against a common view, but that view was changing. Public interest in

---

[41] Adolph Murie, *The Wolves of Mt. McKinley*, Fauna Series No. 5, National Park Service, Department of the Interior (Washington, D.C.: U.S. Government Printing Office, 1944). On the importance of this work see the dedication of L. David Mech's *The Wolf: The Ecology and Behavior of an Endangered Species* (Garden City, N.Y.: Natural History Press, 1970). Durward Allen, Mech's teacher, confirmed these views (personal communication).

[42] Olaus's colleagues in predator and rodent control thought he had not sufficiently emphasized coyotes' appetite for mutton. Park Service officers who thought that "innocent" animals needed protection wanted to fire Adolph. Box 264, Miscellaneous P file, Murie Papers, Conservation Center, Denver Public Library, Denver, Colo., and Box 83, W. L. McAtee Papers, Manuscript Division, Library of Congress, Washington, D.C., have comments on Olaus's paper. On Adolph's troubles see Olaus Murie to Harold E. Anthony, 5 December 1945, Murie File, Department of Mammalogy, American Museum of Natural History, New York; Olaus Murie to Carl L. Hubbs, 30 July 1946, Correspondence, Box 361, Folder 7, Adolph Murie Papers, Conservation Center, Denver Public Library, Denver, Colo. See also File 720, Entry 7, RG 79, National Archives, Washington, D.C.

nature, if it had not reached the level of defending predators, was at least expanding. Peterson's field guide, launched at the low point of the Depression, became a steady seller and the parks were becoming more important for wildlife. In 1934 Congress authorized Everglades National Park, which certainly did not meet the standards of monumental scenery that had defined the system, and in 1940 it added Isle Royale, the largest island in Lake Superior. This was an even greater shift from the ideal. The island was a maze of cutover, disturbed areas, and there were summer cottages scattered through the forest and commercial fishing docks along the shore. It lacked even the massed displays of wildlife that made the Everglades an attraction. It could, though, be managed for wilderness and wildlife, and it was.[43]

Canadian authorities watched events in the United States, but they had to adapt what they learned to their situation. Wildlife, for example, was still abundant north of the border, and while Americans reduced bag limits and ended market hunting entirely in the interests of sport, Canadians did little more than discuss such measures. In the North, where subsistence hunting was the rule and the fur trade an important part of the economy, the lessons of sport hunting were hardly even applied. In the United States large predators were rare. In Canada they had been eliminated in the livestock-growing areas, but north of the agricultural regions they still existed in significant numbers.[44] The Survey's continuing campaign against predators had no counterpart there. The government of the Northwest Territories hired hunters to kill wolves to save the caribou, but that was a minor project. In 1923 it resulted in 3 wolves trapped, 2 shot, and 130 poisoned. It also accounted for 15 foxes and 4 Indian dogs. The next winter the tally was 110 wolves at a profit of $2,331.54. (There is no indication in the records of where this figure came from.)[45] In the late 1930s, when caribou herds seemed to be in steep decline, there were inspections and reports, but only in the late

---

[43] The island became an important wildlife research area when wolves colonized it in the early 1950s. See L. David Mech, *The Wolves of Isle Royale*, Fauna Series No. 7 (Washington, D.C.: U.S. Government Printing Office, 1966), and Durward L. Allen, *Wolves of Minong: Their Vital Role in a Wild Community* (Boston: Houghton Mifflin, 1979). Rolf Peterson, at Michigan Technological University in Houghton, is currently directing work on the island.

[44] Box 1, Acts and Legislation, RG 109, Public Archives Canada, Ottawa; Robert McCandless, *Yukon Wildlife: A Social History* (Edmonton: University of Alberta Press, 1985). There are parallels with Australia, where wildlife officials began debating the advisability of commonwealth regulation of interstate trade in game only in the late 1940s. See *Conference of Authorities on Australian Fauna and Flora* (Hobart: Government Printer, 1949). New Zealand has never attempted such measures.

[45] Lewis, "Lively," 39.

1950s did the Dominion government begin a serious study of caribou and wolves and a program of poisoning.[46]

Policies in the national parks were closer to those in the United States, for the two countries' park systems had developed for many of the same purposes, under similar conditions, and faced many of the same problems, including wild life populations. Like Americans, Canadians had saved the buffalo by putting them in a park and killing their predators. In the early 1920s the Dominion Park Service had sought advice and even bought poison from the Bureau of Biological Survey, which was helping out the National Park Service as well. As in the United States this produced too many animals. In 1923 authorities slaughtered 250 buffalo at Wainwright National Park (which had been established to save the plains buffalo) because there were too many for the range. To avoid a continuing program of this kind it was proposed that the surplus animals be shipped each year to Wood Buffalo National Park in northern Alberta. This brought protests, centered in the Field Naturalists' Club. Opponents argued that the plains buffalo would interbreed with the rarer woods buffalo, whose only herd was in that park, and that the subspecies would eventually be bred out of existence. Since the animals at Wainwright had tuberculosis, shipments would also spread that disease to the still-healthy herd. Official response was prompt, but not based on the merits of the case. Hoyes Lloyd, superintendent of wildlife protection, was the editor of the *Canadian Field Naturalist*, and he and his assistant, Harrison Lewis (also active), were, Lewis recalled, "notified that we could either resign our respective positions with the Field Naturalists' Club and its magazine or be expelled from the Department of the Interior."[47] A total of 6,000 plains buffalo were sent to the southern part of Wood Buffalo Park. They probably brought in tuberculosis and brucellosis, and they certainly swamped the existing herd of wood buffalo. Only an isolated (and until 1960 undiscovered) herd of some 200 animals preserved the strain.[48]

No more than their colleagues to the south did Canadian officials understand what was going wrong. J. B. Harkin, director of Dominion Parks, pointed to excess elk in Jackson Hole, Wyoming, and the starving deer in the Kaibab National Forest in Arizona, where "predatory animals were systematically destroyed" and the surplus animals were unable to migrate. This showed that there was "a special need for a policy

---

[46] Files on wolves, WLU200, Boxes 33, 76, 77, RG 109, Public Archives Canada, Ottawa.

[47] Lewis, "Lively," 113, 109–14. On continuing action see 275, 289, 304. See also Box 50, BU2, RG 84, Public Archives Canada, Ottawa.

[48] Disease continues to be a problem, and there is concern that it will spread to domestic stock. See Federal Environmental Assessment Review Office, *Northern Diseased Bison* (Ottawa: Minister of Supply and Services Canada, 1990).

of preserving the balance of nature, exercising only the control necessary to maintain that balance against such human interference as may
unavoidably occur."[49] He provided no hints as to what control was necessary. A year later, explaining his agency's predator control policy, he
said that "[w]ild life is given absolute protection with the further exception that war is waged on predatory animals to a reasonable extent in
order that the safety of the remainder may be made more secure. . . .
[W]hile it is not the desire that predators be abundant they are not
being exterminated, which would upset the balance of nature."[50] Just
what this meant was not explained.

As in the United States, there was opposition to even this much protection for "vermin." In 1936 the *Calgary Herald* replied to a complaint
about killing mountain lions in the national parks by saying that humans
had taken over their role as checks on the population of deer and elk.
"Moreover, there is a pretty general notion that the parks are intended
for people, the animals being preserved there only to provide interest
for visitors. And in the capacity of entertainers the flesh-eaters have
certain clearly defined limitations." Bears were qualified attractions,
since they robbed food caches. "And as for being entertained in the long
stilly watches by the snarls of mountain lions prowling around the tent
– well, really."[51] Even within the Park Service, sentiment was split. In
1934 game protectors in Wood Buffalo National Park were still using
strychnine to kill wolves, and a year later the staff at Banff warned that
sheep, deer, and goats in the park "would be faced with extinction in
the course of a few years" unless measures were taken against predators.[52] In 1946 A. W. F. Banfield reported that park policy toward predators "is not clearly understood by the majority of members of the Parks
staff. . . . It was found that in the majority of parks, carnivores . . . were
looked upon with disfavor by the staff and every opportunity was seized
to shoot them on sight."[53]

Other currents were running, though. In 1930 the superintendent at
Waterton Lakes National Park had warned that efforts to restore the

---

[49] Announcement of 3 October 1928, Harkin to Superintendents, Box 36, U300, RG 84,
Public Archives Canada, Ottawa.
[50] Harkin to Miss W. B. Conger, Newcomb College, New Orleans, La., 24 September 1929,
Box 36, U300, RG 84, Public Archives Canada, Ottawa.
[51] *Calgary Herald*, 9 January 1936, copy in Box 36 U300, RG 84, Public Archives Canada,
Ottawa.
[52] Memos, 1934–9, and letter, Supervisor, Algonquin, to Williamson, Controller, Dominion Parks, 31 January 1939, v. 3., Box 157, U266, RG 84; Memo, 29 May 1935, Box
157, B261, RG 84, Public Archives Canada, Ottawa.
[53] A. W. F. Banfield, "Report on Wildlife Conditions in the Mountain National Parks,
1946," in Box 39, U300, RG 84, Pubic Archives Canada, Ottawa.

mountain sheep should be examined closely. "It must be borne in mind that there is a limit to the number of sheep that can, with safety, be accommodated within a park." Once that was reached the animals would deteriorate from lack of food and become subject to diseases. "So really there is no good in trying to obtain large numbers."[54] A long memo of 1940, "Predatory Animals in the National Parks," declared that a "significant development of modern scientific thought has been the ecological approach. . . . The individual must be looked upon as part of the whole, and subordinate to the larger social system." It spoke of the pyramid of numbers, the balance of nature, the carrying capacity of land, and the insecurity of a surplus population – terms from wildlife management and animal ecology. Predator elimination had been practiced to protect wildlife; that had "reflected the scientific opinion of the day": Now we realize that predators "were of essential value and deserved equal rights with all other animals."[55]

As in the United States there were arguments for preserving wilderness, though the impetus came not from forestry but from park officials, and they spoke less of pioneer recreation or identification with nature than of science. In 1926, echoing the mammalogists' arguments, Harkin had said that "it is thought to be important . . . to preserve parts of Canada in their original conditions" so that scientists could "study the inter-relationships of one species upon another. Although this is not particularly important at present because much of Canada outside the National Parks is still wilderness, it will become increasingly important as the wilderness outside the parks is affected by the operations of man."[56] In 1940, in a report titled "Sanctuaries of the Northwest Territories," C. H. D. Clarke said that "the enormous expansion of the aircraft industry now in progress [makes it] easy to see that the days of inaccessibility of the Thelon Game Sanctuary are numbered." The area, he warned, "may have to play host to tourists."[57] Three years later, in response to a piece in the *Calgary Herald* called "Are the Parks for Animals or Humans?" the Dominion Park Service said that the parks,

---

[54] Knight, acting superintendent, Waterton Lakes, to Harkin, 6 March 1930, Box 162, W240, RG 84, Public Archives Canada, Ottawa.

[55] W. E. D. Halliday, Dominion Forest Service, Lands, Parks, and Forests Branch, Department of Mines and Resources, "Predatory Animals in the Parks," Box 39, U300, RG 84, Public Archives Canada, Ottawa.

[56] Harkin to W. F. H. Mason, secretary of the Northern Alberta Game and Fish Protective League, 15 March 1926, Box 36, U300, RG 84, Public Archives Canada, Ottawa.

[57] Box 38, U300, RG 84, Public Archives Canada, Ottawa. On postwar developments see Box 39. On earlier concerns about aircraft, not related to the parks, see Lloyd to H. H. Hume, chairman of the Dominions Lands Board, Department of the Interior, 15 February 1932, U266, Box 157, RG 84, Public Archives Canada, Ottawa.

besides serving people, provided places "to conduct fundamental bio-logical studies on the primitive fauna and flora", which were "becoming impossible" elsewhere due to "changes wrought by man."[58] In 1939 a twenty-four-page memo outlined a reorganization that would make the Dominion Park Service's Division of Wildlife Protection into a "Bio-logical Survey of Canada," with a section devoted to scientific studies of animals and their relationship to the environment.[59] The war put these plans on hold, but in 1944 wildlife officials in the Service said that "the very large mass of wildlife management projects that it is proposed to develop in Canada as a part of the post-war reconstruction" would justify the appointment of specialists along the lines followed in the United States.[60]

New Ideas for Society

One of the penalties of an ecological education is that one lives alone in a world of wounds. Much of the damage inflicted on land is invisi-ble to laymen. An ecologist must either harden his shell and make believe that the consequences of science are none of his business, or he must be the doctor who sees the marks of death in a community that believes itself well and does not want to be told otherwise.

Aldo Leopold, 1949[61]

Leopold did not harden his shell. Even as he was building a new pro-fession in the midst of the Great Depression he was engaged in an even bolder venture – applying ecology's perspective to the relations between people and the land. He sought an ethical framework for policy grounded in ecology's description of nature's processes and humanity's ties to them. The ideas that, twenty years after his death, would make him an environmental guru and *A Sand County Almanac* an environ-mental bible, were the slow growth of a lifetime. His interest in nature dated to his youth, and hunting was part of his introduction. "[M]y ear-liest impressions of wildlife and its pursuit," he said, "retain a vivid sharp-ness of form, color and atmosphere that half a century of professional wildlife experience has failed to obliterate or to improve upon."[62] Hunting also provided him his "first exercise in ethical codes." When

---

[58] Memo in reply to a piece in *Calgary Herald*, reprinted in *Ottawa Morning Journal*, 30 June 1943, clipping and reply in Box 38, U300, RG 84, Public Archives Canada, Ottawa.
[59] Harrison Lewis, memo titled "Proposed Biological Survey of Canada," Box 38, U300, RG 84, Public Archives Canada, Ottawa.
[60] Lewis to Smart, 22 June 1944, Box 38, U300, RG 84, Public Archives Canada, Ottawa.
[61] Leopold, *Sand County*, 197.   [62] Ibid., 28.

his father allowed him to hunt alone, he forbade him to shoot a sitting bird, but at his age and with his experience a wing-shot was "hopeless." "Compared with a treed partridge, the devil and his seven kingdoms was [*sic*] a mild temptation."[63]

His education and experience as a young adult were in conservation. He received a master's of forestry degree from the Yale Forestry School in 1909, at the high tide of conservation, in a school that was a training ground for the Forest Service and Gifford Pinchot's crusade. Leopold joined the agency and was posted to the Southwest, where he was pitched into the actual business of conservation. Far from headquarters and superiors, and responsible for vast acres, he and his cohort had to translate ideals into action suited to the land and the people who depended on it. This was part of his education, but he also sought other kinds in these years. He read widely, and he also came into contact with people working on wildlife issues and problems from other perspectives.[64] In 1917 he began submitting notes from his hunting trips to scientific journals. One of these was *Condor*, published at the Museum of Vertebrate Zoology. In 1922 a note on roadrunners brought a tactful but firm challenge to his ideas. He reported shooting a roadrunner to see what it was eating and finding it was a quail chick. He ended the note by urging that roadrunners be put on the blacklist. Associate editor Harry Swarth gently reminded him that roadrunners doubtless ate other things that might counterbalance an occasional quail chick. Leopold agreed and asked that the last part of his note be dropped before publication.[65] He began a correspondence with Joseph Grinnell that ended only with the latter's death in 1939. By then Leopold's oldest son, Starker, was studying for his Ph.D. at the Museum.

He also learned from the land. It was in the Southwest that he came to see the consequences of conservation. He had arrived committed to the gospel of utilitarian conservation and with conventional ideas about game and its protection, and as late as 1918 he had been urging New Mexico's sportsmen to get rid of the last varmint in the state. He said little about the Kaibab at the time, but it is the subtext of "Thinking Like a Mountain," the essay he wrote twenty years later summarizing his ideas about the relationship between deer, wolves, and the land. Looking back he said:

---

[63] Ibid., 129.

[64] Kurt Meine, *Aldo Leopold* (Madison: University of Wisconsin Press, 1988), 87–228, describes this period in detail.

[65] Author's interview with Starker Leopold, 16 June 1981. Aldo Leopold File, Correspondence of the Museum of Vertebrate Zoology, Berkeley, cited with the permission of David Wake, Director.

I was young then, and full of trigger-itch; I thought that because fewer wolves meant more deer, then no wolves would mean hunters' paradise. . . . Since then I have lived to see state after state extirpate its wolves. I have watched the face of many a newly wolfless mountain, and seen the south-facing slopes wrinkle with a maze of new deer trails. . . . I now suspect that just as a deer herd lives in mortal fear of its wolves, so does a mountain live in mortal fear of its deer.[66]

In 1914 he suffered an attack of nephritis that nearly killed him. He used the time spent recovering reading and reflecting, and when he returned to work he began to concentrate more strongly on issues that engaged him. By the time of his transfer to Madison he was focusing on wildlife and wilderness. He was also starting to organize his thoughts. In an unpublished foreword to *A Sand County Almanac* (later printed in *Companion to "A Sand County Almanac"*) he said the transfer inspired his essays. He became associate director of the Forest Service's Forest Products Laboratory, and "[t]he industrial motif of this otherwise admirable organization was so little to my liking that I was moved to set down my naturalistic philosophy."[67] This is stretching it. In the Southwest he had been arguing, though inside the Forest Service for the most part, that wilderness be considered a land use. A year before he went to Madison he spoke to an Albuquerque civic group, and the text of "A Criticism of the Booster Spirit" shows him already at odds with conventional ideas of progress – including his own of a few years before.[68]

In 1928 he left the Forest Service to make his interest in game a profession, but he took along his developing critique of American attitudes toward the land. In his textbook, *Game Management* (1933), he said that while the game manager produces more birds and animals for hunters, what "he really labors for is to bring about a new attitude toward the land." He stands between those who see only economic value and believe the "food-factory" has "the right to be as ugly as need be, provided only it was efficient," and those who see "economic productivity as an unpleasant necessity" and want it kept out of sight. The manager is part of a third, much smaller group that sees the ugliness of development as

---

[66] Leopold, *Sand County*, 138–9. This account of Leopold's thought relies heavily on Flader, *Thinking Like a Mountain*, and Meine, *Aldo Leopold*.

[67] Leopold, Unfinished manuscripts, FOREWORD, Box 17, Leopold Papers, University of Wisconsin Archives, Madison. Quote appears in Callicott, *Companion to "Sand County"*, 285.

[68] Meine, *Aldo Leopold*, 177–8, 194–228; David E. Brown and Neil B. Carmony (editors), *Aldo Leopold's Southwest* (1990; reprinted Albuquerque: University of New Mexico Press, 1995). The Albuquerque paper is printed in Susan Flader and J. Baird Callicott (editors), *The River of the Mother of God* (Madison: University of Wisconsin Press, 1991), 98–105.

neither "the inevitable concomitant of progress" nor a "necessary compromise" but as "the clumsy result of poor technique, bunglingly applied by a human community which is morally and intellectually unequal to the consequences of its own success." The test of civilization was the capacity to live in high densities without destroying the environment, and the "practice of game management may be one of the means of developing a culture which will meet this test." It served "a motivation – the love of sport – narrow enough actually to get action from human beings as now constituted but nevertheless capable of expanding with time into that new social concept toward which conservation is groping."[69]

By the late 1930s he was writing the essays that would make the "land ethic" a central part of the environmental movement's ideas. He was, he realized, in new territory in applying science to people's view and treatment of the land. In "Round River," he said that ecology was as yet "an infant . . . engrossed with its own coinage of big words. Its working days lie in the future. Ecology is destined to become the lore of Round River, a belated attempt to convert our collective knowledge of biotic materials into a collective wisdom of biotic navigation."[70] His own guide was an interplay of science, ideas from the culture, and field observations, each informing and checking the others. His understanding of the balance among deer, wolves, and the mountains shows this. His experience in the Southwest had shaken his convictions, and game management research further undermined old ideas. Trips to two different areas provided new food for thought. In 1935, in Germany, he saw a managed forest with a managed deer herd. The next year, hunting on the Rio Gavilan in Mexico, he saw a deer herd largely controlled by natural forces. These extremes became the touchstones of his thought, though when he came to write "Thinking Like a Mountain," his mature formulation of this complex relationship, he harked back in memory to what he had seen in the Southwest.

Leopold came, ultimately, to see the land as community and humans as citizens. He measured human action by science, but went beyond science. "A thing is right," he said, "when it tends to preserve the integrity, stability and beauty of the biotic community. It is wrong when it tends otherwise."[71] This is a moral declaration and an act of faith. So too are his forceful arguments that there are no useless parts in nature, that everything is essential to the system, and that diversity means sta-

[69] Leopold, *Game Management*, 422–3.
[70] Leopold, *Sand County*, 189.
[71] Ibid., 262.

bility.[72] That leap, that kind of wisdom, was necessary. The problems of land use and wildlife preservation were not only scientific; they involved the values people placed on nature. They were not just theoretical; they had to be addressed in the "here and now." The land ethic is useful today because Leopold was willing take a stand in science and experience and then move beyond and point beyond.

Conclusion

These bold ideas seem an unlikely outcome of such practical work, particularly in light of the scientific ethos, which stressed detachment and objectivity, did not encourage scientists to think in philosophical terms, and discouraged their thinking about society. Nor did the institutions that supported this work – the Bureau of Biological Survey, the Wildlife Division of the National Park Service, and departments in land-grant or professional schools – seem the kind to support this line of thought. All true, but science had always been involved with society and it had always in one way or another been applied. One reason scientists spoke of, or invented, the idea of "pure" science was to fend off people who kept asking them to do something practical rather than letting them work on what interested them. This involvement was particularly close in the study of visible nature. People had as far back as we know taken this as a model for humans and an indication of the order of the world, and the Europeans' formal knowledge had always been deeply involved in the culture. Natural history had developed against the background of European expansion and colonialism, with it, and as part of it, and it had had a long-standing alliance with natural theology. Ecology had come of age with the industrial age, and one of the discipline's first professional concerns had been to save remnants of nature from that great machine. Leopold's own views developed in the context of dwindling game, the Depression, and the Dust Bowl.

Leopold's great contribution was not to provide a moral critique of industrial civilization in the new science but, with the land ethic, to lay out a practical guide for action. Few saw its value at the time. His essays found an appreciative but small audience; almost everyone else overlooked *A Sand County Almanac* when it appeared in 1949, a year after

---

[72] In 1976 R. W. May described the association of diversity with stability as part of the "folk wisdom of ecology"; cited in T. F. H. Allen and Thomas B. Starr, *Hierarchy: Perspectives for Ecological Complexity* (Chicago: University of Chicago Press, 1982), 188. See also Daniel Goodman, "The Theory of Diversity–Stability Relationships in Ecology," *Quarterly Review of Biology*, 50 (September 1975), 237–64, esp. 261; Frank Golley, *A History of the Ecosystem Concept* (New Haven, Conn.: Yale University Press, 1993), 100.

his death. It would be twenty years before the accumulating problems of pollution, vanishing wildlife, and environmental destruction would bring the book a large audience. Only as environmentalism grew did Leopold become a prophet. In 1970 Ballantine Books issued an expanded version, incorporating conservation essays from a later collection, *Round River*, into their "Style of Life" series. It has now gone through more than twenty printings and can be found in bookstores in all the settler lands. The land ethic has become a guide and point of departure and a basic argument for environmentalists defending nature.[73] How the ideas of ecology spread to scientists and the public and have begun to reshape policy and practice is the rest of our story.

[73] The best introduction is J. Baird Callicott, *In Defense of the Land Ethic* (Albany: State University of New York Press, 1989).

SECTION FOUR

NEW KNOWLEDGE, NEW ACTION

The great story of the settlers and their lands in the past forty years has been the impact of ecology's knowledge and perspective on the settlers' understanding. Assessing that now is a bit like writing about natural history's shaping of Victorian culture in 1840. It has only begun. Still, enough has happened that we can speak of new directions.

For twenty years after World War II, both scientists and citizens learned about ecology and applied its ideas for conventional ends, but the two streams of learning were separate. Scientists worked within the linked infrastructures of the scientific community and applied the knowledge they generated to establish programs for managing nature. The public learned less formally and systematically from education and the movies, books, and lectures that were the apparatus of nature entertainment, and it was concerned mainly with parks and a few endangered species. All of this occupies one chapter of this section. The other describes the flood that followed those years. In the past thirty years the union of scientific knowledge and popular interest has begun to radically alter our ideas about the land and to change our treatment of it. The settlers have begun a new kind of conversation – shaped by ecology, the conditions of the land, and settler history – about what it means to live with the land and become native to it.

9

NEW IDEAS, OLD PROBLEMS
THE DIFFUSION OF ECOLOGY, 1948–1967

The uproar over Rachel Carson's *Silent Spring* marked the beginning of the public environmental movement, but also the end of a generation of quiet scientific and public education. In the interwar years a small group had laid the foundations of an ecological perspective on humans and the land. After the war scientists in all the settler countries, then the general public, began learning those lessons. The modern defense of visible nature, now caught up in the larger agendas of the environmental movement, is the result of this quiet shift.

The postwar years did not appear to be a time of reflection, particularly about human dependence on nature, or even seem particularly suited to it. There was, as in the 1920s, a renewed enthusiasm for the conquest of nature, fueled by new technology and a prolonged economic boom. DDT was going to bring "victory" in our "war against the insects," 2,4-D would destroy weeds, Compound 1080 the "varmints," and pumps and bulldozers would bring water to the desert and transform it into field and pasture.[1] The settler societies had emerged from World War II with their industrial plants and farmlands intact, and for a generation postwar relief, reconstruction, pent-up domestic demand, and then the rearming that accompanied the Cold War brought prosperity and expansion. North American farmers again moved out onto the western Great Plains, this time with pumps to tap the accumulated groundwater on the eastern edge of the Rockies. In Australia bulldozers and tractors smashed down mallee and brigalow for new wheat farms.

The public's interest in nature ran in conventional channels. People went to national and state parks in unprecedented numbers and flocked

[1] Edmund P. Russell, "'Speaking of Annihilation': Mobilizing for War Against Human and Insect Enemies, 1914–1945," *Journal of American History*, *82* (March 1996), 1505–9; Thomas R. Dunlap, *Saving America's Wildlife* (Princeton, N.J.: Princeton University Press, 1988), chap. 8.

to beaches, lakeshores, and mountains. They showed, however, little concern for the problems they found there.[2] The products of industrial production and masses of people showed up on beaches and in parks, but memories of depression and unemployment made these seem annoyances or the price of progress, and there seemed to be other areas to go to. Only the most obvious destruction of natural beauty – or threats to established national parks – brought much protest, and that was from a small group of nature lovers. If the general public thought about wilderness, it was in vague terms of "untouched" areas that were assumed to be "out there" for those who wanted to explore. Nor did it pay much attention to environmental issues. William Vogt's *Road to Survival* and Fairfield Osborn's *Our Plundered Planet*, both of which appeared in 1948, made strong cases for a long-term view of the land and for taking the earth as a whole. Though Vogt's was a selection of the Book-of-the-Month Club neither his nor Osborn's work produced alarm or even a very visible debate.[3] A few hailed *A Sand County Almanac* when it appeared the next year, but it would not become popular for another twenty years.

Beneath that air of complacency and optimism, though, concern was growing. In part this was because people were learning more about their connections to the world. Ecology, if not always by that name, became part of the school curriculum and popular nature entertainment – books, movies, television. It also came home. Radioactive isotopes from hydrogen bomb tests in the central Pacific appeared in food, including strontium-90 in that quart of milk nutritionists and the milk industry urged mothers to give their children every day. Newspapers, magazines, and television provided what amounted to a short course in how the movement of the atmosphere, food chains, and bioconcentration brought these products of modern science to you. By 1960 banning nuclear tests in the atmosphere was a political cause and a diplomatic issue. There were more stories, as well, about threats to nature and a new wave of concern about endangered species. Closer to home, the more observant noticed dead birds on their lawns, victims of concentrated insecticide sprays, fewer fish and frogs in local streams.[4] In the

---

[2] On the postwar park boom see Leslie Bella, *Parks for Profit* (Montreal: Harvest House, 1987), 105–27; Alfred Runte, *National Parks: The American Experience* (2d ed., Lincoln: University of Nebraska Press, 1987), 173–9; David Thom, *Heritage: The Parks of the People* (Auckland: Lansdowne Press, 1987); Sarah Bardwell, "National Parks in Victoria, 1866–1956" (unpublished Ph.D. dissertation, Monash University, 1974).

[3] William Vogt, *Road to Survival* (New York: William Sloane, 1948); Fairfield Osborn, *Our Plundered Planet* (Boston: Little, Brown, 1948).

[4] Thomas R. Dunlap, *DDT: Scientists, Citizens, and Public Policy* (Princeton, N.J.: Princeton University Press, 1981), chap. 5.

aftermath of *Silent Spring* more biologists would concern themselves with public policy, but only at the end of the decade would a significant number of them make common cause with alarmed and angry citizens. This chapter, then, tells two stories. One is of the diffusion of the ecologists' vision from a few centers to institutions and scientists throughout the settler nations, the other of the public's education.

Ecology's spread, like natural history's, was from metropolitan centers to the periphery, but both the centers of innovation and the channels of communication were different. The United States was now part of the scientific center, the place of higher education and the development and testing of new ideas, and information went not to semipublic institutions like natural history societies but to small groups of experts. It often went through the new programs in mammalian pest control. After World War II all of the settler countries undertook new or more aggressive action to control "vermin," and – for the first time – they based this work in science. Using science against pests was, in fact, the raison d'être of the government ecological agencies established in Australia and New Zealand in this period. (In Canada, wildlife management loomed much larger.) The public had less formal and organized avenues of learning – newspaper columns, magazines, books, lectures, and two new media, movies and television. In the United States, the largest and richest country, nature films were a big business, and periodicals and nature programs national in scope. In the antipodes local lecturers, amateur movies, and newspaper columns carried more of the load. This was, though, a matter of proportion, not an absolute difference. The Audubon Society held camps and sponsored local lectures, while Australian television produced some excellent nature films (whose "exotic" subjects helped sell them in the large North American market).

Agencies, Ideas, and Programs

After World War II all three Commonwealth countries formed agencies to apply science to wildlife problems – the Canadian Wildlife Service (CWS), the Wild Life Survey Section of the CSIRO (the CSIR's successor), and the Ecology Division of New Zealand's Department of Scientific and Industrial Research (DSIR). Beyond that common mission and a strategy of using the resources of the larger Anglo community to their advantage, the bureaus were shaped by their countries' history and biology. The Canadian one was the best established. It had begun life in 1912 as the Wildlife Division of the Dominion Parks Branch of the Department of the Interior, and agency officials had been planning the new agency for a decade. In 1939 they had suggested a "Biological Survey of Canada." World War II put that on hold, but agency officials used their time to compile a list of people who might be useful after the war, including current employees with specialized training, ones who could be trained after the war, and expatriate Canadians who might be lured home by a job in their field.[5] The CWS took over the Wildlife Division's responsibilities, so it had at its beginning a well-defined mission. More than the others, it was concerned with wildlife rather than pests. Canadian parks were now seen as wildlife sanctuaries and research areas. A 1944 memo said the agency was trying to preserve "representative examples of natural or primitive conditions." That required scientific information, and "[g]ood technical advice of well-trained wild-life scientists is available now and extensive research by such men ... will result in improved advice in the future."[6] In 1946 Harrison Lewis said that plans for postwar work were on file; "the chief difficulty at present is the scarcity of suitably trained scientific personnel."[7] That summer Dominion Parks hired A. W. F. Banfield, returning from military service, to survey wildlife conditions in the parks.[8] Two years later the new agency was set up. It was headed by a Canadian and staffed by Canadians, though it drew scientists from around the world.[9] In 1959 the University of British Columbia established a graduate program in wildlife science, headed by Ian McTaggart Cowan, who had worked for the agency, and Canadians now could train at home.

The Australian and New Zealand agencies were established as a result of immediate problems. Killing rabbits had been a low priority during the war and they had multiplied. The situation was approaching a crisis. Legislators turned to science in part because of its successes during the war, but also because there were scientists on hand, in both countries immigrants who campaigned for a new agency. In Australia it was Francis Ratcliffe, in New Zealand Kazimierz Wodzicki. Wodzicki's story is an indication of how large a role chance and individual initiative

---

[5] List, HFL in upper corner, January 1944, Box 38, U300, RG 84, Records of the Dominion Park Service, Public Archives Canada, Ottawa.

[6] Unsigned memo, from internal evidence probably Spring 1944, Box 30 U300, RG 84, Public Archives Canada, Ottawa. Other documents in the file support this view.

[7] Harrison Lewis, "Wildlife Research in Canada," memo of May 1946, Box 38, U300, RG 84, Public Archives Canada, Ottawa.

[8] Box 162, U300–1, RG 84, Public Archives Canada, Ottawa.

[9] Harrison Lewis, "Lively: A History of the Canadian Wildlife Service," unpublished manuscript, RG 109, Records of the Canadian Wildlife Service, Public Archives Canada, Ottawa, 263–9. On recruiting see Box 300, W300, RG 84, Public Archives Canada, Ottawa. One letter, Lewis to de Vos, 16 December 1946, suggests some of the difficulties. Lewis wrote to one prospective employee that it was understandable that a young woman used to the Australian climate might not want to live in the Northwest Territories. He suggested the candidate give his fiancée's views "great weight" and consider them with "great care." Box 39, as above.

played in these small communities.[10] He had been professor of animal anatomy at the Agricultural University in Warsaw when World War II broke out. Joining the Polish government in exile in London, he had been posted to New Zealand as counsel-general. As a member of that government and of the Polish nobility, he may have decided to stay in New Zealand for his health. In any event, he had been doing scientific work in the country and in January 1946, "at the suggestion of the Prime Minister," he was given a temporary job surveying the state of New Zealand's wildlife. His report, delivered in early 1948, was the foundation for and first publication of the new division.[11] For staff both men looked to Britain, and particularly Oxford. Wodzicki seems to have been impressed by its reputation; for Ratcliffe the connection was literally the "old school tie."[12] This was not "colonial science" in the classical sense – second-class citizens taking on menial chores for their scientific betters. The antipodean agencies were using institutions they could not afford to build themselves.

The great attraction for both was Elton's Bureau of Animal Population, which was becoming in the postwar years an international training ground and crossroads. It was one of the few graduate training programs in terrestrial vertebrate ecology, and Elton was as interested in practical applications as in theory. For the agencies it was a place to find new people, train their own, and learn the latest. Ratcliffe was there during 1948, and even as the Wild Life Survey Section was being organized it offered Peter Crowcroft, from the University of Tasmania, the choice of continuing his work on fish parasites or studying animal ecology at Oxford.[13] The first New Zealander, Gordon Williams of the New Zealand Wildlife Service, arrived in 1950. J. S. Watson, who had spent 1941–9 at the Bureau of Animal Population, returned in 1955 on leave from the DSIR. J. S. Tener of the Canadian Wildlife Service was there in 1952–3. People also came from established programs and agencies. Thomas Park, a member of the animal ecology group at the University of Chicago, sent three students. Frank Pitelka spent a sabbatical year there on leave from the Museum of Vertebrate Zoology. Eugene Odum, one of the postwar leaders in American ecology, made

[10] Boris Schedvin, in researching his history of the CSIR to 1949, *Shaping Science and Industry* (Sydney: Allen & Unwin, 1987), found so little wildlife work that he deferred all discussion to the second volume, which would deal with myxomatosis. On New Zealand see Ross Galbreath, *Working for Wildlife* (Wellington: Bridget Williams, 1993), 30–40.

[11] Galbreath, *Working for Wildlife*, 51–5.

[12] On Wodzicki, author's interview, 19 June 1990, with John Flux, a rabbit biologist for the Section from the early 1950s to 1991. See also Galbreath, *Working for Wildlife*, 51–5.

[13] Peter Crowcroft, *Elton's Ecologists* (Chicago: University of Chicago Press, 1991), 53–4, 73.

a long visit. Marston Bates, working for the Rockefeller Foundation, visited for two days in 1950 to "discuss human ecology problems." The next year Victor Cahalane, of the U.S. National Park Service, stopped off after a tour of African game reserves.[14] The Bureau was part of the invisible college of ecologists as well as the visible structure of British academe and science.

Postwar Pest Control

Agencies and academic programs provided theory, but in each country people had to apply it to their own situation. This was as true of work on mammalian pests as anything else. Canada largely escaped the frenzy – though there was increased attention to predators in the North – but the others succumbed. In the United States the coyote was the main target, in Australia the rabbit (with the dingo in second place), and New Zealand ran separate but intense campaigns against rabbits and deer. Science had a different role in each situation. Controlling dingoes and coyotes was largely a matter of spreading poison bait, and scientists worked out better ways of doing that. Rabbit control took different paths in Australia and New Zealand, and called on science in different ways. In Australia investigations into disease produced startling results. The Wild Life Survey Section established a rabbit disease, myxomatosis, which virtually wiped out the animals over a large area of the continent. Myxo, as it was called, did not take hold in New Zealand. The Ecology Division aided, but also criticized, a successful conventional program. Science had the least impact on deer control (though the struggle of conventional ideas with new ecological ones within the bureaucracy is important). Deer were brought under control largely by people intent on making money shooting deer and using technology that enabled them to do the job with an efficiency far greater than the government's deer cullers. This review begins with the least ecologically sophisticated program – work on dingoes and coyotes – proceeds though New Zealand deer control, and ends with the work on rabbits in Australia and New Zealand.

"A Wholesale Abandon"

About three million coyotes were taken in the United States in the thirty-year period 1916–1946. In subsequent years the use of "coyote-getters" and poisoned bait stations has made it impossible to count the

---

[14] Ibid., 65–6, 158–61, 87.

"take," but it is generally conceded that these newer methods of control are more effective than prior methods.

Robert Rudd, 1964[15]

This was American coyote control – lots of poison and many dead animals – and dingo control was much the same. The programs continued on this basis because the public cared little about these animals, and stock owners a lot, and they saw the matter in simple terms. Predators were bad and ought to be killed. Their perspective was based on common sense, economics, and a visceral hatred. Sheep owners, whose charges were particularly vulnerable, were the most vocal. They had no good idea how many animals they lost to predators, or how these losses compared with those from disease, internal parasites, poisonous plants, or the incredible stupidity of sheep, but death by coyote or dingo was understandable, these animals could be killed, and doing it was much more satisfying than rooting out poisonous plants or running the sheep through a chemical dip to kill parasites. So across the Anglo world ranchers carried rifles and poison and set traps, and they hung the carcasses or skins of "outlaw" canids on fences and barns.[16]

Fueling the postwar programs was a new poison, Compound 1080, the result of wartime research on rodent control. Scientists had found that sodium fluoroacetate (1080 was its laboratory test number) was deadly not only to rats and mice but to canids as well. Early postwar tests showed that two mouthfuls of dead horse, treated with 1.6 grams of 1080 per hundred pounds of meat, were enough to kill a coyote. It was much less lethal to other forms of life, and optimism ran high. One article was titled "New War Born 1080 Coyote Poison May Kill All Predators."[17] That it fit existing programs was another point in its favor. In the 1930s the Biological Survey (which would become the Fish and Wildlife Service in 1940) had begun work on "bait stations" to kill coyotes through the winter and allow graziers to move to high-country pastures earlier in the spring. The technique consisted of poisoning a quarter or half a dead horse in the fall and staking it out before the snow fell.

---

[15] Robert Rudd, *Pesticides and the Living Landscape* (Madison: University of Wisconsin Press, 1964), 54–5.

[16] Dunlap, *Saving America's Wildlife*, chaps. 4 and 8; Roland Breckwoldt, *A Very Elegant Animal: The Dingo* (Melbourne: Angus & Robertson, 1988), 92–105.

[17] Lewis Laney, "New War Born 1,080 Coyote Poison May Kill All Predators," *New Mexico Stockman* (May 1948), 75; copy in 1080, Articles and Publications, General Records, Division of Wildlife Services, RG 22, Records of the U.S. Fish and Wildlife Service, National Archives, Washington, D.C. On the general climate of opinion see Dunlap, *DDT*, chap. 5; On 1080, idem, *Saving America's Wildlife*, chap. 8.

Compound 1080 seemed just what was needed, but it was very toxic and water-soluble as well. To reduce the dangers, the agency wrote stringent guidelines. It was manufactured under license and sold only to the Fish and Wildlife Service and certified pest control operators (who used it in urban rat control). For coyotes it was to be used only in bait stations, which were to be placed far from homes or inhabited areas, and only as a last resort. Pressure from stock interests soon made these precautions a dead letter. In 1949 E. R. Kalmbach, head of the Service's Denver research lab, complained to his superiors in Washington that the agency was "promoting the use of 1080 far beyond the limits that have been recommended through adequate research." Two years later he showed that in more than half the West a coyote had a poison station in its range and that 91 percent of Idaho range-land, 83 percent of Utah, and 71 percent of Nevada were covered. This had been done without reducing other methods of control and only four years after the Service had approved 1080 "primarily on acute preda-tion areas where other method have not gained the desired degree of control."[18]

The pattern in Australia was much the same, though there was more enthusiasm and less caution. Two scientists with experience in the field said that their countrymen had "always used poisons with a wholesale abandon that horrifies visitors from other countries," and their use of 1080 certainly qualified.[19] The Fish and Wildlife Service shied away from aerial baits; Australians began using them in 1946, and against rabbits, which requires the use of small-dose baits. The persistence and toxicity of 1080 soon led to a new poisoning program. Graziers began collect-ing dead rabbits and throwing them out of planes over rugged country near their pastures, hoping for a secondary kill of dingoes. The gov-ernment followed. A "newsreel film of the day shows baits being shov-eled out of a DC-3 accompanied by an earnest voice reading a dramatic script on the farmer's fight against the killer warrigal [dingo]."[20] As in the United States people continued to use other methods. Postwar prosperity revived barrier fences that had been abandoned during the Depression. In the late 1940s work began on a new one to isolate the Centre from grazing lands closer to the coast. It eventually stretched in a 8,614-kilometer arc from the Great Bight on the southern coast to

---

[18] E. R. Kalmbach to C. C. Presnall, 14 February 1949, in 1,080 Correspondence Instructions to Regions (ADC), and memo of 10 January 1951, in Poison 1,080 – Studies of. Both in General Records, Division of Wildlife Services, RG 22, Records of the U.S. Fish and Wildlife Service, National Archives, Washington, D.C.

[19] Frank Fenner and Francis N. Ratcliffe, *Myxomatosis* (Cambridge University Press, 1965), 30.

[20] Breckwoldt, *A Very Elegant Animal*, 239–40, quote on 240.

Queensland. In 1980 it was shortened to 5,614 kilometers, which is still "3,374 km longer than the Great Wall of China."[21]

Until the late 1960s there were few protestors, and criticism, as in the 1920s, was largely confined to the professional and scientific communities.[22] The situation was not like the 1920s in that the most vocal opposition came from E. R. Kalmbach and Clarence Cottam, who were career employees of the Fish and Wildlife Service. That they kept their jobs was in part due to their tact and bureaucratic skills, but also because they had more evidence and they worked in a changed organization. Olaus Murie had written privately to his superiors, concerned that he might be a "black sheep" for his view on poisoning. Kalmbach sent official memos. Murie had spoken as if his passion for nature was an odd taste; this generation took it for granted that preserving wildlife was part of the mission. Policy, though, would change only when the public became concerned, and that would wait for the rise of environmental consciousness in the late 1960s.[23]

From Guesswork to Measurement

Ecological ideas did not shape New Zealand's deer control program, but their effect on policy was more visible than in the United States. Here the science came from abroad, fully developed, and its assumptions and conclusions clashed sharply with conventional wisdom. A dispute in the late 1940s over wapiti in Fiordland National Park brought the change. The animals had been introduced early in the century for sport, but now groups interested in the forests called for their elimination. They were, critics charged, destructive aliens. New Zealand sportsmen sprang to their defense. Neither side in this dispute was quite what it had been a generation before. The forest's defenders were more clearly articulating a vision of a primeval landscape, and the New Zealand Deerstalkers Association (formed in the southern town of Invarcargill in 1938) was championing not British deerstalking but an American version of the sport. It depicted deerstalking as open to all and emphasized skill in the woods and with a weapon, the joys of the outdoors and the rugged life, and deerstalking as a way for fathers to help form their sons' characters.[24] A group was formed to investigate the situation, and it invited

---

[21] Ibid., 195–210, quote on 195.
[22] On protests in Australia see Folder 4, Box 1, and Folder 54, Box 6, Serventy Collection, National Library of Australia, Canberra.
[23] Dunlap, *Saving America's Wildlife*, chap. 8.
[24] On the organization see Philip Holden, *The Deerstalkers* (Auckland: Hodder & Stoughton, 1987), 12–32.

Olaus Murie, now working on elk in Wyoming, to join them. (He came as New Zealand's first Fulbright fellow, a part of the new connections America was forming to the rest of the world as Cold War divisions hardened.)

The group's conclusions challenged the unspoken assumptions that had guided thinking about deer and forests. It said there was no need to eradicate the wapiti; the forest could support a moderate population. Murie went further. In his own report to the minister of internal affairs he suggested that the policy of extermination be reconsidered and that more fieldwork be undertaken to find out what deer actually were doing to the various forests.[25] Shortly thereafter the Department of Internal Affairs hired an American biologist, Thane Riney. New Zealanders, historian Ross Galbreath said, had known that wildlife populations could be managed by "methods based on ecological principles," but "estimates which had been made of deer numbers, of their impact on plant growth or erosion, or of the effectiveness of deer destruction operation had all been based on qualitative, subjective observations – in other words, on guesswork." For twenty years there had been no real measure "made of how effective [the deer destruction campaign] actually was in terms of reducing either deer numbers or the forest damage or erosion the deer were said to be causing."[26] Now they had someone who "came from the fountainhead – or near it" and who could show them how it was done. Riney had just finished his master's degree at the University of California, Berkeley, under the direction of A. Starker Leopold (Aldo's eldest son).[27]

Graeme Caughley, who began as a deer shooter on the deer campaign and went on to become a wildlife biologist, said that Riney's ideas and approach were "entirely alien to Major Yerex" (who directed the program) and that Riney got "in hot water with the Department [of Internal Affairs] because he had scant respect for holy writ [the department's ideas about deer and forests] and set about examining these assumptions as if they were hypotheses."[28] Some of his techniques were straight from ecology and game management. He checked changes in population density by counting deer droppings along transects (lines drawn through an area). To check the age and composition of the herds he asked the deerstalkers for information on animals they shot. That got him a sample, even if it was not a representative one. It also involved the

[25] Galbreath, *Working for Wildlife*, 68–9.   [26] Ibid., 70.

[27] Ibid. For Riney's own early views, see Thane Riney, "New Zealand Wildlife Problems and Status of Wildlife Research," *New Zealand Science Review, 10* (March 1952), 26–32.

[28] Graeme Caughley, *The Deer Wars* (Auckland: Heinemann, 1983), 70.

hunters in work on "their" animals, an aspect of public relations that game managers in the United States regarded as part of their professional practice. The New Zealand bureaucracy was not enthusiastic. Riney went beyond this, though, to examine policy. Using maps he showed there was little overlap between the areas of high deer density, where the Wildlife Branch concentrated its shooters, and those areas the Forest Service and the Soil Conservation and Rivers Control Council considered at risk of erosion. Stationing observers with high-powered telescopes on ridges to observe hunters in the valley below, he showed that hunting was not all that effective. On a day when a government shooter sighted four deer and shot one, watchers noted twenty-three occasions on which an animal had simply moved out of his way.[29]

These conclusions were not entirely welcome. It was bad enough that Riney suggested the program was not working. Worse was his conclusion that it was based on a misconception. Deer, he said, were changing the forests but they were not eating them to the ground. Cockayne's nightmare of denuded slopes, raging waters, and cities swept into the sea was just that – a bad dream. Worst of all, Riney showed that in one watershed sheep, not deer, were causing erosion. He perhaps complicated matters by being an American at a time when anti-American sentiment was high. Although Riney "mastered the intricate ritual of the New Zealand tea ceremony," Caughley said, "he floundered on submerged reefs in the poorly charted channels of the New Zealand psyche" and left in 1958, having "singlehandedly advanced New Zealand wildlife research by about twenty-five years."[30]

In his history of the New Zealand Wildlife Service, Ross Galbreath provided a more measured contrast between Riney and Yerex, and by implication between the old program and the new. Yerex, he said, had a New Zealander's "feeling for the native bush and the importance of retaining it unspoiled; Riney, as an American and a scientist, took a more detached view." The two men also had "very different views of the place and role of a scientist in the organization." Riney sought to answer fundamental questions about animal populations, "and for validation of his work ... looked to the wider scientific community. ... Yerex, on the other hand, expected deference to his authority and loyalty to the organization." The scientist's role was "to investigate specified problems and to provide information to the administration, which would then set priori-

[29] Galbreath, *Working for Wildlife*, 70–1, on Riney. The *Journal of Wildlife Management* for these years describes techniques.

[30] Caughley, *Deer Wars*, 71. I am indebted to Graeme Caughley, who died in 1993, for two very interesting interviews about his research and the state of wildlife science in English-speaking countries.

ties and decide upon action."[31] Scientists were, in the jargon, to be on tap, not on top. Beyond Galbreath's analysis was a deeper division – in how to see nature. Yerex, though he might not have phrased it in just this way, acted as if nature could be grasped directly and understood by common sense – as witness the organization of the extermination campaign. Riney constructed nature from measurements interpreted by theory, and from this standpoint questioned the "commonsense" links among deer, forests, and erosion and between individual kills and population reduction.

That he was an outsider challenging an established policy and bureau- cracy from a perspective his superiors did not share reduced Riney's influence, but the crucial element was that policy did not follow from scientific findings. It was the result of political pressures from interested groups. This was most apparent in coyote poisoning, which remained a matter of mass poisoning until a significant portion of the public wanted to save predator populations as part of saving the environment. It was also the case in New Zealand as well, except that the key was not envi- ronmental sentiment but economics. Rising prices for venison, velvet (the covering of the growing antler), and deer organs (used in tradi- tional Asian medicine) encouraged hunters to turn what had been sport or a source of small change into a business. In the late 1950s hunters began taking jeeps and trucks into the backcountry on weekends. The next step was to fly to makeshift strips deeper in the mountains. Killing peaked in the late 1960s with the development of the helicopter gunship team. A pilot flew; a shooter leaned out the right side with a scope- sighted semiautomatic rifle; a third man at a staging point gutted the animals; and a fourth stayed in the valley below, minding the portable cooler and refueling the copter. In the early days, when deer were thick and unwary, a team could take more than a hundred in a single day, and on occasion tallies were double that. It was the mechanized equivalent of the Great Plains buffalo slaughter. Foot-bound government hunters became an embarrassment to the department. As demand for deer prod- ucts rose, farmers demanded a cut. Parliament obliged by making deer farming legal, and there was a brief flurry of live-capture hunting. Deer are now raised behind fences like cattle, and while control operations continue in a few areas eradication is not even a distant goal.[32]

"A Spectacular Epizootic"

Rabbit control provided the richest mix of politics, science, and views of nature, as well as the clearest examples of the difficulties of turning

[31] Galbreath, *Working for Wildlife*, 72.
[32] Caughley, *Deer Wars*, 85–105; Carolyn M. King (editor), *The Handbook of New Zealand Mammals* (Auckland: Oxford University Press, 1990), 454–8.

ecological knowledge into effective action. In the late 1940s, after seventy years of conferences, commissions, poisons, patent remedies, and failure, both Australia and New Zealand decided to try science. In Australia it was successful beyond all expectations. The Wild Life Survey Section set off "a spectacular epizootic which for scale and speed of spread must be almost without parallel in the history of infections."[33] In less than a decade myxomatosis reduced the continental rabbit population by roughly 90 percent. In New Zealand science was an important adjunct to a program of conventional controls that reduced rabbits to a local nuisance. These were not, though, simple tales of science triumphant. Far from it. There was dogged scientific work – over several decades in the Australian case – but luck and organization played crucial roles.

Ratcliffe's analysis of the rabbit problem during World War II showed that there was a new view even before the Australian government established the Wild Life Survey Section. People had commonly spoken of rabbits as a single thing: the "rabbit menace" or "the gray blanket." Ratcliffe pointed out that the interaction of rabbit populations, the land, and human activity in Australia created three quite different situations. In well-watered areas near the coast, rabbits thrived, but cultivation was intense, and normal farming operations, some purposeful plowing up of their warrens, and a bit of trapping, poisoning, and hunting kept them under control. Because production per acre was high, farmers could afford these measures. In the very arid pastures surrounding the interior desert there was no need for rabbit control. Frequent, harsh droughts kept numbers down. It was in the intermediate area, where the climate was mild enough for rabbits to flourish but rainfall too low to support intensive farming, that the problems lay. In many years farmers and graziers needed control but could not afford it.[34] This land, unfortunately, was where much of the wheat was grown and the sheep raised.

There is a certain irony in Ratcliffe's making this analysis. It implied that Australians should learn to live with rabbits rather than search for a magic bullet, but Ratcliffe's bureau produced the most spectacular magic bullet in pest control history. Though myxomatosis burst into public consciousness with the epidemic that began in 1950, its story goes well back. It was discovered in 1898, when scientists traced an epidemic among European rabbits being used in a South American laboratory to a previously unknown disease of New World rabbits. Among them it produced few symptoms, but it was highly lethal to the related Old World species. In 1919 Dr. Henrique de Beaurepaire Aragao of the Instituta Oswaldo Cruz,

---

[33] Fenner and Ratcliffe, *Myxomatosis*, 276.

[34] Francis N. Ratcliffe, *The Rabbit Problem* (Melbourne: Commonwealth Scientific and Industrial Research Organization, 1951), 3–4.

Rio de Janeiro, suggested its use to Australian authorities. Brute force methods were not keeping rabbits down, but the same fears that had curtailed earlier suggestions for disease limited this one. Besides, there was now a rabbit industry, and CSIR officials pointed out that the disease would spread far more readily among caged animals than in free populations. H. R. Seddon, director of veterinary research in the New South Wales Department of Agriculture, ran laboratory tests on samples Aragao sent him in 1926, but he could not get permission for field trials.[35]

Matters rested there for eight years, until Dr. Jean Macnamara, a Melbourne polio specialist, learned about the disease (but not, apparently, Seddon's studies) while working in London. She recommended it to the high commissioner for Australia, who passed the suggestion on to the Commonwealth government. Quarantine authorities objected, but CSIR officials were now less concerned about the rabbit industry. They arranged for tests in Britain, and then for field trials on an island off South Australia (a concession to quarantine), then in an arid part of the state. The results were disappointing. The disease did not infect common domestic or native mammals, and it did kill rabbits, but it did not reliably spread in the wild. Where there were stickfast fleas, Australian insects that had adapted to the rabbit, it did, but these were not present over the rabbit's range. L. B. Bull told his superiors in the CSIR that the disease would do some good, but probably would not be useful under most natural conditions. In 1943 the agency abandoned the project, and the states ignored its suggestion that they take it up.[36]

Bull's discouraging verdict was hardly surprising. Starting an epidemic required luck or knowledge, and Bull had no luck and there was not much knowledge. The ecology of disease was in its infancy. Entomologists, ecologists, and doctors were just working out the ways in which insects, other mammals, and humans interacted to produce malaria and yellow fever in South and Central America. That had taken years of work by teams of scientists, and the species involved were, except for the mosquitoes, relatively well understood. Bull had a small budget and was working with a system no one knew much about. Even the distribution of rabbit parasites over the continent was not known, much less how they spread disease and under what conditions. As it turned

---

[35] Fenner and Ratcliffe, *Myxomatosis*, is the standard reference (see introductory chapter on early history), supplemented, more recently, by Frank Fenner and John Ross, "Myxomatosis," in Harry V. Thompson and Carolyn M. King (editors), *The European Rabbit* (Oxford: Oxford University Press, 1994), 205–39.

[36] This work is summarized in L. B. Bull and M. W. Mules, "An Investigation of *Myxomatosis cuniculi* with Special Reference to the Possible Use of the Disease to Control Rabbit Populations in Australia," *Journal of the Council on Scientific and Industrial Research*, 17 (1944), 79–93. Fenner and Ross, "Myxomatosis," 205–6, provides a short summary.

out, transmission varied dramatically in different environments. When myxomatosis was introduced into Europe in the 1950s (by a French landowner who had heard of Australia's success), British scientists found out that fleas were the main carriers and that they were more efficient in transmitting moderately virulent strains than either very deadly or highly attenuated ones. In Australia, mortality patterns from the mosquito-borne epizootics that had been responsible for the initial kill were very different from those that followed the introduction of rabbit fleas in the 1960s. This, as it turned out, was because of the weather and time of year when the different insects did their work.[37]

Bull did not stumble on the right conditions, but the Wild Life Survey Section did. In 1950 political pressure from Macnamara and other enthusiasts forced Ratcliffe, against his better judgment, to begin field trials. Initial results seemed to bear out his pessimism, and crews were breaking camp in December when reports began trickling in of dead rabbits outside the test areas. By February the disease had spread over an area 1,100 by 1,000 miles. More than mosquitoes were at work; as word spread graziers began driving to hard-hit areas and getting sick animals to put in their pastures. By September state authorities were convinced. They started inoculating rabbits with the virus. Tasmania tried to save its rabbit industry, but too many people smuggled sick rabbits across Bass Strait. Compared with myxomatosis the Black Death had been mild. Over much of the continent rabbits almost disappeared. Plants flourished. By 1953, even before the first wave of infection had burned through the country, the wool clip had risen by some 40,000,000 pounds, worth $A60,000,000 – economic evidence of ecological change. The epidemic died out in the late 1950s, when rabbits were too few and scattered to support the chain of infection, many were resistant, and the most virulent strains had died out. The disease remained endemic, flaring up when conditions were right. Only in the late 1980s did rabbit populations again reach alarming levels (and rabbit calcivirus disease is now moving through the continent).[38]

---

[37] Fenner and Ross, "Myxomatosis," 216–17, 227. On the disease in Europe and Britain see N. W. Moore, *The Bird of Time* (Cambridge University Press, 1987), 121–41. For the application of these dynamics of mammalian diseases to humans see René Dubos, *The Mirage of Health* (New York: Doubleday, 1959), 78–81.

[38] Fenner and Ratcliffe, *Myxomatosis*, provides details. See also on the later period Fenner and Ross, "Myxomatosis." Only in the 1990s did growing rabbit populations again cause serious concern. Researchers have begun working on new strains of myxomatosis, and in late 1995 there was an unexplained outbreak of a new disease, calcivirus, which was being tested offshore. Author's conversations and papers at the Sixth International Theriological Congress, Sydney, August 1993. Calcivirus publicity statements by the CSIRO and the Minister of Primary Industries can be found on the Internet.

Success brought the Wild Life Survey Section a rush of funds for the study of the disease. Its early spread, up the Murray–Darling River system, suggested that an insect that required water was the primary agent, and the scientists began looking at mosquitoes. They also tested the efficiency of the ticks and fleas that parasitized rabbits around the country. The disease itself came under scrutiny. The deadliest strains killed their hosts before the disease could be passed on. Did that mean the disease would rapidly attenuate to a nonlethal form? It did not, but the study of virulence became a long-term research interest. Wider questions about the effects on native vegetation and wildlife populations as the rabbits disappeared received short shrift. Even myxo research ground down as the rabbit population fell. "After 1956, when most of the intensive field studies had been terminated and the novelty of myxomatosis had worn off, the collection and collation of field data on an Australia-wide basis virtually ceased."[39] Still, for the next twenty years the Section "dined out," as one ecologist put it, on the success. In the early 1960s, for example, the Australian Meat Board supported a decade-long study of the ecology of dingoes that included work on predator–prey relations, the role of fire in Australian ecosystems, and hybridization between dogs and dingoes – subjects that went well beyond the graziers' immediate concerns.[40]

"The Knowledge of Animal Biology"

Across the Tasman Sea, New Zealanders were discovering science, though the effects were less dramatic. Surveying rabbit legislation in 1950, Kazimierz Wodzicki commented dryly, "Until the last few decades it was thought that legislative measures alone without the knowledge of animal biology would achieve the destruction of pests."[41] Beyond this, the Australian and New Zealand stories diverge sharply. Myxomatosis did not spread in the islands, and the postwar campaign employed traditional methods. The difference was that, this time, they were systematically applied and guided (though not directed) by science.[42] In 1947 Parliament established a new national Rabbit Destruction Council and a goal of wiping out rabbits. While local boards retained the power to tax stock, trained people carried out the actual control measures, and

---

[39] Fenner and Ratcliffe, *Myxomatosis*, 298.

[40] Author's interviews with Alan Newsome, Division of Ecology and Wildlife, CSIRO, 28 May and 2 June 1992.

[41] Kazimierz Wodzicki, *Introduced Mammals of New Zealand* (Wellington: Department of Scientific and Industrial Research, 1950), 127.

[42] John A. Gibb and J. Morgan Williams, "The Rabbit in New Zealand," in Thompson and King, *The European Rabbit*, 166.

action was coordinated across the country. To remove any incentive to keep rabbits, the rabbit industry was abolished; after a grace period, trade in fur or meat became illegal. The national board frowned on keeping rabbits as pets, and it was said to have "wished to prohibit the sale of stuffed rabbits as cuddly toys, but this was never mentioned in its Annual Reports."[43]

In the early years scientists were deeply involved. "Work on rabbits was the bread and butter of research in Ecology Division . . . in the 1950s."[44] Much of it involved formulating and distributing poisons or testing techniques. Research showed, for example, that while night shooting produced impressive piles of dead rabbits, it reduced populations only briefly. Some studies were more radical. The scientists were from the first skeptical of the possibility of eradicating rabbits, but they were reluctant to say so. As with the wapiti, they left it to an outsider, Walter Howard, a biologist from the University of California, Davis, who spent 1958–9 in New Zealand as a Fulbright fellow, to raise the issue. The Rabbit Destruction Council and the local boards were not receptive, and the Ecology Division set out to test the proposition. Its most dramatic demonstration came in the mid-1960s, when it convinced the Wairarapa Rabbit Control Board to suspend all operations on a 1,000-hectare tract for three years. At the end of the period the rabbit population was unchanged. Indeed, it might have gone down. The board was "extremely reluctant to accept the implications of this trial." (Its skepticism continued. This particular local group was among the hold-outs when, in 1988, authorities recommended cutting back control measures.)[45]

This kind of coordinated national action was possible because New Zealand farming and grazing were relatively intense, providing a good return per acre, and the program used poison, the least expensive method. In the first years older chemicals were used, but by the mid-1950s Compound 1080 was the mainstay. Aerial baiting was used as

---

[43] Ibid., 163–4, quote on 167. The basic law was the Rabbit Nuisance Act, 1928, No. 8; Amendment of 1947. Laws were consolidated by the Rabbits Act, 1955, No. 28; the next general statute was the Agricultural Pests Destruction Act, 1967, No. 147. An outside report on the effectiveness of the organization is Walter E. Howard, *The Rabbit Problem in New Zealand*, DSIR Information Series No. 16 (Wellington: Department of Scientific and Industrial Research, 1958).

[44] John Gibb, "A Brief Review of Rabbit Research in Ecology Division since the 1950s," unpublished paper prepared for the Ecology Division, New Zealand Department of Scientific and Industrial Research, March 1983. Library, Ecology Division, DSIR, Lower Hutt, New Zealand. With the breakup of the Division in 1992 these papers may be elsewhere.

[45] Ibid.; Gibb and Williams, "Rabbit in New Zealand," 166.

early as 1949, but most baits were distributed from the ground. Results were quickly apparent, and in the early years, increased production as rabbit populations fell paid for control; only after 1960 did costs exceed immediate benefits. By the late 1980s control was cut back in most areas, and a 1994 report said that "the rabbit is now stabilized at low densities almost everywhere, primarily by natural processes; and pasture production is threatened only in the 'semi-arid' tussocks grasslands of the South Island."[46]

All these efforts show that while science had considerable influence on "practical" matters it did not set policy. Vocal and interested groups did that. In part this was because it was only rarely able to make a decisive difference – myxomatosis being the obvious case – in part because any policy change affected some people adversely (New Zealand's stringent postwar campaign was a political coup as startling in its way as myxomatosis). The short-term view, however, is not the only one. Science's greatest impact was not the guidance it gave policy-makers but the perspective in which it framed the questions, which would, in the long run, affect administrators and the public alike. In North America, for instance, acceptance of this new point of view brought protection for predators in the 1970s. This shift in public ideas is the second element of change in these years. It was not, in form, much like the scientists' debates. There were no institutions, policies, or research. People learned from books, magazines, films, and lectures. Theories and data were incidental, illustrations of a much deeper argument, which people grasped much less systematically and consciously than the scientists had their ideas. The latter learned about nature as a complex, dynamic, interconnected whole, one in which humans were involved and which they were changing in important ways.

Popular Understanding of Nature

"The Contamination of Man's Total Environment"

Scientists became familiar with ecology through professional learning and practice, the public through the apparatus of public education that had been developing since natural history had been a popular

---

[46] Gibb and Williams, "Rabbit in New Zealand," 164, 167, 163. A useful midterm report is John Gibb and John Flux, "Mammals," in Gordon R. Williams (editor), *The Natural History of New Zealand: An Ecological Survey* (Wellington: Reed, 1973), 358–9. A more current one is John Parkes, "Rabbits as Pests in New Zealand: A Summary of the Issues and Critical Information," Landcare Research Contract Report LC9495/141, Lincoln, New Zealand. My thanks to Landcare Research for this report.

recreation. The public learned that biological systems were intricately interconnected and that the growing human population and industrial development threatened many species. What gave these concerns point was the more troubling knowledge that humans were personally involved. It was one thing to know that pollution and population were destroying natural beauty, another to find that human ingenuity threatened humans, even oneself and one's family. Such, though, were the revelations that radioactive fallout was spreading around the world as it fell with the rain and accumulated in the milk children drank and poured on their breakfast cereal. In 1962 Rachel Carson's *Silent Spring* crystallized public fears. In an eloquent book backed by appeals to scientific research, Carson argued that we had "put poisonous and biologically potent chemicals indiscriminately into the hands of persons largely or wholly ignorant of their potential for harm." With the problem of nuclear war the "central problem of our age" was "the contamination of man's total environment" with chemicals that disrupted ecosystems, killed animals, and even humans. It could "even penetrate the germplasm to shatter or alter the very material of heredity upon which the shape of the future depends."[47] Carson proposed a radical remedy. We had to abandon our ideas of "conquering" nature and learn to live as what we were, citizens of a biological community.

This was the trigger for the popular environmental movement, for it presented the issue – chemical contamination of the environment – around which formed the campaign to save ourselves. Concern with contamination also fueled the campaign to save nature, but this also drew on popular nature programs that gave the public a vivid, if vicarious, experience of life in the wild. Nature photography had shown wildlife and the land as spectacles or events at a distance. Postwar equipment allowed people to watch a hawk stooping on its prey and mice giving birth in their burrows, follow the slow unfolding of a flower, and see night-living animals going about their lives. Television created a new and wider forum for the nature lecturer. Marlon Perkins in the United States was perhaps the most successful of the first generation. He spent almost two decades chasing through the "Wild Kingdom" for the Mutual of Omaha insurance company.

Much of what people saw was "gee whiz" science, isolated facts about odd creatures or disconnected but thrilling events, but they also learned about the world, for the films had an undercurrent of ecology. The science was an organizing device, allowing filmmakers to make snippets of film into a story. The varied shots of awesome landscapes and spectacular sunsets, time-lapse sequences of vegetation, the miracle of birth,

---

[47] Rachel Carson, *Silent Spring* (Boston: Houghton Mifflin, 1962), 23, 18.

cute baby animals, exciting chases, and mating dances were all parts of the life of the land, the living desert, or the vanishing prairie. In the United States this was big business. Walt Disney produced full-length features for movie theaters, which were recycled in pieces on television. Vincent Serventy, Australia's most prominent postwar nature popularizer and educator, said that "probably there has been no greater force for nature conservation." Americans were not the only ones in the business, though. Francis Ratcliffe, working on popular nature education for the Australian Conservation Foundation in the late 1960s, found that nature films were among the British Broadcasting Corporation's most popular products.[48]

Films and television supplemented rather than replaced older parts of this apparatus of nature education and entertainment, particularly in the smaller societies. Serventy's career is a good example. He was from Western Australia, and one of the few national figures up to that time who had not been involved in the circle of Melbourne journalists that had formed around Donald Macdonald and continued after his death. Serventy had a B.A. in science and education and worked in the Western Australia school system. In the 1960s he was in charge of the Department of Education's Nature Advisory Service. He was also an active field scientist. He did zoological work on the Australian Geography Society expedition to the Recherche Archipelago, a natural history survey of the Great Victorian Desert, studied mutton birds for the CSIRO, and spent six months on the Great Barrier Reef. He was also personally active in nature education. From the 1940s he edited a column of newspaper nature notes. He made a film about koalas with the Department of Information and produced nature shows for the Australian Broadcast Commission and science telecasts for Western Australia. He lectured widely; he gave one talk to the Blavatsky Lodge of the Theosophical Society in Sydney, not a venue commonly associated with nature in Serventy's sense. He became a television personality; in 1970 the Australian Broadcasting System did a series, *Around the Bush with Vincent Serventy*.[49]

His message and themes came from natural history, but his concerns were modern. *A Continent in Danger* (1966), written as consciousness of the environmental crisis was rising, shows both strains.[50] The organiza-

[48] Unsigned and unaddressed letter to editor on criticism of Disney's films, Folder 13, Box 2, Serventy Collection; Ratcliffe to Garfield Barwick, 9 September 1965, Box 5, Ratcliffe Papers, Australian National Library, Canberra.

[49] The Serventy Collection provides this information in scattered boxes. The first two give a good idea of the range of his work and interests. The files testify to Serventy's consuming interest in nature, not filing systems. The paperback published in conjunction with the series was *Around the Bush with Vincent Serventy* (Sydney: Australian Broadcasting Commission, 1970).

[50] Vincent Serventy, *A Continent in Danger* (Sydney: Ure Smith, 1966).

tion of the book was reminiscent of natural history, beginning with the formation of the continent and in successive chapters taking up in turn the various forms of Australian life: egg-laying mammals, marsupials, "The Modern Mammals," birds, and reptiles. "The Ecology of Fire" introduced a more modern concern, though the word "ecology" was not in the index. After this "Man Arrives," and we consider "The Present Position." Like many of his predecessors, Serventy lamented the destruction caused by the Europeans. With them begins "the destruction of the continent's unique assembly of plants and animals," which "is still going on today."[51] The final chapter, "A Plan for Action," is conventional conservation and preservation, but it ends with a discussion of *Silent Spring*. The year the book appeared Serventy became editor of *Wildlife*, a magazine founded a few years before by the Wildlife Preservation Society of Queensland. He also became involved in the growing number of protests against development that would coalesce into an Australian environmental movement.

Serventy, moving from natural history and education to ecology and action, had his counterparts elsewhere. In the aftermath of *Silent Spring*, scientists had been reluctant to defend Carson or to advocate action, but by the late 1960s a growing number were challenging the ideal of detachment.[52] In 1964 Robert Rudd (who had worked at the Museum of Vertebrate Zoology) published *Pesticides and the Living Landscape*, a field scientists' *Silent Spring*. The next year, Raymond Dasmann described, in *The Destruction of California*, the impact of modern irrigation, farming, highways, and suburbs on the Golden State.[53] In 1966 a group of Australian biologists and naturalists presented the environmental case for saving nature in A. J. Marshall's *The Great Extermination: A Guide to Anglo-Australian Cupidity, Wickedness, and Waste*. In prose that was at times as blunt as the title the authors lamented the destruction of Australian plants, wildlife, and reef communities, and called for saving all wildlife, even "our heritage of native fishes" (not a central concern for nature lovers up to that point).[54]

---

[51] Ibid., 18.

[52] Dunlap, *DDT*, chap. 4. Early studies that were couched in the "proper" form include Robert Rudd, *Pesticides: Their Use and Toxicity in Relation to Wildlife*, Bulletin 7 (Sacramento: California Department of Fish and Game, 1956), and John George, *The Program to Eradicate the Imported Fire Ant* (New York: Conservation Foundation, 1958).

[53] Raymond Dasmann, *The Destruction of California* (New York: Macmillan, 1965). Robert Rudd, *Pesticides and the Living Landscape* (Madison: University of Wisconsin Press, 1964).

[54] J. S. Turner, "The Decline of the Plants," 134–55, and David Pollard and Trevor D. Scott, "River and Reef," 112, both in A. J. Marshall (editor), *The Great Extermination: A Guide to Anglo-Australian Cupidity, Wickedness, and Waste* (Melbourne: Heinemann, 1966). Though not as thoroughly backed by ecological knowledge as some of the American works, *The Great Extermination* made a decisive break with the past in several respects.

Beyond Pests: Endangered Species

Concern about the "environment," which developed after *Silent Spring*, built on an earlier defense of the parts of nature. After World War II advocates campaigned to expand national parks, make new ones, and provide better access to them. In the United States wilderness became a cause, achieving legislative status with the Wilderness Act of 1964. There was also a rising concern with endangered species. In the postwar decades the disconnected initiatives and emergency campaigns began to coalesce into continuing programs. Efforts to save the whooping crane in North America illustrate the process. Like other favored species in these years – buffalo, antelope, and koala – the crane was one people could see and that they associated with the land. Adults stood about five feet (1.5 meters) tall and had wingspans of around seven feet (2.1 meters), and their calls could be heard for miles across the marshes where they fed. Though they had ranged over much of central North America, they had never been common, and the hunting and habitat destruction that accompanied Anglo expansion reduced them to a remnant by 1900. The Migratory Bird Treaty of 1916 set a ten-year closed season, which was extended, but the end seemed near. When, in 1922, the birds abandoned the last known nesting site, Mud Lake in Saskatchewan, there were obituaries (premature, as it turned out).[55]

Even so, little was done. Fred Bradshaw, chief game guardian of Saskatchewan and head of the Provincial Museum of Natural History, made the birds his project from the early 1920s until his retirement in 1935, writing to people from near the Arctic Circle to the northern coast of South America to trace their migration and speaking to groups throughout the province, asking for information and pleading for protection. South of the border T. Gilbert Pearson, head of the Audubon

---

James Barrett (editor), *Save Australia: A Plea for the Right Use of our Flora and Fauna* (Melbourne: Macmillan, 1925), had drawn on experts in resource management and had spoken of the continental economy and appealed to efficiency. Biologists and amateur naturalists wrote *The Great Extermination*. They were not concerned with the economy or short-term returns from business as much as with the ecosystems that supported everything and their long-term health. *Save Australia* had not thought it possible or useful to save the native fauna, repeating the belief that it could not compete with European introductions. Marshall's contributors thought it could be saved and said it should be, as a "national heritage."

[55] This account draws on Thomas Dunlap, "Organization and Wildlife Preservation: The Case of the Whooping Crane in North America," *Social Studies of Science, 21* (1991), 197–221. On obituaries see Hamilton M. Laing, "Exit the Whooping Crane," *Manitoba Free Press* (5 January 1922), typescript in Whooping Crane Files, 1931, Saskatchewan Museum; Hal C. Evarts, "The Last Straggler," *Saturday Evening Post, 196* (14 July 1923), 48.

Society, located the cranes' wintering ground on the Texas Gulf coast, but only in 1938 did the U.S. government create Aransas National Wildlife Refuge to protect the area. In 1940 a storm destroyed all but one of a nonmigratory Louisiana flock, and that bird died when Fish and Wildlife Service agents attempted to capture it. The Aransas flock was all that was left, and counts on the refuge showed there were less than two dozen birds. The Service could do little more than administer the refuge, for no one knew where the birds nested.

There were, though, stirrings that would develop into a strong campaign after World War II. In 1938 the National Audubon Society started a program to help endangered species. It was small and it had a rocky start. The staff consisted of Robert Porter Allen, an enthusiastic college dropout, and World War II soon put an end to his work. He returned in 1946, however, and made the crane one of his first projects. He studied the birds at Aransas in the winter and spent summers looking for the nesting site, somewhere north of southern Saskatchewan. In the early 1950s Fred Bard, who had been Bradshaw's assistant in Saskatchewan and was now director of the provincial museum, joined him. With the cooperation of the Fish and Wildlife Service and the Canadian Wildlife Service they built a network of spotters from Texas to Saskatchewan and publicized the cause. By the late 1950s newspapers were following the birds' progress each spring and fall, and the whooper was a symbol of endangered wildlife. It would become one of the first protected under the American endangered species program, written into law between 1966 and 1973, and the subject of a U.S.–Canadian agreement.

New Zealanders' interest went through a similar pattern. There had been concern about vanishing native birds, and the government had set aside a few refuges, but action was spotty in the years before World War II. The government had pigs killed to protect the only known nesting grounds of Buller's shearwater, and it supported private efforts to protect royal albatrosses nesting near Dunedin, but these, like the campaign to help the whooper, were the result of a few enthusiasts pressing for action. There was more interest after the war, and in 1948, in a move that was "as it turned out, very well timed," the Department of Internal Affairs announced the formation of a Native Birds Preservation Committee.[56] Two weeks before its first meeting a population of takahe, a native flightless rail, was discovered in a remote valley in Fiordland National Park. This was news, reported even in the *Times* of London.

---

[56] Galbreath, *Working for Wildlife*, 81–4, quote on 85. This account is also based on the files of the Department of Internal Affairs relating to the takahe, Wildlife 47/48/3, National Archives, Wellington.

The takahe had been written off long ago. In 1898 the government had paid a considerable sum for what was thought to be the last stuffed specimen. The area was quickly closed to visitors, and the Minister of Internal Affairs announced that the new committee would consider further measures.[57] In his history of the New Zealand Wildlife Service Ross Galbreath sees this as part of a larger shift. In 1952 Parliament passed a new national park act and in the next four years added several new areas to the system. It established a "forest sanctuary" in the Waipoua kauri forests. A new wildlife act formally reversed the earlier presumption about wildlife, declaring that all species were protected unless they were declared game and a hunting season set. The survival of certain ones – the takahe, kiwi, and kotuku (a native heron) – became "a matter of national pride."[58]

Certainly the issue did not go away. In 1952 the secretary of the Department of Internal Affairs (the senior career official) told his minister that, although there had been "severe losses of trained officers" in the Wildlife Division and there were more takahe than had at first been thought, the project was still "an essential responsibility of this Department and one which is expected of it by the general public to do all possible to ensure preservation of the birds."[59] The first question was whether anything should be done. The Forest and Bird Protection Society argued that the takahe should be left alone. Most members of the Native Birds Preservation Committee, on the other hand, favored active measures, with captive breeding as the most popular choice. The need for research put a decision off for a few years, but in the end officials judged it "scientifically [and] politically safe to proceed." In 1957 authorities, with full publicity, took four eggs.[60] None hatched. The next year the department did not make its plans public, partly to avoid another public relations fiasco, but also because Southland residents were against removing any of "their" birds.[61]

[57] Galbreath, *Working for Wildlife*, 85–6.
[58] Ibid., 91, 93. The settlers were using Maori names for the native birds, another indication of the wave of nature nationalism. In 1958, when the Wildlife Branch got a distinctive seal and badge, it featured the kotuku.
[59] Secretary, Department of Internal Affairs, to the Minister, 24 October 1952, DIA files, Wildlife 47/48/3, National Archives, Wellington; File 762, Ms. papers 444, Royal Forest and Bird Protection Society Papers, Turnbull Library, Wellington. Galbreath, *Working for Wildlife*, chaps. 6 and 8, covers these efforts and describes the philosophical splits between the active and passive camps, which was a feature of all the programs. This was a general concern among those interested in rare and endangered birds. See letter of A. H. Chisholm, the Australian nature writer, to Perrine Moncrieff, 9 December 1942, in Folder 1, Ms. papers 4723, Royal Forest and Bird Protection Society Papers.
[60] Galbreath, *Working for Wildlife*, 88.
[61] DIA files, Wildlife Division, 47/48/3, New Zealand National Archives, Wellington.

In 1954 the accidental discovery of the whooping crane's nesting ground presented North American wildlife authorities with the same dilemma: whether to allow the birds to recover on their own or to take active measures. If the wildlife agencies did not act and the birds became extinct, they would now, surely, be blamed. If, on the other hand, they did something and the birds died out, the same thing would happen. Here too sentiment was split. The Audubon Society was strongly against any active measures. People, Robert Allen argued, knew too little to take effective action, and there were too few to birds to risk experimenting on them. He also felt that the crane was part of wild America and extinction preferable to survival as a handful of caged birds. Others, particularly a group in the Fish and Wildlife Service, were more sanguine about humans' ability to help and more worried about natural disasters. There was the bad example of the attempted capture of the lone survivor of the Louisiana flock, but the danger was still extreme. A few trigger-happy hunters, another bad storm, or a disease could push the small population over the edge. They also rejected the idea that the birds were better off dead than in captivity. Live birds could always be reintroduced into the wild.[62]

Like their New Zealand counterparts North American authorities concluded that they had to take action and that captive breeding was the least risky course. They acted, though, on a much larger scale.[63] The New Zealanders had taken a few eggs and given them to an amateur aviculturist. Later programs, such as the rescue of the black robin in the 1980s, were more elaborate but still relied on a few generalists. These were commonly recruited from the ranks – some of the biologists began as deer shooters – trained on the job, and expected to turn their hand to anything that came along.[64] American helicopters flew Canadian Wildlife Service biologists into the breeding grounds to gather eggs, which were rushed by jet across the continent to a specially built whooping crane breeding unit in Maryland. There they were cared for by a team that had developed its techniques and expertise by practicing on the related but abundant sandhill crane.[65]

---

[62] Dunlap, "Organization and Wildlife Preservation."

[63] The United States carried the main burden. It spent some $2,000,000 a year on crane research in this period against Canada's $50,000. Graham Cooch, author's interview, 5 June 1989.

[64] Galbreath, *Working for Wildlife*, chap. 6; David Butler and Don Merton, *The Black Robin: Saving the World's Most Endangered Bird* (Auckland: Oxford University Press, 1992).

[65] Dunlap, "Organization and Wildlife Preservation." Practicing with a related species became common. New Zealanders would practice on the South Island robin before trying captive breeding on the endangered black robin. Butler and Merton, *The Black Robin*, describes the project. See also Galbreath, *Working for Wildlife*, chap. 8.

The shift from emergency actions to save a few species to continuing programs for many was most pronounced in the United States, which eventually extended legal protection to all native species above the level of microscopic life. There, as nowhere else in the settler countries, this work became the focus of major environmental battles. Action elsewhere was not so sweeping, but the focus did shift from emblematic species to many, and from narrow action to increase the population to broader programs to save the systems on which they depended. Science's role was, in contrast to pest control work, to develop rather than criticize, and it was at the center of these efforts. It was not, however, all, and its circumstances point to limits on its usefulness. Some parts of the recovery programs depended directly on theory-driven science, but others required something more like natural history's attention to a mass of facts gathered by observation. In the end the critical element often turned out to be something else, the mass of practical knowledge accumulated by experience. Aviculturists' appeal to the Fish and Wildlife Service for whooper eggs to raise pointed out that breeding birds in captivity was a matter not of science but of technique. Nor was it always clear that sophisticated actions were the most important. While taking whooper eggs for captive breeding Canadian scientists also removed the infertile ones. This led the birds to lay more, giving the wild population a small but continuing boost that may have played the decisive role in its recovery. Certainly captive breeding, though it established a safety net, did not result in the reintroduction of cranes into the wild in substantial numbers.[66] Nor were resources, beyond some minimum, decisive. The New Zealand Wildlife Service managed, on what its American counterparts would have considered pocket change, to carry out successful programs, and the methods it worked out to safeguard offshore islands from introduced pests were quite sophisticated, if simple in their use.[67]

Differences in programs reflected differences in the nations' wealth, but also the roles that wildlife had in the society. In North America it was becoming a visible sign of the invisible grace of the natural world. The wolf was a totem, its presence the guarantee of ecosystem integrity and the possibility of the wilderness experience. In Canadian and U.S. parks there were wolf-howling junkets (which relied on the fact that wolves will respond even to bad imitation howls). Their pictures

[66] Dunlap, "Organization and Wildlife Preservation."

[67] Galbreath, *Working for Wildlife*, chap. 6; Butler and Merton, *The Black Robin*. On these projects see also A. F. J. Warren, "Management of Endangered Species: Policy Influences and Development in New Zealand" (research project for master of science degree, University of Canterbury, 1987), 17–38.

appeared on posters, mugs, sweatshirts, and cribbage boards. There
were programs to adopt a wild wolf.[68] In New Zealand native birds were
a national heritage and a treasure, but there was less feeling that the
primeval islands were a lost Eden, and where it was present the sentiment inhered, as in Australia, in the vegetation and the landscape.
Everywhere, though, endangered species appeared in a new context.
The issue was less saving a creature than preserving the workings of a
precious natural world.

Conclusion

The settler societies took in ecology as they had natural history, but not
in quite the same way. Both systems embodied authoritative knowledge,
and both were universal ideas that had local application. Taxonomy was
everywhere the same, the creatures in each land different. Every land
had ecosystems, but its own. The extreme case, perhaps, was Australia.
Graeme Caughley, who worked for the CSIRO for some years, claimed
that Australian ecosystems were so different that a biologist from abroad
needed a couple of years simply to become familiar with how they
worked.[69] The production of theory and the training of experts were
even more centralized than they had been, and the participation of the
smaller societies was limited. The smaller countries worked out institutional answers. The CSIRO and DSIR were not simply modern analogs
of the "birds of passage" that had marked colonial science in the nineteenth century but solutions to the need to get scientific knowledge
under these circumstances.

   Ecology, though, was not a complete guide to living on the land. Its
theories were still new and the range of its expertise limited. Even when
it could provide important information and a new perspective, it was, as
Riney's experience shows, not decisive. But the fundamental problem
was that it did not, could not, say what ought to be done. That was the
next phase, one that came in the late 1960s. It was marked by an expan-

---

[68] Thomas R. Dunlap, *Saving America's Wildlife* (Princeton, N.J.: Princeton University Press,
   1988), chap. 10. Good wolves in literature begin with Seton, but the wolf's entry into
   popular culture on a grand scale dates from the environmental movement. This
   conclusion is based on the author's extensive, if casual, reading in nature catalogs
   and observing life in the United States. On wilderness as an environmental religion
   see William Cronon, "The Trouble with Wilderness," in William Cronon (editor),
   Uncommon Ground (New York: Norton, 1995), 61–90, reprinted in *Environmental
   History, 1* (January 1996), 7–28, and Thomas R. Dunlap's comments on Cronon's
   essay, "But What Did You Go Out into the Wilderness to See?" *Environmental History, 1*
   (January 1996), 43–6.
[69] Author's interview, 4 June 1992.

sion of interest. People sought to save not species but environments, and they began a fundamental reexamination of ideas about humans' relationship to nature. This, the impact of an environmental perspective on the defense of visible nature and its value to settler societies, is the last, or latest, chapter in the settlers' search for a place in the land and its place in them.

10

THE NEW WORLD OF NATURE
THE DEBATE TRANSFORMED

The knowledge of how, and how closely, we are connected to the world around us and what human population and technology are doing to it form the background of the modern concern with nature in the settler lands. It demands a new kind of action and regards nature in a new way. All is to be saved. Areas, plants, and animals that earlier generations disregarded or took for granted have become precious elements of national heritage or essential elements of a system that have to be saved whole or not at all. The remnants of wild land have become places of spiritual retreat, national treasures, or benchmarks of a healthy state of the land. Ideas have changed as well. "Progress" and "development," which were once unalloyed goods and the basis of settler dreams and economies, have come under attack as dangerous illusions. We have in the past forty years begun a new kind of conversation about, and with, the land. This chapter deals with that, the beginning of what promises to be a transformation in settler ideas and action.

Each settler society took its own course in adapting environmental ideas as it adopted them, for each now had a considerable history, but the process ran roughly along common lines. In each there was a national movement, a coalition of interests and groups whose agenda became part of the political landscape. Everywhere it drew for support on the educated, on professionals – whose children were the most prominent protestors in the early phase. Science was central to the arguments. At times it was more useful as metaphor and authority than as knowledge (ecologists often did not have the detailed information needed to direct policy), and it did not always shape decisions, but it was the ground of discussion. Each country saw roughly the same stages at the same time; environmentalism was, even more than earlier responses to nature, a common Anglo movement. In the early 1960s environmentalists organized. Late in the decade, riding the wave of public concern, they demanded action, and in the next few years new laws were passed and new agencies established to administer them. By the mid-1970s mass public demonstrations were fading, but the environmentalist defense of nature continued to transform activists, policy, and the public. We are still grappling with the implications of the insights of ecology, learning to see the world and ourselves in its perspective.

Environmentalism, like the science of ecology that supported it, had its start and early center in the United States. This was not a tribute to American wisdom. It was the result of American circumstances. That country had the largest population and biggest industrial plant, which meant its destruction of nature was the most extensive and visible. It also had the largest scientific community, which meant that destruction was analyzed. In addition, nature had been for more than a century an important part of American nationalism and culture, and that gave the American debates a rich context. As a result, the United States was, particularly in the early years, a source of information and ideas. As a New Zealand textbook of environmental law put it, American legal decisions were, for New Zealand courts, "of persuasive value only," but there was much to be learned from the "broad and definitive body of decisional law" that American environmental litigation had generated.[1] The same was true of every aspect of the movement, from administration to the value of wilderness. Persuasive value, though, was the Americans' largest contribution. The other societies, living in different lands, had to work out their own solutions.

A description of developments in even one country would occupy a book (or several). This chapter traces trends and suggests the range of variation. Many of the illustrations come from Australia, for the same reason Goldilocks settled on the little bear's chair, porridge, and bed; it was just right. The United States was too large and populous to be much affected by the other settler countries. To focus on it would be to miss much of the mixture of national sentiment and common Anglo ideas that, even more than earlier reactions to nature, marks the environmental debate over the defense and value of visible nature. Canada's cross-border problems and connections obscure its distinctive national characteristics. New Zealand had too few people to support a strong internal dialogue. Australia had enough people and institutions to foster a strong local movement and local concerns, yet was not so large that Australians did not look abroad for experts and ideas.

---

[1] D. A. R. Williams, *Environmental Law* (Auckland: Butterworth of New Zealand, 1980), ix. Appeal to the United States was not new; the legal opinion upholding the Migratory Bird Treaty in Canada appealed to the American decision *Holland* v. *U.S.* See Boxes 1–3, Acts and Legislation, RG 109, Records of the Canadian Wildlife Service, Public Archives Canada, Ottawa.

Organized for Action

[Environmentalists are] subversives who, for personal gain or a lust for power, are desirous of breaking down what is left of our "free enterprise" system entirely; those latter people, whose numbers are swelled by a great mass of unwashed, unspanked, dole-bludging dropouts, are threatening the lives and fortunes of the Australian community as it has never been threatened before.

Australian, 1978[2]

This angry Australian mine owner – who had his counterparts elsewhere – was reacting to the protestors who were the environmental movement's foot soldiers in the early years. They were the oldest of the baby boomers, many of them enjoying the increased access to higher education that had come after World War II, and in addition to enthusiasm brought a new element to nature protection. Environmentalism, much more deeply than earlier campaigns for nature, was part of social reform. It marched with opposition to the Vietnam War and nuclear weapons, and its rhetoric included anticapitalism, several varieties of socialist thought, and local communitarian traditions. Activists often had "life-styles" that outraged their elders. They did not, of themselves, threaten "the lives and fortunes of the Australian community," but the mine owner's instinct was sound. The ideas behind the environmental movement were, and are, a fundamental challenge to the conventional wisdom of the settler societies.

Young activists stamped the movement with their energy, ideals, and rhetoric, but it was their parents who gave it political strength. Samuel P. Hays has characterized environmentalism in the United States as a part of the culture of consumption, and that description fits the other countries as well.[3] There were radical implications in its stands, but it was a conservative movement: educated, middle-class city or suburban dwellers seeking political action to preserve or better their lives. In the 1950s they had demanded improvements in the national parks and protested pollution. After *Silent Spring* they worried about pesticide residues, and they began to see all these problems in terms of the "environment." G. M. Bates, in his *Environmental Law in Australia*, spoke of "enforcing the public interest in the maintenance of environmental quality for both present and future generations. . . . [The law] mirrors

---

[2] *Australian, 10* (April 1978), quoted in J. M. Powell, *An Historical Geography of Modern Australia* (Cambridge University Press, 1988), 249.

[3] The following discussion draws on ideas in Samuel P. Hays, *Beauty, Health, and Permanence* (Cambridge University Press, 1987).

public acceptance that non-economic as well as economic factors are important in enhancing that 'quality of life' to which we all aspire."[4] He spoke for the vast majority of the group that gave the movement its political power.

Environmentalism was strongest in cities and in regions dominated by cities. (In the United States Hays found it more closely associated with well-established cities than with those that had sprung up since World War II.) Rural areas tended to be less supportive, and where the local economy depended upon primary production, opposition was open and sometimes fierce. This is as true of Australia as the American and Canadian West. The two self-governing states outside the Southeast – Queensland and Western Australia – lagged in environmental enthusiasm, and Queensland's premier in these years, Joh Bjelke-Peterson, achieved a stature among environmentalists something like James Watt's in the United States.[5] In each country, though, the distribution of power among levels of government shaped the movement. National policies were most important in New Zealand, which had no provinces or states, and the United States, which had a strong central government and a large federal public domain. The major organizations had offices in or a connection to the capital, and national policies quickly became the center of concern. In Canada and Australia constitutional limits on federal power meant the movement developed more at the provincial or state level. This was particularly true in Australia, for Canada's border with the United States and concern with the Arctic gave the federal government a strong role.[6]

Existing organizations and programs, the base on which the movement could build, also varied. In the United States there were national groups, most formed in the battles of the Progressive period or the New Deal years. In Australia and New Zealand the bushwalking clubs that were the main defenders of nature were a legacy of the 1920s. Organizational

---

[4] G. M. Bates, *Environmental Law in Australia* (Sydney: Butterworths, 1992), 6.
[5] Hays's *Beauty, Health, and Permanence* provides a regional analysis of the United States. On Queensland see Judith Wright, *The Coral Battleground* (Melbourne: Nelson, 1977); on Western Australia, K. R. Newby, "The Fitzgerald River National Park, Western Australia: Some Conservation Issues," in J. G. Mosley and J. Messer (editors), *Fighting for Wilderness* (Fontana: Australian Conservation Foundation, 1984), 83–95. Conclusions are necessarily tentative; a good history of environmental organization in any of the Commonwealth countries remains to be written.
[6] On constitutional differences see Bruce W. Hodgins et al. (editors), *Federalism in Canada and Australia* (Peterborough: Frost Centre, 1989); one of its chapters treats environmental matters: Donna Craig and Ben W. Boer, "Federalism and Environmental Law in Australia and Canada," 301–16.

activity in Canada is harder to measure, for Canadians often joined American organizations, and some groups, even large ones, were tied to American counterparts. The Canadian Audubon Society, for example, was not established as a separate organization with its own magazine until 1948.[7] Programs drew on ideas and proposals of a generation before and were framed in terms defined in the 1890s – national parks, outdoor recreation, and amenities. Interest, though, grew through the 1950s, and the number of groups and members rose more quickly in the next decade. In the late 1960s and early 1970s there was a virtual explosion of new organizations, and established groups like the Sierra Club doubled in size every few years. The rate of formation and the rise in membership slowed by the mid-1970s, but public interest did not flag.[8]

Because conservation and nature organizations were an established force in the United States before the environmental era, they provided the base for the new movement and helped lead it. Representatives of the Sierra Club, the Audubon Society, and the National Wildlife Federation testified before congressional committees and lobbied agencies.

---

[7] Harrison Lewis, "Lively: A History of the Canadian Wildlife Service," unpublished manuscript, RG 109, Records of the Canadian Wildlife Service, Public Archives Canada, Ottawa, 498.

[8] There is no accurate guide to the formation of private nonprofit groups – anyone with a mimeo or copying machine could form one, and many had only a short life – but a good rough indicator is the National Wildlife Federation's *Conservation Directory*, issued annually (Washington, D.C.: National Wildlife Federation). Victor Scheffer, *The Shaping of Environmentalism in America* (Seattle: University of Washington Press, 1991), 113, suggests a rate of growth in the 1960s and 1970s that is six times that of previous decades. John Dargavel, *Fashioning Australia's Forests* (Melbourne: Oxford University Press, 1995), 143–9, estimates there were less than fifty conservation organizations in the country in 1950 and sees a rapid growth similar to that shown here for the United States. The first register of Australian groups, the Australian Conservation Foundation's 1973 *Conservation Directory* (Melbourne: Australian Conservation Foundation, 1973), lists 206 organizations in the most populous state, New South Wales. Of those giving the date they were founded, 7 were in the 1940s, 20 in the 1950s, 13 in the first half of the 1960s, 33 in the second half of the decade, and 22 between 1971 and 1973. All lists are necessarily biased toward groups that survived, and the ACF's included many – the Catholic Bushwalking Club in New South Wales, for example – that had other than environmental aims, but it matches evidence from other sources. A more recent one is the Foundation's *Green Pages, 1991–1992* (Melbourne: Australian Conservation Foundation, 1992). On New Zealand see Records of Environmental and Conservation Groups of New Zealand, Inc., Record 92-071-1, 92-107-4, Turnbull Library, Wellington. Other rough guides are the publications and records of the environmental organizations listed in the card catalogs of the national libraries in Ottawa, Wellington, and Canberra and the state libraries of Queensland, New South Wales, Victoria, and Tasmania. On parallel developments in Great Britain see Philip Lowe and Jane Goyder, *Environmental Groups in Politics* (London: Allen & Unwin, 1983), 15–17.

Their magazines championed environmental issues and rallied people to the cause.[9] In turn the new issues reshaped them. The National Wildlife Federation, founded in 1936 as a sportsmen's organization, was by the early 1970s a coalition of hunters, bird-watchers, and people interested in humane treatment of animals. Making policies and leading the group required some diplomacy.[10] The Sierra Club, long associated with California's economic elite, was as late as the mid-1960s still quite traditional. "If, in the mid-1960s, three quarters of the members joined for the outings program and a only a quarter for the conservation program, by the early 1970s the ratio was reversed. Members were paying their dues to see an effective conservation program."[11] The club split over the aggressive tactics and programs of its executive director, David Brower, and in 1969 the board forced him to resign. The Audubon Society kept its disagreements out of the newspapers.

Elsewhere new leaders and new organizations were more prominent, but in Australia early environmental action was a mixture of old and new. The Australian Conservation Foundation, for example, had its origins in Prince Phillip's hope, expressed in 1964, that Australia would be a member of the World Wildlife Fund when he arrived on a royal visit the following year. Prime Minister Menzies promptly got a group of scientists and businessmen to form an organization that could apply for membership and the government provided 1,000 pounds in seed money. This was in the established Australian tradition of elite leadership (which coexists with a fierce strain of egalitarianism and contempt for authority). There were no public members, and the president was Sir Garfield Barwick, chief justice of Australia's High Court. Francis Ratcliffe, the senior scientist, saw it as a collection of wise men who would boost important research and advise ministers from behind the scenes.[12] This was reminiscent of the Boone and Crockett Club or the National Audubon Society early in the century. The social arrangements that supported this fusion of personal influence and official position had faded

---

[9] Michael P. Cohen, *The History of the Sierra Club* (San Francisco: Sierra Club, 1988); Frank Graham, Jr., *The Audubon Ark* (New York: Knopf, 1990); Thomas B. Allen, *Guardian of the Wild: The Story of the National Wildlife Federation, 1936–1986* (Bloomington: Indiana University Press, 1987).

[10] Author's interview with Thomas Kimball, President of NWF, 19 December 1979.

[11] Cohen, *History of Sierra Club*, 436. See also Allen, *Guardian of the Wild*. The author also relies here on the Kimball interview. Graham, *Audubon Ark*, gives information on that organization. The same is true for New Zealand, though less obviously. See, e.g., the Forest and Bird File, Record 92-107-2, Environmental and Conservation Organizations, Inc. Collection, Turnbull Library, Wellington.

[12] Box 5, Francis Ratcliffe Papers, Archival Collections, Australian National Library, Canberra.

in the United States. They were doing so in Australia even as the ACF was formed.

In Queensland activists were rallying a wider public constituency for nature protection. In 1962 they established the Wildlife Preservation Society of Queensland and took the unusual, and bold, step of publishing a magazine. Americans had been doing this since the turn of the century, but the Australian market had not supported this kind of thing. Crosbie Morrison, the great interwar nature popularizer, had edited *Wildlife* from 1938 to 1954, but it had survived on the publisher's goodwill.[13] The first years of the new *Wildlife* suggested that things were not much different. The editor begged for articles and scrambled for money to pay the printer; local enthusiasts persuaded shop owners to stock it; the staff found or built a distribution network; and everyone publicized the venture. The situation, though, was different. In 1966, the year the society hired Vincent Serventy as editor, circulation was around 2,000. That was also the year *The Great Extermination* appeared. In 1967 activists at the University of Queensland formed the Queensland Littoral Society to fight mining and other development on the Great Barrier Reef. By 1970 the magazine's circulation was 11,500.[14]

A clearer indication of what was going on was the analysis Dominic Serventy, Vincent's scientist brother, offered the ACF in 1970 in response to its proposal for a new magazine. He was, he said, opposed to a "high-quality" magazine. That was preaching to the converted. He thought the ACF should try to reach two other audiences, by other means. One was the people who made policy, a small group of government ministers and senior civil servants. They should be given something full of facts and documentation, for they were "immune to glossy magazines and nice pictures." The other group was the angry but disorganized general public. In Western Australia, he pointed out, it had been a mass movement of voters, not "the orthodox natural history and conservation societies" that forced creation of the Ministry of Conservation. To reach the people in the street the ACF should have a press bureau that would work to put environmental issues into newspapers, periodicals like *Australian Women's Weekly* and television programs like

---

[13] Graham Pizzey, *Crosbie Morrison* (Melbourne: Victoria Press, 1992), 199–203. On the American scene see Frank Luther Mott, *A History of American Magazines* (5 vols.; Cambridge, Mass.: Harvard University Press, 1930–68).

[14] Figure from memo of 9 September 70, Folder 103, Box 13, Serventy Collection, National Library, Canberra. See Boxes 12, 13, and 16 for records of the magazine from its founding. Materials are scattered. See also Box 82, MS 11437, Papers of the Royal Australian Ornithological Union, State Library, Victoria, Melbourne.

Four Corners and *Today Tonight*.[15] Serventy was pointing to a new social reality, the breakdown of the political and social order that had supported saving nature as an amenity within an industrial society and economy. A network of groups and interested individuals (some of them scientists), representing an interested segment of the population, had sought political action to preserve areas or species. A mass movement was developing that would demand that nature be a central concern and that its need affect, if not guide, economic development. This exaggerates the break between the old and the new, but only by a little. The change that was coming – the one we are now going through – involved a fundamental shift in perspective and values.

Starting with Species

The new view of nature in terms of processes and relationships did not change policies or even approaches overnight. The kind of change environmental ideas brought was apparent in the change in endangered species protection. This was an issue that concerned nature's pieces, and it drew on existing laws and agencies, but in the first years of the public crusade the new perspective reshaped legal protection. Things went furthest in the United States, where a coalition of nature lovers, environmentalists, and those seeking humane treatment of animals lobbied Congress for successively stricter laws with a broader approach. Three acts, passed in 1966, 1969, and 1973, shifted policy and protection from a few species to all, and rather than just forbidding hunting and keeping people away from limited areas where the species rested or bred, these sought to preserve the ecosystem that supported the species and they (potentially at least) limited or outlawed many human activities over large areas. If they did not favor ecosystems over economic development, they at least weighed them against each other. The United States also worked for and ratified the 1973 Convention on International Trade in Endangered Species.

Elsewhere legislation was less comprehensive. Canadians were as interested in wildlife as Americans were, but they did not enact a comprehensive law, and while New Zealanders made saving their native birds a priority, the campaign was not the occasion for popular outrage or a catalyst for other environmental action. In Australia alarm about the larger kangaroo species did not lead to a general law, but it did play a role in mobilizing public sentiment and legislative action on the environment. The situation began with an increased kill of kangaroos in the

---

[15] D. L. Serventy to R. D. Piesse, 26 March 1970, Folder 19, Box 2, Serventy Collection, National Library, Canberra.

mid-1960s, a drought over large parts of the inland, and rising interest in wildlife and the environment. People had been killing kangaroos since the days of Captain Cook, and there had been sporadic attempts to regulate the killing since the nineteenth century, but now an industry seemed to be developing, a high-tech version of buffalo killing. Hunters were going out at night, shooting kangaroos by spotlight, and packing the carcasses in mobile coolers. The hides went for shoe leather, the meat for pet food.

The coalition that emerged to protest the killing included people who liked wildlife, others who advocated humane treatment of animals, and a much larger group that wanted to see kangaroos when they went for a Sunday drive or just to know the animals were out there.[16] The states took no action (Queensland took the same stance it had toward koala killing in the 1930s), and the protestors turned to the Commonwealth. Between March 1968 and May 1970 Parliament received eighty-four petitions protesting the commercial harvesting of kangaroos. That they received a sympathetic hearing shows how much had changed. In the late 1940s the Commonwealth had refused to help the states save the koalas, saying it was a matter for the states. Now speakers rose in Parliament to point with alarm. "We all know," one member said, "what happened to the koala bear. We were very lucky that the slaughter of the koala was stopped in time. We were also fortunate to save the platypus."[17] Another member, E. C. M. Fox, used the uproar to press for a survey of native wildlife and its ecology as a guide to saving it.[18] In the United States the passenger pigeon had become extinct, and the buffalo had been reduced to a remnant. Was Australian wildlife to suffer the same fate? There was little to save it. New South Wales, he pointed out, had only two people to enforce the Fauna Protection Act and only six to

[16] Boxes 1822, 1823 Australian Conservation Foundation Papers, State Library of Victoria, Melbourne.

[17] The first extended comments on the subject in the Australian Parliament, House of Representatives, appear in the *Hansard*, v. 56 (26th Parliament, 1st session), 20, September 1967. Hereafter cited as *Hansard*.

[18] On kangaroos see H.J. Frith and J. H. Calaby, *Kangaroos* (Melbourne: F. W. Cheshire, 1969), 161–9, 185–97; on the protests, Parliament, House of Representatives, Select Committee on Wildlife Conservation, *Wildlife Conservation*, Report from the House of Representatives Select Committee (Canberra: Government Publishing Service, 1972), 7 (hereafter Fox Report), and Box 1822, Australian Conservation Foundation Papers, State Library of Victoria, Melbourne. Fox apparently made his interests well known. In March 1968, the attorney general, Mr. Bowen, replied to a question about the possibility of federal wildlife protection by saying that he was "aware of the honourable member's interest in the subject, some correspondence has passed between us." *Hansard*, 58, 13 March 1968. See also Folder 308, Box 36, Serventy Collection, National Library, Canberra.

oversee laws protecting animals, wild flowers, and native plants – this for an area of 309,443 square miles. Western Australia had seven wardens and a cadet to cover 1,000,000 square miles. Fox called for a meeting of state ministers, an embargo on exports of kangaroo products, the establishment of a national wildlife agency, and education for all on the value of "our wonderful heritage of native flora and fauna."[19] He suggested a study of the adequacy of existing reserves to protect wildlife, an investigation of the impact of pollution, pesticide contamination, and feral animals on native species, and a review of the Commonwealth's role in wildlife protection.

The government agreed, as did the opposition, and in May 1970 the House appointed a seven-person Select Committee on Wildlife Conservation, commonly called the Fox committee.[20] By now the environment was a political issue across the Anglo world. A month earlier 250,000 students, demonstrators, and interested citizens had rallied in Washington, D.C., for the first Earth Day, and schools and towns across the United States had put on their own programs. Canadians held similar meetings. In New Zealand local groups would, within a year, form a national coalition around the campaign to save Lake Manapouri from hydroelectric development. In Australia local and state groups had organized to fight beach mining, clear-felling (in North American English, clear-cutting), and woodchipping (turning entire forests in small pieces of wood for processing). The threatened flooding of Lake Pedder in the Tasmanian mountains was becoming a national issue.[21]

The Fox committee's report, delivered in October 1972, fit this militant mood. Though it built on early measures for nature protection, it charted a new course. It proposed studies on everything from park expansion to the environmental impact of industry and agriculture. In the meantime, it said, Australia should follow the Americans' lead in reg-

---

[19] *Hansard, 58*, 2 May 1968.

[20] *Hansard, 67* (27th Parliament, 1st session), 14 and 15 May 1970, 2036–7, 2227–8.

[21] On North America there are a number of histories and a mass of material. On New Zealand, Roger Wilson, *From Manapouri to Aramoana: The Battle for New Zealand's Environment* (Wellington: Earthworks, 1982), provides what is probably the most available account outside the country. See also David Thom, *Heritage: The Parks of the People* (Auckland: Lansdowne Press, 1987), 187–9. On Australia the popular literature shows trends. A set of comments, including information on school projects following Earth Day, is in Folder 312, Box 36, Serventy Collection, National Library, Canberra. An early report on woodchipping and its potential is K. W. Cremer, "Effects on Forestry of the Increasing Demand for Wood Chips," *Newsletter*, Institute of Foresters of Australia (December 1968), 9–13. Copy in Folder 1, Box 1, Judith Wright Papers, National Library of Australia, Canberra. See other materials in this folder as well. Woodchipping continued to be a subject of interest to environmentalists. See material in Folder "Corresp. 1980," Box 51, Serventy Collection.

ulating DDT, saving endangered species, and protecting parks. It called for a national policy of protecting all habitat areas by including them in national (i.e., state) parks, with Commonwealth grants to help the states acquire areas "which are of national significance." It laid out a new role for the Commonwealth. Australia was the only technically advanced nation in the world without a national institute for faunal studies. There should be a biological survey office charged with collecting the now-scattered federal specimen collections, "undertaking on a continuing basis surveys of birds, mammals, and reptiles and their ecology and . . . establishing a national collection of wildlife species." Australia also needed a parks and wildlife agency to administer preservation programs in Commonwealth Territories and take responsibility for international treaties on migratory and endangered species. It would also, "in cooperation with the States," survey park needs and wildlife populations, set national guidelines, and monitor endangered species.[22]

The states were already taking action. Between 1966 and 1977 all but Victoria (which overhauled its agencies) created new, comprehensive parks and wildlife bureaus. Earlier legislation had focused on scenery or particular species and put parks where no one else wanted the land. Now laws spoke of a "national heritage," called for habitat preservation, and gave the parks responsibility for preserving examples of all natural areas within the state. Parliament joined in, creating the Australian National Parks and Wildlife Service (now the Australian Conservation Agency), the Biological Resources Study, and the Council of Nature Conservation Ministers. It did not, though, set up a national program. The park agency could only advise the states, and the Council, composed of the heads of the Commonwealth, state, and territorial wildlife and conservation authorities, was even more directly a mechanism for coordinating state efforts rather than launching federal ones. The limits of federal action were clearest in the issue that triggered this action, the killing of kangaroos. The Fox committee proposed bold plans for federal action on other fronts, but here gingerly suggested that the Commonwealth "approach" the states "with a view to obtaining greater uniformity of laws relating to the taking of kangaroos."[23] Things have since gone slowly.

---

[22] Fox Report, 1, 1, 3.

[23] Fox Report, 3. Australian Environmental Council, *Guide to Environmental Legislation and Administrative Arrangements in Australia, Second Edition*, Report No. 18 (Canberra: Australian Government Publishing Service, 1986), 183–4, summarizes this shift. Analysis based on a reading of state statutes, most from the National Library, Canberra, and Australian National Parks and Wildlife Service annual reports and its 1989 review, Department of the Arts, Sport, the Environment, Tourism and Territories, *Report on the Review of the Australian National Parks and Wildlife Service* (Canberra: Australian Government Printing Service, 1989).

The first thing needed was research, for even the distribution of the various species was poorly understood, and although the most common complaint about kangaroos was that they competed with sheep for pasture grass, the CSIRO reported that there had "been less work done on the food eaten by kangaroos than on most other aspects of their biology."[24] To remedy this the New South Wales Fish and Wildlife Service worked out techniques to count and monitor populations, then invited the CSIRO to participate in a study of kangaroo ecology in sheep grazing lands. Much was learned, but the 1988 Senate Select Committee on Animal Welfare called for more resources for these projects, and it did not seek Commonwealth authority over kangaroos.[25]

Going to the Land

Wildlife was important for the growing environmental defense of nature, but land was central. This was a matter partly of human perception, partly of law. Wildlife was interesting but elusive. The land was always there, and it appealed to the senses in a variety of ways and offered various pleasures. People could walk over it, admire views, look for wildlife, and smell the flowers. They could do all this with friends. Photographs (the modern equivalent of landscape painting) could spread to a wide public a vision of nature as Eden or spectacle. Land was also built into society in ways animals were not. The latter were *ferae naturae*, no one's property, or the property of the sovereign until reduced to individual possession by killing in a lawful manner. Land was "real estate," a term that testified to its importance and solidity. Wildlife law was thin, land law complex.

For environmentalists, the important point was that land law included a century of precedent, categories like "national park" and "nature reserve," which could be adapted for ecosystem protection. This use was most obvious in Australia, where ideas about noneconomic uses of land had not been well developed and saving wild lands was a key element in the environmental defense of nature. The controversy over development in Victoria's Little Desert in the late 1960s shows the early stages.[26] The

---

[24] Frith and Calaby, *Kangaroos*, 9, 74, 83, 51–5.

[25] On the ecology of these inland pastures see Graeme Caughley, "Ecological Relationships," in Graeme Caughley, Neil Shepherd, and Jeff Short (editors), *Kangaroos: Their Ecology and Management in the Sheep Rangelands of Australia* (Cambridge University Press, 1987), 159–87. The rest of the volume describes the project. On later work see Parliament of Australia, Senate Select Committee on Animal Welfare, *Kangaroos* (Canberra: Australian Government Publishing Service, 1988).

[26] This account draws heavily on Libby Robin, "Of Desert and Watershed: The Rise of Ecological Consciousness in Victoria, Australia," in Michael Shortland (editor), *Science and Nature* (Oxford: British Society for the History of Science, 1993), 115–49.

minister of lands, William MacDonald, could hardly have expected that plans to subdivide and settle this area would cause any concern. The state had carried out many such projects since World War II, and the Little Desert seemed no different. It was not particularly attractive, being covered by mallee, a scrubby plant that put up many stems from a single base, and lacking in conventional "views." It was the home of the malleefowl, a turkey-like bird that built nest mounds to incubate its eggs, but natural history curiosities had never been an issue. Protest, though, was immediate and widespread, and it had new elements and arguments. In her history of this debate Libby Robin pointed out that the "motto of the Victorian National Parks Association, established in 1952, [was] 'For all the people, for all time.'" Now there was more "prominence given to scientific values of National Parks."[27] Marshall's *The Great Extermination* was "undoubtedly important in giving the scientific values of National Parks a high profile. . . . Following soon after the internationally acclaimed *Silent Spring* (1963 to Australian audiences), [its] local content made it particularly relevant and disturbing. The book was read and cited by active conservationists."[28]

Scientists, rather than lobbying discreetly from behind the scenes, took an active role, working across disciplinary lines and crossing between academia and government. One ecologist at Monash University in Melbourne "actively involved senior scientifically trained resource managers in the teaching of his honours students. The lectures given by these public servants were responsible for the enthusiasm of one group of young activists, Monash University Biological Students Society, about the Little Desert issue."[29] Robin attributes the scientists' activity to the small size of their communities in Australia. American ecologists could retreat to the laboratory. There were too few Australian scientists for the professional community to serve as refuge. One result was "unexpected alliances. If there are few experts, then barriers between university and other scientists cannot be maintained."[30] This was a factor, but not a complete explanation. Everywhere the sense of crisis and the recognition that ecological problems cut across disciplinary boundaries were breaking down walls between fields and between scientists and the public. Within the scientific community there were calls for professional action, and Paul Ehrlich, Barry Commoner, and Garret Hardin were becoming public figures. In 1962 few scientists had been willing to speak out in

---

[27] Ibid., 137, 136.   [28] Ibid., 137, 138.

[29] Ibid., 141, 137–42. For a similar phenomenon in the United States, see Thomas R. Dunlap, *DDT: Scientists, Citizens, and Public Policy* (Princeton, N.J.: Princeton University Press, 1981), 156.

[30] Robin, "Of Desert and Watershed," 142.

Carson's defense. Seven years later the group seeking to ban DDT in Wisconsin had more experts willing, and wanting, to testify than it could use.[31]

Protests over the Little Desert showed that national parks were becoming a much more important issue. They had once been the concern of small groups. Delegations from bushwalking clubs had "waited upon" the minister and "proposed" plans. Now a coalition of bushwalking and conservation societies, the Save Our Bushlands Action Committee, held public meetings and rallies. MacDonald offered to compromise; the area least suitable for farming would be made into a park. That activists might oppose this was to be expected. That legislators did was more surprising – and an indication that nature was an issue important enough for political maneuvering. In October 1969 the Opposition in the Legislative Council (the upper house of the legislature) formed a committee of inquiry, and in December the Council refused to fund development of the area. Premier Bolte spoke about conservation and promised a new land resources council (which was ultimately established as the Victorian Land Conservation Council). In the election of 1970 the government lost two seats. One of them was MacDonald's.[32]

Leadership, though, remained what it had been in the early postwar years. Bushwalking clubs were central, and the head of the Save our Bushlands Action Committee was a former chairman of the State Soil Conservation Authority and president of the National Resources Conservation League. The people in charge were established figures, and "it was their contemporaries, and in some cases friends, who held the reins of power within the government." The "young activists . . . were conspicuous by their absence."[33] An upheaval in the Australian Conservation Foundation showed that this genteel, "top-down" tradition was passing. By 1970 most of the organizers had come to believe that the organization could not function as a small, self-selected group overseeing research, and they set about changing the ACF into a national group with a public membership and the mission of educating people about the environment. Ratcliffe, charging that it had strayed from its original purpose, resigned.[34] In October 1973 dissidents among this expanded membership used parliamentary procedures to stall a general meeting until their opponents left, put their own people on the Council, deposed

---

[31] Dunlap, *DDT*, 155–7; author's interview with Victor Yannacone, 10 December 1973.
[32] Robin, "Of Desert and Watershed," 125–9.
[33] Ibid., 133.
[34] Box 5, Ratcliffe Papers, National Library, Canberra.

Sir Garfield Barwick, and elected a new director. Seven scientists on the ACF Council resigned in protest and sent a report to the membership, "How the ACF Was Taken Over."[35]

If the Little Desert was the first stage, the battle to save Lake Pedder was the second. It was more visible, and national, and it provided a rallying point that raised consciousness and helped organize the movement. A small mountain lake on an island off the continent's south coast seems an unlikely focal point. Tasmania was the country's smallest state in both area and population. It had, though, things that were rare in the rest of the nation – rugged mountains, dense forests, and wild rivers – and it was only a short plane ride from major population centers. It had in the postwar years become a popular nature vacation spot for mainlanders.[36] The lake itself was a gem. Photographs, which played a major role in publicizing it, showed a small blue lake, its shore sculpted by wind and wave, on a grassy plain against a backdrop of rugged mountains.[37]

The groundwork for the battle was laid in the 1950s when a coalition of Tasmania's bushwalking clubs mounted a successful campaign to have Lake Pedder declared a national park. In the early 1960s, with interest in the area growing, they formed the South-West Committee to lobby for the park's expansion. The state's Animals and Birds Protection Board joined, asking that the Southwest be declared a faunal reserve. It was, in 1966, but that year the state also began a study of the region's hydroelectric potential. The investigating committee included representatives from the Hydro-Electric Commission and three departments concerned with development – Lands and Surveys, Mines, and Forestry. There were none from the Scenery Preservation Board, Tasmania's park agency, and that body seems not to have had a chance to comment on the committee's report before it was sent to Parliament. These arrangements may have been intended to avoid controversy, or administrators may have assumed there would be no objections. Tasmania had for years pinned

---

[35] Minutes of meeting and pamphlet in Box 1823, Australian Conservation Foundation Papers, State Library of Victoria, Melbourne. Dr. Patricia Mather, one of the resigning members, provided details on the meeting and the scientists' protests (author's interview, 5 August 1990).

[36] On Lake Pedder see Dick Johnson, *Lake Pedder* (Lake Pedder Action Committee of Victoria and Tasmania and the Australian Union of Students, 1972), and records of the Tasmanian Wilderness Society, now the Wilderness Society, in the State Library of Victoria, Melbourne. A useful retrospective on tactics is Bob Brown, "Wilderness v. Hydro-Electricity in South West Tasmania," in Mosley and Messer, *Fighting for Wilderness*, 59–68.

[37] On the importance of photographs see Brown, "Wilderness v. Hydro-Electricity," 60–1. An American comparison would be the Sierra Club's use of photos as a reminder of what was lost when the Glen Canyon Dam was built.

its hopes for prosperity on cheap electricity, and development had always taken precedence over parks. However, when it came out that the planned dam would flood Lake Pedder and the plain around it, opponents formed a Save Lake Pedder National Park Committee and gathered 10,000 signatures (from a population of roughly 350,000) on a petition opposing the dam. As plans went forward opposition grew, and there were protests and demonstrations not only in Hobart but in Melbourne and Sydney.

The ensuing battle was another mix, not quite the same as that over Little Desert. The protestors took a conventional stand, though one not often honored in Australia. Parks were to be inviolate and valued for things other than money. Here the arguments sound much like those John Muir had made for the Hetch-Hetchy valley sixty years before (though with less religious imagery). There was, though, new language, reflecting new concerns. In 1972, as defeat loomed, one advocate asked, "If a national park as important as Lake Pedder can be sacrificed where in Australia is the environment safe from destruction?"[38] The cause was mixed with popular social reforms. Activists also protested the Vietnam War and nuclear weapons, and their program had ideological strains that ranged from conventionally Marxist to ideals of community based on small-scale production. These had not usually been associated with the defense of national parks.[39] Parks had been defended as places around which progress might flow, islands in the stream of development. Now development itself came under criticism. The Hydro-Electric Commission, which had been the herald of prosperity and the vanguard of progress, found itself attacked as "the electric Kremlin," making secret deals that destroyed Australians' national heritage.[40] The leaders now comprised a different group. There were fewer of the older generation, who could appeal to "their contemporaries, and in some cases friends, who held the reins of power within the government."[41] Younger people were conspicuous.

The park's advocates lost, which was not unusual. But while American reaction, in both the Hetch-Hetchy battles around 1910 and the fight over Echo Park in the early 1950s, had been to redouble lobbying for parks and wild nature, the Australians took another tack.[42] In late

---

[38] Johnson, *Lake Pedder*, 15.

[39] Records of the Tasmanian Wilderness Society, now the Wilderness Society, in the State Library of Victoria, Melbourne, are useful here.

[40] Ibid.

[41] Robin, "Of Desert and Watershed," 133.

[42] On the Echo Park battle, the great postwar dam fight before environmentalism changed the rules, see Mark Harvey, *Symbol of Wilderness* (Albuquerque: University of New Mexico Press, 1994).

1972, as the growing reservoir submerged Lake Pedder, the shore, and the plain beyond, its defenders formed the United Tasmania Group and entered electoral politics with a program to restructure the island's economy and society to take account of what they saw as ecological realities.[43] This, and a party formed in New Zealand in the wake of the Lake Manapouri fight (a group that, like the Tasmanian one, claimed to be the first Green political party in the world), were visible evidence that the defense of nature was becoming part of a larger attempt to fit humans into the world of nature.

This would require fundamental social and economic change, and questions about the extent, nature, and speed of change have become the subtext of environmental battles, or the text itself. Forest policy has been one of the most hotly contested areas. Not only are forests an important part of the land and humans' ideas of it, they are, almost alone among environments people care deeply about, economic resources that humans use and often (particularly under industrial use) destroy. As concern in this area changed from general interest and the defense of a few spectacular types (such as the sequoias of California) to campaigns for the preservation of all forest types, the stakes grew, and they were increased by the coincidence of greater pressure on forests and greater public interest in preserving them. Often they involved the older and more spectacular areas, what became known as old growth and ancient forests (both appear in the various literatures). This interest was hardly surprising in New Zealand, where native forests had long been a center of preservation interest, or the United States, where forests had been a symbol of the primeval since the beginning of the frontier myth, but it appeared even in Australia, where forests had been much less a part of national concern for nature. At the turn of the century the settlers had made the eucalyptus a national symbol, but they had not sought to save the forests. Now they did.

The campaigns drew public interest from rising environmental sentiment but also because the world market and modern technology were taking a large and obvious toll. In the early 1960s the expansion of the Japanese economy and new technologies that made it profitable to process eucalyptus brought firms to Australia seeking contracts for woodchipping. These operations involved cutting everything over a large area and reducing it to small pieces of wood, which were processed or exported for processing. Even people who were not concerned about

---

[43] Boxes 2669–71, Records of the Tasmanian Wilderness Society, now the Wilderness Society, in the State Library of Victoria, Melbourne, and the Wilderness Society's *Wilderness News*. Partial collections are in the Melbourne records and the Tasmanian National Parks and Wildlife Agency Library, Hobart.

the environment or proud of their country's unique biological heritage were dismayed by denuded hillsides and slash piles. The concentration of jobs in large firms and the use of new technology displaced many rural people who depended on relatively low-tech labor in the woods for part or all of their income.[44]

Besides forest preservation and environmental considerations there was wilderness. Australians had seen the country as bush and had gone to it for mateship. Now they were finding (or admitting they found) something else. In 1974 Dick Johnson, a veteran of the Lake Pedder battle, appealed to wilderness in calling for a Victorian alpine park (part of the plans from the Dunphy era for mountain reserves). He spoke of its "essential utility [which] lies in its emotional and spiritual attraction to man." In examining the "characteristics of untouched land" he leaned heavily, he admitted, "on the American experience, since it is in this country that the wilderness idea has been most effectively articulated." He was quite clear as to why Australians had not developed the concept.

> The crucial spiritual components of wilderness have been ignored in that horror of emotional articulation which is the national gaucherie. In attempts to justify our stance we grope about for utilitarian explanations which simply don't exist. And for want of someone to say that Australians love their wild places, we stand about in embarrassed silence while our masters rip the living guts from the wilderness that is left.[45]

Embarrassed silence, though, had had its day.

The Politics of Winning Wilderness

> The Franklin issue was not only on the political agenda – it has dominated the political life of Tasmania for almost three years and in late '82–'83 became a major issue at the Federal level. It has been argued that the campaign has fundamentally changed the politics of winning wilderness.
>
> Penny Figgis, 1983[46]

---

[44] Dargavel, *Fashioning Australia's Forests*, 83–110. A discussion of the social aspects of these battles, the pitting of rural forest workers against urban environmentalists, is Ian Watson's *Fighting over the Forests* (Sydney: Allen & Unwin, 1990).

[45] Johnson, *Alps*, 16, 19. See also Judith Wright McKinney, "The Individual in a New Environmental Age," *Australian Journal of Social Issues, 8* (1973); copy in Folder 310, Box 36, Serventy Collection, National Library, Canberra.

[46] Penny Figgis, "Out of the Wilderness for the Wilderness Issue?" in Mosley and Messer, *Fighting for Wilderness*, 228.

The fight for the Franklin River, a second phase of the defense of Tasmanian wild lands, brought Australian environmentalism to a new level.[47] Like the battle over Lake Pedder, this one was waged over a proposed dam, but where discussions of Pedder had focused, overtly at least, on conventional topics like scenery and parks, this one moved wilderness to the center, and advocates spoke more openly and clearly about its value and place in society. Lake Pedder had seen some protests on the mainland, but it had been more a Tasmanian battle. Before the Franklin River controversy was settled, national politicians had taken stands and the issue had twice come before the High Court, where the justices considered the Commonwealth's responsibilities and Australia's adherence to UNESCO's standards for World Heritage Sites.

The Southwest was the center of Tasmanian wilderness. It had been, from the early days of settlement, a place to avoid. It was steep and rugged country, and the forests, nourished by the abundant rain brought by the trade winds, dense and tangled. There were no good harbors along the coast, and the rivers were narrow and rocky. There had been some lumbering but little settlement, and the area was not covered by the state's electric power grid (and still is not). Even bushwalkers had hardly penetrated much of it. The first Anglos to see the full extent of the Franklin River were a group who took rubber boats down it in 1976, the year the Tasmanian Wilderness Society was formed. The river became an issue in 1979, when the government proclaimed the South West Conservation Area and the Hydro-Electric Commission recommended building a dam on the Gordon River. The reservoir behind it would extend up the lower Franklin, and activists immediately protested. The Franklin was the longest wild river left on the island. Mainlanders quickly joined in; Sydneysiders formed the South-West Tasmania Committee, and Victorians a similar committee in Melbourne. When the Tasmanian group called for a national march, committees were organized for Hobart, northwestern and western Tasmania, New South Wales, Victoria, South Australia, Western Australia, and Canberra. The ACF assigned a full-time staff member to the campaign.

The government suggested shifting the dam site to a point on the Gordon upstream of the Franklin's confluence. The Legislative Council (the upper house of the state legislature) balked, and in December 1981 the state held a referendum to choose between the plans. Environmentalists wanted a "no-dam" choice on the ballot, and when they did not get it they organized a write-in campaign that drew more than a third

---

[47] On Australian wilderness sentiment compare the papers in Mosley and Messer, *Fighting for Wilderness*, with a collection from the Sierra Club conferences, William Schwartz (editor), *Voices for the Wilderness* (New York: Ballantine, 1969).

of the vote.[48] The government revived the original scheme; the Wilderness Society took to the streets. Rallies across the state featured the campaign's mascot, a thirty-foot, inflatable platypus, "Franklin." (The platypus later became the society's emblem.) A Labour Party conference at a resort along the Derwent River was the occasion for guerrilla theater. A car pulled up and five people wearing rubber masks of Tasmanian Premier Robin Gray got out. Later a figure in a suit and one of the masks, carrying a harpoon, took to a boat while "Franklin" rafted down the river. There was a spirited battle, and in the end the platypus swallowed the "premier" – to the delight of news photographers.[49]

In December 1982 protestors set out to block construction at the dam site by a nonviolent protest. The tactic seems to have been inspired by an incident at Terania Creek in December 1979. People opposing logging of the rain forest attempted to stop the construction of a logging road by nonviolent physical obstruction (the first such environmental protest in Australia). Bob Brown, one of the Tasmanian organizers, said that planning for the Franklin protest began in 1980, when it seemed likely that construction would proceed, and was stepped up after the vote. The ideology was international – fliers quoted Ghandi and Martin Luther King – but everything else was local.[50] Organizers were thorough. To avoid unfavorable publicity they warned protestors to avoid activities, like skinny-dipping, that might alienate residents. All volunteers had to attend a workshop on nonviolent resistance before they were sent out.[51] The indoctrination seemed to have taken hold. The protestors swamped the Tasmanian prison system but cooperated with one another and the authorities.[52] The action was less successful, at least in its immediate objective. It did not stop construction. However, the publicity it generated helped make the Franklin a national issue.

[48] Newsletter of the Tasmanian Wilderness Society, Tasmanian Fish and Wildlife Agency Library. A summary can be found in Geoff Law, "The 1980s: A Great Decade for Wilderness," *Wilderness*, *11* (March 1990), 5–6. Various sources report different percentages of the "no-dam" vote, but 35–45% is the range.

[49] Wilderness Society, *Newsletter*, 7 (August 1986). Geoff Law, who played Gray, found wearing a rubber mask over a beard while doing hard exercise quite uncomfortable. Author's interview, 17 July 1990.

[50] Bob Brown, "Wilderness v. Hydro-Electricity," 64. On Terania Creek see Dargavel, *Fashioning Australia's Forests*, 184–5. The ongoing clash over forest policy in the North American Pacific Northwest illustrates these same themes, including the class divisions and the dominant role of industrial corporations.

[51] Law interview, Boxes 2669–71, Tasmanian Wilderness Society Records, State Library of Victoria, Melbourne. Brown, "Wilderness v. Hydro-Electricity," also stresses this point.

[52] David Biles and John Howe, "Tasmania and the 'Greenies': Research Note on Prison Crowding," *Australia and New Zealand Journal of Criminology*, *17* (March 1984), 41–8; copy in Box 2671, Tasmanian Wilderness Society Records, Melbourne.

What also helped was that the political climate had changed. The Commonwealth was willing to intervene, and it had ways to do it. In 1975 the Australian National Commission Act had set up the Register of the National Estate, and in 1980 the Commission had placed the South West Conservation Area on it. The next year the three Western Tasmanian Wilderness National Parks were nominated for World Heritage listing, a designation developed by UNESCO to give protection to natural and cultural treasures around the world. When the Tasmanian Parliament passed the dam act in 1982 it asked the Commonwealth to withdraw the nominations. The Commonwealth refused – something it would not have done a decade before, when development was still an unchallenged good. The areas were listed just as the blockade began.

Federal interest, though, was no guarantee of federal supremacy. The Australian constitution, written to safeguard state interests, put severe limits on what the government of the whole could do in one part, and it was not clear how a declaration of heritage rights or a public interest in the environment would affect these. The Franklin controversy promised to drag them all in; a textbook on Australian natural resources law began its chapter "The Conservation of South-West Tasmania" with a declaration: "Many of the constitutional impediments to Commonwealth involvement in conservation may be examined by reference to the proposal to develop areas of wilderness in Tasmania for hydro-electric generation."[53] Legal impediments have commonly been the occasion for legal ingenuity, and this was no exception. The Wilderness Society (in the course of the fight for the Franklin, it dropped "Tasmanian" from its name) came forward to argue that the Australian Loan Council's granting of funds for construction of the dam was illegal. The Council, as an agency of the Commonwealth, was by this action helping to destroy the Southwest wilderness and so violating the National Heritage Act and Australia's ratification of the UN World Heritage Convention. This was not a terribly far-fetched claim as such things go – recall Congress's appeal to its powers over interstate commerce to set water fowl hunting seasons – but the High Court was not sympathetic.[54] It rejected the Wilderness Society's claims in every respect. The next year the issue came back, and "the legal background had changed."[55] The Southwest parks were now on the World Heritage list; the governor-general had proclaimed regulations for their management as a World Heritage Site; and the Commonwealth had enacted the World Heritage Properties Conservation Act. Protecting the site had become part of the

---

[53] D. E. Fisher, *Natural Resources Law in Australia* (Sydney: Law Book Co., 1987), 223.
[54] On the basis of Australian federal power see Bates, *Environmental Law in Australia*, 54–66.
[55] Fisher, *Natural Resources Law*, 224–6; quote on 226.

country's international obligations. Public sentiment, too, had changed, as shown by Prime Minister Bob Hawke's coming out in public against the dam. The High Court now ruled that construction of the dam required Commonwealth consent. This was a victory for environmental values, but a decidedly ambiguous one. The justices split four to three and reached their decision only after a lengthy analysis of constitutional language on external powers, corporations, and the Commonwealth's power to make laws for people of any race (since the case involved the flooding of Aboriginal sites). The decision, as the textbook on natural resources law put it, was a "long and complicated series of judgments [that] rendered valid in certain respects the refusal of the Common- wealth Government to consent to the development."[56]

Australian interest in wilderness ran parallel to a general concern with saving or restoring areas that were not visibly affected by humans, or at least by industrial civilization. The American interest in wild nature as spiritual refuge and evidence of pioneer history had fostered the first debates, and by the environmental era there was a history of thought and action stretching back some fifty years. The U.S. Forest Service's reser- vation of the Gila Wilderness in 1923 began an exploration of definitions and administrative strategies (which informed Dunphy's plans). The establishment of the Wilderness Society in 1935 provided a forum for dis- cussion. In the late 1940s the Sierra Club established regular conferences on the subject, and the Wilderness Society began the campaign that cul- minated in the 1964 Wilderness Act. The United States now has some 500 wilderness areas covering some 90 million acres. They are not part of a single system in the way the national parks are. About half are in the National Parks, and the Forest Service, the Fish and Wildlife Service, and the Bureau of Land Management administer the rest. Protection in the other countries is not as strong. Both Canada and New Zealand allow areas to be established or opened to development by administrative action, and in Australia the states, which retain power in this area, did not begin to explore the subject until the 1980s.[57]

---

[56] Ibid., 226–34; quote on 226.

[57] Things are hardly uniform. C. J. Mobbs (editor), *Nature Conservation Reserves in Aus- tralia*, Occasional Paper No. 19 (Canberra: Australian National Parks and Wildlife Service, 1989), lists thirty-four different names for reserves on land, and twelve for those in water or wetlands. See also summary from Land Conservation Council, *Wilderness: Special Investigation, Descriptive Report* (Melbourne: Land Conservation Council, 1990), 20. On New Zealand reserves see C. Bassett and K. H. Miers, "Scientific Reserves in State Forests," *Journal of the Royal Society of New Zealand, 14*, 1 (1984), 29–35, and Robert Cahn and Patricia L. Cahn, "Reorganizing Conservation Efforts in New Zealand," *Envi- ronment, 31* (April 1989), 18–20, 40–5.

American ideas and the country's legal example were influential. The Victorian Land Conservation Council, for example, declared that it "was the Americans who pursued the idea [of wilderness] and brought it to maturity," and as background to its work traced the intellectual line from Emerson and Thoreau to Aldo Leopold and Bob Marshall and the legal one from the Forest Service's work in the interwar period to the 1964 Wilderness Act.[58] Australians are reprising the U.S. Forest Service debates of the interwar period – Leopold's original proposals, for instance, incorporated pioneer ranching as a land use within wilderness areas – but in the context of Australian ideas about the land.[59] The initial focus was on bush wilderness, areas within the mental horizon of settlement. The Centre, in many ways not as well known or as thoroughly studied, was not part of the early debates. Here too scientists seem to have known things well before the public. John Calaby, with the Wild Life Survey Section, recalled forty years later the thrill of being the first to make scientific collections in areas of the Far North, near what is now Kakadu National Park, in the early 1950s.[60] In 1958 Alan Newsome, funded by the Division of Animal Industry of the Northern Territory, followed Finlayson and found more "evidence of the past history of the life of the country" slipping "into obscurity . . . [and more of] the original plan [being] lost."[61] Foxes and rabbits had taken over more of the land. That seems to be changing. Not only are some interior areas now protected as wilderness, Kakadu and the North have become part of the national agenda, and the Centre, Ayers Rock (Uluru), and Alice Springs have become national (tourist) symbols. Whether the Centre will become a part of the mental landscape, and what value Australians will give it, are open questions. Anglo Australia still hesitates before the vastness and harshness of the "great red landscapes, wide open to the splendour of the sun."

The legal wrangling over the Southwest showed another aspect of the new drive for nature protection. The law had an indispensable role, but

---

[58] Land Conservation Council, *Wilderness*, 12, 12–13.

[59] See, e.g., "The Wilderness and Its Place in Forest Recreational Policy" (1921) and "A Plea for Wilderness Hunting" (1924), reprinted in David E. Brown and Neil B. Carmony (editors), *Aldo Leopold's Southwest* (1990; reprinted Albuquerque: University of New Mexico Press, 1995), 146–63, and editor's note preceding them. Mosley and Messer, *Fighting for Wilderness*, particularly the senior editor's introduction, 1–6, offer a snapshot of sentiment in the mid-1980s. For more recent studies see Land Conservation Council, *Wilderness*.

[60] Author's interview, 2 June 1992.

[61] H. H. Finlayson, *The Red Centre* (Sydney: Angus & Robertson, 1935), 16; author's interview with Alan Newsome, 28 May 1992.

its categories were not adapted to environmental action.[62] It had established its concepts when nature was seen as pieces to be divided up for private use, and it presumed that economic development was not only the highest good but the highest for any area.[63] National parks had not challenged this; they were areas no one else wanted, serving the social purpose of providing rest and recreation by contact with nature. The limits of this view were becoming clear even before the environmental movement raised the question of ecosystems and biological integrity. Drawing lines on the land did not separate nature and humans. As one Canadian park manager said, "It is impossible to train elk and deer to sally forth from National Parks in just the right numbers and at just the right times to suit the sportsmen and yet do no damage to agricultural and ranching interests."[64] Not only did environmentalism put this conflict in a new light, it raised new kinds of cases that forced the development of new categories. It also raised new questions about society's needs, as distinct from those of individuals, ones the law was equally unprepared to deal with. Anglo law, except for criminal law, was largely concerned to safeguard individuals from injury and regulate people's relations with one another. It said little about the diffuse rights of many people, none of whom, usually, had any interest as the law recognized it in a particular area of country or a particular ecosystem.

New Thoughts for New Areas

These problems came up in virtually every major environmental dispute, but they were particularly important – and obvious – when environmentalists sought to save something in nature that earlier generations had taken for granted or not seen as a particular thing at all. Australia's Great Barrier Reef is perhaps the best example of the redefinition of nature that environmentalism forced and of the legal problems this involved. The Reef, now a World Heritage Site, is a chain of some 1,500 coral formations and 200 continental islands stretching for some 1,200

---

[62] The Environmental Defense Fund's successful action against DDT in Wisconsin (see Dunlap, *DDT*) helped make environmental litigation popular in the United States in the early 1970s, but suits were often too expensive, drawn-out, and inconclusive for the tactic to become popular. See Joseph Sax, *Defending the Environment* (New York: Knopf, 1971), for the United States, and a later Australian appraisal, Brian J. Preston, *Environmental Litigation* (Sydney: Law Book Co., 1989).

[63] For a historical analysis of the law's role in allocating resources and the impact of social attitudes on the law see J. Willard Hurst, *Law and Economic Growth* (Cambridge, Mass.: Harvard University Press, 1964).

[64] Harrison Lewis to Smart, 24 February 1947, Box 39, U300, RG 84, Records of the Canadian Wildlife Service, Public Archives Canada, Ottawa.

miles along the Queensland coast. It was built by myriad coral animals working over thousands of years, their efforts shaped by the currents, the rise and fall of sea level, and the continent's slow drift into the tropics. All but a fraction is under water, and individual formations lie from twenty to eighty miles offshore, separated from each other by narrow passes or large stretches of ocean. It is the largest thing built by living creatures, but only astronauts have seen it as an entity and all at once – and then only by a feat of mental construction.

Into the twentieth century it was little known or studied. It had fascinated naturalists since the days of Captain Cook, but just getting there, much less staying for any period of time, was difficult. Ships, understandably, gave it a wide berth, and until the sugar boom of the 1880s there were few Anglo settlements on the coast. In 1920 the University of Queensland began research in marine biology, but the program was small and poorly funded. Only help from British scientists carried it through the Depression.[65] After World War II the southern end of the Reef became a popular tourist destination, and by the early 1960s resort construction and shell collecting were having a significant impact. There were other threats as well. Increased sugar production and new pesticides meant more residues, and new kinds of residues, washing off the fields into the ocean. Cane growers wanted to mine it for lime; companies looked at mineral deposits, including some of the beach sands; and oil companies scouted for places to drill. Some concerned Queenslanders suggested making it an underwater national park, but this was a long stretch. The concept was novel and the Country Party government that came to power in 1957 favored economic development and had little interest in nature protection.[66]

Reaction to a population explosion of crown-of-thorns starfish in the mid-1960s showed that the public was becoming concerned. The starfish, which lived on coral, now seemed to threaten large parts of the Reef. Twenty years earlier the nature notes columns would have covered this. Now it appeared in the regular news. People spoke of it as a plague, and the pessimistic feared it would damage whole formations or kill entire sections of the Reef. The CSIRO, which was supporting reef research, began investigating, and the ACF announced a national

---

[65] Dorothy Hill, "The Great Barrier Reef Committee, 1922–85, Part I, The First Thirty Years," and "Part II, The Last Three Decades," *Historical Records of Australian Science*, 6, Nos. 1 and 2; copies courtesy Dr. Patricia Mather, curator, Queensland Museum, South Brisbane. First part was unpaged, second is pages 195–211. A sketch of the Reef's history and of the environmental battles is James Bowen, "The Great Barrier Reef: Toward Conservation and Management," in Stephen Dovers (editor), *Australian Environmental History* (Melbourne: Oxford University Press, 1994), 235–56.

[66] Wright, *Coral Battleground*, 3; Bowen, "Great Barrier Reef," 238–41.

seminar on reef problems and research needs.[67] Development, or the threat of it, also brought action. In 1967 applications to mine the Reef and drill test wells for oil prompted a group at the University of Queensland to form the Littoral Society of Queensland and print what were claimed to be the first bumper stickers made in Australia – SAVE THE BARRIER REEF. It was the start of a campaign that would continue until the proclamation of the Great Barrier Reef National Park in 1975, an epic that has been described as "the greatest, most sustained environmental issue yet witnessed in Australia."[68]

It was great and sustained but also fragmented and confused. There was no one battle for the Reef but rather a series of actions against various plans for mining, drilling, or building a resort, each of which threatened some set of formations, islands, and beaches. Arguments were almost as diverse; advocates appealed to everything from natural beauty to money from tourism to national heritage and international significance. At the heart of their defense, though, was the idea of the Reef as a living system, to be saved or lost as a system. Defenders described it as a complex set of communities interconnected by currents, tides, food chains, and the life cycles of its species. Currents would carry oil spilled in one area to others. Mining dead coral would disrupt natural processes of recycling over a much wider area. Chemicals cane growers spread on their fields would wash into rivers, killing mangrove swamps and wetlands vital to the Reef.

This argument was in part an act of faith, for no one knew enough about the Reef as an ecosystem or set of ecosystems to predict with certainty the effects of any of these developments. It was, though, grounded in all the science activists could find. To fight the first mining application they recruited Australian biologists and two Americans working on the Reef. (The Queensland government, in turn, hired an American geologist, H. S. Ladd, to survey the area and help judge mining applications.)[69] They used environmental disasters elsewhere to point out what might happen on the Reef. The first stages of the fight over oil drilling coincided with the *Torrey Canyon* oil spill, and the Queensland government was reviewing applications for test wells on the Reef when an oil well in the Santa Barbara channel blew out. California's tragedy,

---

[67] Hill, "Great Barrier Reef Committee," Parts I and II; D. C. Potts, "The Crown of Thorns Starfish – Man-Induced Pest or Natural Phenomenon?" in R. L. Kitching and R. E. Jones (editors), *The Ecology of Pests: Some Australian Case Histories* (Melbourne: Commonwealth Scientific and Industrial Research Organization, 1981), 55–86; Bowen, "Great Barrier Reef," 244–5.

[68] Bowen, "Great Barrier Reef," 239.

[69] Wright, *Coral Battleground*, 5, 12, 22–34. On the relative lack of Australian experts see Bowen, "Great Barrier Reef," 249, 251.

one of the area's defenders said, "was the Reef's good fortune. It made the best possible publicity."[70] One of the Americans working on the Reef helped education and the cause by giving a public lecture on the Santa Barbara oil spill.[71]

Battles on and over the Reef showed the inadequacies of existing law. They were considerable, beginning with the question of who owned the area or, to be more precise, which Australian authorities had what rights in what places. The Reef's defenders were "naturally interested in this question, since our attempts to challenge state ownership had been brushed aside in the Ellison Reef case. . . . Over the years . . . [we] consulted several firms of solicitors in an attempt to find out what the legal situation in Reef waters really was." The topic was of general concern; the ACF's seminar on the Reef and its problems included a session devoted to it.[72] The "legal situation" is best described as uncertain. Property law mainly concerned land; rights in the sea were not well defined. The Queensland government tried, in the Mineral Resources (Adjacent Submarine Areas) Act (1964) to cut the knot by applying its mining laws to the adjacent continental shelf. This, as a textbook on natural resources law dryly put it, "effectively claimed for the State a jurisdiction unknown to a former colony before federation."[73] In 1967 the states and the Commonwealth agreed on a joint administration of offshore oil drilling, but did not (perhaps deliberately) resolve the issue of federal versus state rights. The next year the Commonwealth Continental Shelf (Living Natural Resources) Act declared that the nation had rights over Reef organisms. It did not, however, claim all Reef waters, only those beyond Queensland's territorial seas. The Labour government that came to power in 1972 sought more authority. The Commonwealth Seas and Submerged Land Act declared that the federal government had sovereign rights over the continental shelf (though the Commonwealth now concedes to the states development rights within the three-mile limit and a voice in making policy farther out).[74] The government appointed a Committee of Enquiry into the National Estate, whose 1974 report was the basis for the Great Barrier Reef Marine National Park Act (1975). That continued the division of authority. The park board is made up of representatives of the Commonwealth and the state; Queensland is responsible for day-to-day operations; and the federal govern-

[70] Wright, *Coral Battleground*, 51–5; quote on 51.
[71] Ibid., 75. See also D. W. Connell, "The Barrier Reef Conservation Issue," *Search*, 2 (June 1971), 188–92; Connell was president of the Queensland Littoral Society. Copy from Folder 331, Box 33, Serventy Collection, National Library, Canberra.
[72] Wright, *Coral Battleground*, 15; Bowen, "Great Barrier Reef," 248.
[73] Fisher, *Natural Resources Law*, 433.
[74] Ibid., 434–5; Bowen, "Great Barrier Reef," 248.

ment sets overall policy and provides most of the money. Implementation of the new rules has been gradual, and the Commonwealth has sought to educate the public before acting.[75]

Elsewhere the same kind of negotiations went on, and division of authority among levels of government and definitions of new things in nature were the simplest. The more difficult questions concerned who had what stake in natural areas and how these interests were to be balanced. If, as environmentalists contended, the preservation of nature meant the protection of ecosystems, how humans used nature had to change in many ways. Forests, for reasons noted earlier, have been important battlegrounds, and the American debate over logging the ancient forests of the Pacific Northwest a good recent example. The legal lever was the status of the northern spotted owl under the Endangered Species Act, but the central issue was how a social interest in the preservation of ecosystems could be reconciled with the individual and community interests of people who lived in the area, lived by cutting down trees.[76] That is still not settled, and when it is there will be other battles. The environmental perspective is forcing the settler societies to rethink what nature is and what values it has (for whom and over what period of time). Whatever decisions we come to, we will have to take new action that will shape and change our lives. These issues wind up in court because this is the socially acceptable forum for making decisions like these.

Conclusion

When the great wave of environmental enthusiasm ebbed in the late 1970s some people believed, or hoped, that the disappearance of the "great mass of unwashed, unspanked, dole-bludging dropouts" meant the end of the movement. It did not, entering instead a period of consolidation. Since the great burst of legislation and agency building in the late 1960s and early 1970s, environmentalists have been struggling to use these effectively. The same thing happened in politics. Green political parties reached an immediate peak, from which they have

---

[75] Australian Environmental Council, *Guide to Environmental Legislation*, 236–7; Bowen, "Great Barrier Reef," 250–2. Annual reports of the Australian National Parks and Wildlife Service (now the Australian Nature Conservation Agency) trace this relationship.

[76] On the law see Steven Lewis Yaffee, *Prohibitive Policy* (Cambridge, Mass.: MIT Press, 1982). There is a large literature on the owl and the forests. For polemical discussions see Alston Chase, *In a Dark Wood* (Boston: Houghton Mifflin, 1995), and reaction to it. A good survey is James D. Proctor's "Whose Nature? The Contested Moral Terrain of Ancient Forests," in William Cronon (editor), *Uncommon Ground* (New York: Norton, 1995, 269–97.

declined, probably to regroup and return. They were never important in the United States, where a winner-take-all electoral system has historically throttled all third parties. Even in Australia, where they made quick gains, they were important only in the "multi-member electorates of the Australian Senate, Tasmania's House of Assembly, and Australian Capital Territory's Assembly." As a major force they were confined to Tasmania, where they held the balance of power between the Liberals and Labour into the early 1990s.[77] Now they have only a toehold. Environmental groups, though, still affect politics by endorsing candidates and pressing their agenda.[78] Of more importance, the movement's perspective has become the common currency of public discourse. Most politicians, however hypocritically, burn incense on its altars, and opponents pose as principled defenders of a "balanced" position against an elitist and extremist movement.[79]

Environmentalism has been successful enough to attract an organized opposition. In the early 1980s the business communities in these countries, realizing that it constituted a serious, as distinct from public relations, problem, abandoned their ad hoc and occasionally disorganized efforts in favor of an organized and increasingly sophisticated defense and then a counterattack. This too has encountered limits, as shown by the failure of the Reagan administration, whose true believers saw environmentalism as a hangover from the hippie movement of the 1960s, to restore the old legal and social framework. It was forced to fall back on administrative and legal obstruction. There followed an effort, outside the administration, to develop a countermovement at the grass roots, of which the American "wise use" movement is the best example. It has mounted an ideological campaign, appealing to Americans' commitment to individual freedom and private property, depicting itself as the defender of economic opportunity for working people and environmentalists as urban elitists trying to "lock up" land for their recreation.[80] Even here, it might be noted, the opposition accepts many of

---

[77] Dargavel, *Fashioning Australia's Forests*, 221, 222–3.

[78] Hays, *Beauty, Health, and Permanence*, particularly chaps. 4 and 13–16; Dargavel, *Fashioning Australia's Forests*, 223–30.

[79] This appears in everything from political platforms to the examination of dam projects. On changes in public ideas, compare what the 1975 novel *Ecotopia* considered "utopian" awareness with policies and attitudes about recycling or waste disposal. Ernest Callenbach, *Ecotopia* (1975; reprinted New York: Bantam, 1977). Environmentalism has not conquered, but it has reshaped, the mental landscape.

[80] A start here is Richard White, "Are You an Environmentalist or Do You Work for a Living?" and Proctor, "Whose Nature?" in Cronon (editor), *Uncommon Ground*, 171–85, 269–97. Hays, *Beauty, Health, and Permanence*, outlines the national debate. Contemporary political materials and newspapers are also good sources. On Australia see Watson, *Fighting over the Forests*, and Mosley and Messer, *Fighting for Wilderness*.

the movement's terms, wrapping itself in the mantle of conservation and calling for "balance."

Environmentalism has also fostered a vigorous internal dialogue (though this has also been described as confused and demoralized). One element has been the debate over tactics. Greenpeace activists protected whales by interfering with whaling ships or even trying to sink them. In the late 1970s an American organization, Earth First!, formed around a program that included – at least as a last resort – sabotage ("ecotage" in the movement's lexicon) in defense of wilderness.[81] This has dwindled, but the issue is still alive. A second, and more important, line of development has been intellectual. The defense of nature requires justification, and many people have tried to provide it. The most obvious split was between deep ecology and what it saw as a shallow mainstream defense, based in human needs and ends. The new movement, which began in the 1970s around the ideas and example of Arne Naess, called for commitment to nature for its own sake. In the past twenty years philosophers and activists have begun to explore the implications of that stance. There was by the early 1990s an obvious split between those who put wilderness and nature first and those who felt that the immediate conditions of people's lives were the most important thing. The division of Earth First! over this issue was the most visible evidence in the United States, but the same can be seen elsewhere. The movement has also moved beyond its base in the middle and upper middle classes. That was apparent in the United States with the rise of groups seeking environmental justice or fighting environmental racism. They were concerned less with wilderness or wild nature than with conditions of urban life, and they explicitly considered race and class. The change was less apparent elsewhere. In Australia, for instance, labor unions were involved in the defense of the Great Barrier Reef.

Environmentalism has also shaped science, or made more respectable a current within the community. Field biologists, as we have seen, never fully embraced the ideal of detachment when it came to the subject matter of their work, and there was from the late 1940s a continuing current of activism. Scientists helped form nature conservancies in the United States and Great Britain and the International Union for the Protection of Nature (later the International Union for the Conservation of Nature and Natural Resources), which was associated with the United

---

[81] David Foreman, *Confessions of an Eco-Warrior* (New York: Harmony Books, 1991), gives an inside account. A more scholarly approach is Martha F. Lee's, *Earth First! Environmental Apocalypse* (Syracuse, N.Y.: Syracuse University Press, 1995).

Nations.[82] In 1961 a group, most members of which were active in the IUCN or its programs, formed the World Wildlife Fund to act on urgent issues of wildlife protection that the IUCN seemed unable to address. Scientists helped write the Convention on International Trade in Endangered Species and have served on the commissions dealing with its enforcement. In the 1980s, with the ecological crisis deepening, they formed a new discipline, conservation biology, devoted to using science to save species and ecosystems. That meant dealing with the social as well as the biological questions involved in nature protection, and it has incorporated the kinds of social goals and strategies that marked game management.[83]

As in the change from folkbiology to natural history, the addition of ecology's perspective is a messy and incomplete process in which views change as they interact. It has to be this way. People speak of processes and relationships in nature, but they see – can only see – plants, animals, and landscapes. They appeal to the idea of global ecosystems but necessarily live in some particular place (an insight grounding the movement for bioregional living). The old drives, too, are still there. The great days of conquest are gone, but people still want to manage and change the land around them, if only to restore an Eden that existed before they arrived. They also want to have a place on the land. Ecology has not changed this. It has only provided a new vocabulary and benchmarks. Where are we now, as we seek to use ecology to hear the spirits of the land?

[82] On the movement for international protection of wildlife, which is part of this, see Sherman Strong Hayden, *The International Protection of Wild Life* (New York: Columbia University Press, 1942), and Robert Boardman, *International Organization and the Conservation of Nature* (Bloomington: Indiana University Press, 1981). On the United States see Robert L. Burgess, *The Ecological Society of America* (Oak Ridge, Tenn.: Oak Ridge National Laboratory, n.d.), reprinted in Frank N. Egerton (editor), *History of American Ecology* (New York: Arno, 1977); on Britain, Stephen Bocking, "Conserving Nature and Building a Science: British Ecologists and the Origins of the Nature Conservancy," in Shortland, *Science and Nature*, 89–114.

[83] On conservation biology see David Takacs, *The Idea of Biodiversity* (Baltimore: Johns Hopkins University Press, 1996). At the 1993 International Theriological Congress, for example, speakers at the plenary sessions (which presented the latest information on key topics) were instructed to give particular attention to the conservation biology aspects of their subject. This was in marked contrast to attitudes four years earlier. The 1997 Congress confirmed this direction.

EPILOGUE
THE UPSHOT

> When it suits them, men may take control and play fine tricks
> and hustle Nature. Yet we may believe that Australia, quietly
> and imperceptibly (what do a few centuries matter, after so
> long a waiting?), is experimenting on the men as they exper-
> imented on the dogs. She will be satisfied at long last, and
> when she is satisfied an Australian nation will in truth exist.
>
> W. K. Hancock, 1930[1]

The settlers have indeed taken control and played fine tricks, and not
only in Australia. Look back at the landscapes we visited in imagination
at the beginning of this book. In New Jersey the forest is gone. We are
standing in loose, well-worked soil, among rows of dark green, ground-
hugging plants. Potatoes, say the rural people among us. By the sign at
the edge of the field a developer is planning a crop of new houses. The
only sizable trees in sight are a white oak in front of the house across
the road and two Norway maples by the one next door. At our next stop
Ottawa and Hull cover the bluffs and riverbank near our campsite. In
Kansas, since this is June, we are in a field of almost ripe wheat. The
meadowlarks are beyond the fence among a herd of Charolais – big
white cattle introduced forty years ago. In Zion trailers and pop-up tents
fill the floodplain where we camped. On the west coast the trees are
gone, lumbered out in the postwar boom. Our ecologist, looking at the
weeds and grass and kicking a charred stump or two, says the regrowth
burned off two or three years ago.

Across the Pacific there are sheep and fences on the plain where we
saw the Southern Cross. The grass is different and the saltbush is almost

---

[1] W. K. Hancock, *Australia* (London: Ernest Benn, 1930), 239; quoted in J. M. Powell,
"Griffith Taylor and 'Australia Unlimited'" (Brisbane: University of Queensland Press,
1993), 1.

gone. Sheep, says our ecologist, and even the city people among us can see the effects of their hooves. The woods on the southern coast are now Melbourne suburb. The stream is paved over, and while we can still find eucalyptus, this area was subdivided and built up before the enthusiasm for native species hit in the 1970s. Most of the trees are European or North American. Our last stop, in New Zealand, is now pasture. To get to the hot spring we just climb over the barbwire fence (at one of the posts, please) and stroll through the short grass past what our Kiwi host calls "hamburger bulls" (Holstein steers destined for a fast-food restaurant). Someone built a proper pool and bathhouse at the hot spring, but that was years ago. It is long abandoned and rotting away. Earlier we needed our ecologist to point out human effects on landscapes that seemed to our untutored eyes natural. Now we must search for even remnants of what was. This would be true across our countries; even on the mountaintops we can see contrails from passenger jets. Only on the largest scale do the lands seem the same, in the sweep of clouds and the violence of thunderstorms on the plains, the night sky (if we can get away from lights), the scents of flowers and trees, forest and ocean breeze, and bird and insect calls.

Old worlds have indeed been overthrown and new ones are forming, but "the slow re-establishment of a new equilibrium" is as much in the settlers' minds as on the land, and the interaction of ideas, experience, and culture is at least as complex as that of plants, animals, and the land. The earlier peoples had lived on the land so long and had been so isolated that, save for the Maori in New Zealand, they knew of no other home. The Anglos were from elsewhere, knew it, and looked abroad even as they sought a home on the land. They used their old culture, adapting its forms, to express new sentiments. The canvases of Frederick McCubbin and Charles Russell are settler experience and myth set down in the conventions of European art. The same is true of nature writings, from Banks and Bartram's reports, through Seton's animal stories, Guthrie-Smith's memoir of his sheep station, and May Gibb's fairy tales, to Aldo Leopold's essays. There was also, however, a process of exchange that reshaped ideas. They hybridized in ways that species usually do not. The Americans, for example, adopted and adapted Romanticism as a way to view their land, but Romanticism was itself, in part, a product of the idea of North American wilderness in European minds. The national parks of Australia, Canada, and New Zealand draw, in different proportions, on American ideas about preserving vast stretches of wild country and British ones about urban green spaces. Conservation was a hybrid, formed by settler experience in each country, experience in Europe and the European colonies of empire, and an international cadre of experts.[2] Ecology came from and was changed by its own version of this international circuit.

The impulse behind this search was an idea so common and well accepted that the settlers rarely mentioned it. It was that people were related to the land on which they lived. The sentiment took different forms in each generation, but it was a constant. At first it meant resisting the pull of the land, coming to terms with it by making it familiar. Later generations reversed the logic. The new land had made them a new people, and they celebrated their differences from the old country by rejecting the landscapes of "home." (They also took pride in changing the landscape, but intellectual consistency is not a requirement for cultures or individuals.) Industrialization, which seemed to detach people from the land, encouraged the search for connections through play and study, and in the past generation the environmental movement's quest for ecologically sound local economies that will foster real ties to the land has given another meaning to the quest.[3]

Just as important, the settlers knew the land in different ways. Their predecessors had lived in local areas by local knowledge; the Anglos had from the start lived within global networks of exchange and understood their lands within a universal system of knowledge, a system that assumed the human intellect could understand this world from the outside, as an object. Sir Thomas Browne, speaking of "Nature" as "that universall and publik Manuscript, that lies expans'ed unto the eyes of all" expressed an attitude that guided and still guides settler understanding.[4] It was the ground of natural history, then ecology, and is so embedded in the culture that it appears in contexts that seem entirely alien to science and its ethos. Consider Annie Dillard's *Pilgrim at Tinker Creek*, one of the most popular nature books of the past generation. It is a mystical, even spiritual nature journal, centered on individual experience on a small plot of ground and organized by the passage of the seasons. Yet not only is it shot through with science, from natural history to ecology, these perspectives are necessary. They produce the awe at the world around us that leads to a higher understanding.[5] A more prosaic example is

---

[2] Richard Grove, *Green Imperialism* (Cambridge University Press, 1995).

[3] An informative and insightful example of how this is to be carried out is Wes Jackson, *Becoming Native to this Place* (Lexington: University Press of Kentucky, 1994). An example of a less rigorous search is Stephanie Mills, *Whatever Happened to Ecology?* (San Francisco: Sierra Club, 1989).

[4] Thomas Browne, *Religio Medici* (1642; Oxford: Oxford University Press, 1964), 15.

[5] Annie Dillard, *Pilgrim at Tinker Creek* (New York: Harper's, 1974). On the book's structure see Lawrence Buell, *The Environmental Imagination* (Cambridge, Mass.: Harvard University Press, 1995), 237–42.

the nature writing of the American Edward Abbey, no friend of the filing-cabinet mentality of science (taken as formal study). Describing a walk in the mountains, he unselfconsciously speaks of seeing the "larkspur . . . of the species called Subalpine or Barbery (*Delphinimu barbeyi*). Climbing higher, I enter by degrees into the Hudsonian life zone, leaving behind the Canadian with its aspen and Douglas fir."[6] Here are Linnaeus and C. Hart Merriam. Speaking of the canyons of the Southwest, he appeals to the deep time of geology to give significance and awe to the landscape. In everyday life people without formal education in science or interest in nature beyond lawn grass and tomato plants speak of the weather in terms of fronts and jet streams, unconscious of the science behind these ideas. Recently El Niño, a construction from reams of data extending over half the world, has become a figure of casual conversation and popular culture.

This is not to say we have become scientists. Far from it. It is just that our universal "direct experience" of nature and the "essential tangibility of the research material" at this level have kept scientific concepts close to popular ones.[7] A comparison with concepts of the physical world may be useful here. We say (at least if we have advanced education) that we live in a universe described by Einstein's theory of relativity. Pressed, we may admit that the world we experience is really a Newtonian one and that Einsteinian physics applies only at velocities near the speed of light. Nonsense. In daily life we live as if Aristotle were right. We know that thrown objects move in curved lines, fall toward the ground, and slow down and stop unless we continue to apply force.[8] On the level of the physical world, the differences are academic. Physics disdains the world of experience – as a subject, classical mechanics is complete – while ecology still speaks of the world we know, the nature we seek out for recreation and physical challenge, the nature we see as part of our lives.

The ties among various bodies of knowledge have reinforced a common human tendency not to go back to first principles, but to add new ideas on top of old. An additional complication is that the focus of settler interest has shifted from one generation to the next. In the nineteenth century the dominant tendency was to see the land as a whole, a process most apparent in Canada; one historian of science found in

---

[6] Edward Abbey, *Desert Solitaire* (New York: McGraw-Hill, 1968), 222–3.

[7] T. F. H. Allen and Thomas B. Starr, *Hierarchy: Perspectives for Ecological Complexity* (Chicago: University of Chicago Press, 1982), 3. On the necessary ties of scientific and popular views see Scott Atran, *Cognitive Foundations of Natural History* (Cambridge University Press, 1993), 15–80.

[8] My thanks to a former colleague, David Lux, for several conversations on this point and the analogy.

natural history the key to the national vision and titled her book *Inventing Canada*.[9] A later generation moved closer, using nature literature and outdoor recreation to show the land as people saw it and appealing to pioneer life as the formative experience. The stereotypes thus formed permeated the culture as thoroughly as the values and perspective of science. An Australian woman, recalling her childhood reading in the 1950s, said that these books, mostly written in the early twentieth century, convinced her and her playmates that they were tough little bush children, skilled in the lore of the outback. In later years, she went on, they were a bit surprised to discover that they did not know even how to ride.[10] The urban and urbane Canadian novelist Robertson Davies said of one of his characters, a highly cultured man, that he felt awkward around his nephews, for he "had an un-evolved Canadian idea that an uncle ought to teach a boy to shoot, or fish, or make a wigwam out of birch bark."[11]

At the same time local lore and experience shaped smaller loyalties. In the United States and Canada, regions are obvious and contested on levels from popular culture to academic discourse. In Australia, state loyalties are fierce and closely held, and there are smaller divisions within those, based on crops, rivers, and distance from the state capital.[12] Even New Zealand, small as it is, has its own provincial loyalties (never mind that Premier Vogel got rid of provinces as political entities in 1876).

The settlers' focus also shifted from species from "home" to the values of native nature. The national parks, which had begun as islands of scenery or places for recreation, became areas for preserving wild nature, and wilderness areas and special nature reserves chart the growing interest in a pristine land, as it was before settlement or at least white settlement. So does the more conscious remolding of local and

[9] Suzanne Zeller, *Inventing Canada* (Toronto: University of Toronto Press, 1987).

[10] Elizabeth Ward, "A Child's Reading in Australia," *Washington Post, Book World,* 4 (November 1990), 17, 22.

[11] Robertson Davies, *Bred in the Bone* (New York: Viking, 1985), 430. For a more systematic treatment see Margaret Atwood, *Strange Things: The Malevolent North in Canadian Literature* (Oxford: Oxford University Press, 1995). Canadian popular literature and magazines show the same oddities and incongruities.

[12] Ian Watson, *Fighting over the Forests* (Sydney: Allen & Unwin, 1990), as well as John Dargavel, *Fashioning Australia's Forests* (Melbourne: Oxford University Press, 1995), point to this phenomenon as part of the development of, and resistance to, the plans for the environmental preservation of forests. See Richard White, "Are You an Environmentalist or Do You Work for a Living?" and James D. Proctor, "Whose Nature? The Contested Moral Terrain of Ancient Forests and Social Justice," both in William Cronon (editor), *Uncommon Ground* (New York: Norton, 1995), 171–85 and 269–97. On local loyalties a good example and a good analysis is Tom Griffiths's *Hunters and Collectors* (Cambridge University Press, 1996).

even domestic landscapes. In the 1970s Canberra suburbs began to plant native trees in preference to exotics, and Australian nurseries now stock not only Australian plants but regional specialities. In the United States, native plants are a recognized subdivision of the business, and even regular firms offer native grasses and native plants for landscaping.[13]

Now the question of what is truly native is more pressing. Even in New Zealand, where the distinction would seem to be self-evident, ecological reality pulls one way and settler sentiment another. In 1973 two ecologists reported that the Maori rat, the kirre, was now being called "native" and that it had "even crept into the ranks of desirable native wildlife, vying with such elite as the tuatara and saddleback for protection on select island refuges. To what dizzier heights," they asked, "can an introduced rat with but squatter's rights aspire – and how much longer must later introductions await similar recognition? . . . This country," they declared, "will come of age ecologically when Western man and his animal introductions are regarded as part of the natural environment."[14] In 1990 Carolyn King echoed these sentiments in *Handbook of New Zealand Mammals*: "It is time that the native and introduced mammals were treated in practice as resident species of equal status in the scientific sense." To set the example she listed them together. This, she said, recognizes that Europeans live in the country, but also that there exists "a working, evolving community. . . . [that] will continue to evolve according to natural processes largely beyond our control."[15] Not everyone accepts this. When King came to New Zealand in the early 1980s to study stoats and their effects on native wildlife (from, of course, Oxford), she found that the public wanted not study but extermination. It was not simply a matter of predators eating "nice" animals. What enrages the average conservationist, she said, "is that mammalian predators are *alien* intruders into our forest; that they were never 'meant' to be here and that they roar after prey that are doubly defenseless, provided not by God but by man's own bungling and mismanagement."[16]

Australian attitudes show this confusion as well. Some people view rabbits as a foreign plague; others think of them as nice bunnies. To

---

[13] Sara Stein, *Noah's Garden* (Boston: Houghton Mifflin, 1993), is an example of the American literature; see particularly 202–3. Catalogs and directories provide information on the United States. Information on Australia comes from casual conversations with Australians.

[14] J. A. Gibb and J. E. C. Flux, "Mammals," in Gordon R. Williams (editor), *The Natural History of New Zealand: An Ecological Survey* (Wellington, Reed, 1973), 365.

[15] Carolyn King, *Handbook of New Zealand Mammals* (Auckland: Oxford University Press, 1990), 9.

[16] Carolyn King, *Immigrant Killers* (Auckland: Oxford University Press, 1984), 125.

counter the influence of Beatrix Potter, the Anti-Rabbit League of Australia is seeking to put in its sentimental place the bilby, a native marsupial about the same size and with a similar diet. It recently convinced Cadbury's of Australia to make chocolate bilbies for Easter.[17] Then there are the deer that found small footholds in the Southeast. One of their modern defenders said that deer are "to the naturalist's mind . . . [an] introduced species and largely forgotten. Being un-Australian in origin they are considered not quite right for the country." He went on to say that "the introduction of exotics into the wild without scientific study of their possible impact is wrong in principle. That the deer we now have in our bush are a valuable and irreplaceable asset is an opinion likely to be increasingly substantiated." As for those who see the animals as exotics that should be eliminated: "It is sheer humbug today for a white exotic human, yearning for a Dreamtime environment, to point the bone at a particular wildlife species he or she has been told is in the bush, which they are never likely to see."[18] The reader can dissect at leisure the ironies involved in white Australians using metaphors from Aboriginal culture to discuss who or what is native.

In North America excluding exotics has been part of national park policy since the late 1930s, but it has never been a high priority of the public or the agency. The sentiment appears in Canada, but with even less urgency. In 1956 J. B. Munro, head of the Canadian Wildlife Service, said that he was "opposed on aesthetic grounds to cluttering up the North American landscape with European and Asiatic species."[19] He did not, however, recommend action. The debate over native nature has more commonly been framed in terms of preserving or restoring landscapes to their condition before the whites arrived. The desire for purity, though, clashes not only with reality but with the vision of the "natures" the settlers made. Some see feral horses and burros in the West as aliens, a danger to the native vegetation. Others view them as part of the heritage of the "old West" or Spanish settlement.

These debates all draw on science but they go beyond it. In part this is a matter of human limits. We do not know enough to describe the world with certainty. Ecology, to return to the analogy of the biological world around us as a set of card games, knows the most important rules of the various games (or what it thinks are the most important ones),

---

[17] The Anti-Rabbit League's web page has information. My thanks to John Flux for information.

[18] Arthur Bentley, *An Introduction to the Deer of Australia with Special Reference to Victoria* (Melbourne: Koetong Trust Service Fund, Forests Commission, Victoria, 1978), 17, 21.

[19] J. B. Munro to S. C. Whitlock, Department of Conservation, 16 January 1956, Box 43, RG 109, Records of the Canadian Wildlife Service, Public Archives Canada, Ottawa.

but it does not know them all, and it cannot tell us which cards are face down on the table. We always act on incomplete knowledge. At times we draw an ace – myxomatosis was one – but much of our "success" is a matter of not having our bets called. We now hold some losing hands that previous generations thought were winners. Even a perfect science, though, would not answer all the questions, for they go beyond matters of what is to what should be. E. O. Wilson's idea that "[t]he role of science, like that of art, is to blend exact imagery with more distant meaning, the parts we already understand with those given as new into larger patterns that are coherent enough to be acceptable as truth," is incomplete.[20] It is people, not science, who do the blending.

The emphasis here has been on patterns, but within this story of common learning there is, beside the multitude of differences among countries, what seems a fundamental split between the North American nations and those in the South about what constitutes "nature" and how it is related to society. In North America the settlers have come to see these as two separate worlds, with the great goal of nature appreciation being to break down the barrier and re-enter Eden. Thoreau hints at it, and it emerges full-blown in John Muir's ecstatic writings about the Sierra Nevada. It fueled demands for pristine national parks and the preservation of wilderness, and it is part of the environmental vision as well. Consider, for example, Farley Mowat's best-selling *Never Cry Wolf*. Ostensibly the record of the author's research for the Canadian Wildlife Service, it is in fact the tale of a spiritual journey, from hating and fearing wolves to seeing them as part of the life of the land and accepting them as his guides into that world. The climax comes when Mowat, exploring what he thinks is an empty den, meets a wolf and is swept again by fear and hatred. Now the wolf's cry was "a voice which spoke of a lost world which once was ours before we chose the alien role, a world which I had glimpsed and almost entered . . . only to be excluded, at the end, by my own self."[21] No such theme runs through Australian and New Zealand writings. People do not seek God or even transcendence in their national parks or wilderness, nor do they call the wombat (still less the kiwi) brother.

This division appears in the different meanings the settlers give to the word "environment." In the United States and Canada it is commonly

---

[20] E. O. Wilson, *Biophilia* (Cambridge, Mass.: Harvard University Press, 1994), 51.

[21] Farley Mowat, *Never Cry Wolf* (Boston: Little, Brown, 1963), 162–3. Biologists, particularly those in the Canadian Wildlife Service, saw the book as a mythic production. See Wolves, WLU 200, Box 77, Records of the Canadian Wildlife Service, RG 109, Public Archives Canada, Ottawa. This recently received a public airing; see John Goddard, "A Real Whopper," *Saturday Night*, 111 (May 1996), 46–54 et seq.

attached to the "natural environment," and in the early 1990s the environmental justice movement, which sees environment in terms of daily life in the industrial city, baffled mainstream organizations in the United States. This would not have been the reaction in Australia or New Zealand, for the movements in those countries have included these subjects from the start. A New Zealand textbook on environmental law, for example, had no chapter on wildlife or the national parks because these were not among the topics "most likely to be encountered in practice."[22] The first chapter of *Environmental Planning in New Zealand* is on the "urban environment," and wildlife is not even in the index.[23] *Environmental Law in Australia* acknowledges the "intertwining of various 'environments,'" gives no special place to nature, and says that the first purpose of environmental law is "to protect the general public interest in a safe, healthy, and pleasant environment."[24] Even historical studies show this split. In the United States environmental history has focused on the conservation of resources and the preservation of nature. There are many studies on national parks, forests, wildlife, wilderness, and changes in the land, far fewer on what is referred to as the "built environment." Australian environmental historians (the field has been slow to develop in Canada and New Zealand) consider housing and slums along with parks and wild lands, and there is a tacit assumption that people made it all, the natural environment by destruction, their own by construction. This assumption is not always tacit. The subtitle of Geoffrey Bolton's *Spoils and Spoilers* is *Australians Make Their Environment, 1788–1980.* There is also a stronger connection between politics and nature. Some American studies connect the two explicitly, but they are few and lack the passion of, say, William Lines's *Taming the Great South Land*, which declares bluntly, "The British Empire and Australia's tethers to the post-Enlightened industrial world – human constructions all – not nature, created modern Australia." It sees the destruction of nature in terms of some humans exploiting others.[25]

This is the result not of climate or landscape, but of history. The North American settlers arrived in a land that had been for more than

[22] D. A. R. Williams, *Environmental Law* (Wellington: Butterworth of New Zealand, 1980), ix.

[23] P. Ali Memon and Harvey C. Perkins (editors), *Environmental Planning in New Zealand* (Palmerston North: Dunmore Press, 1993). A recent critique of these efforts is Ton Buhrs and Robert V. Bartlett's *Environmental Policy in New Zealand* (Auckland: Oxford University Press, 1993).

[24] G. M. Bates, *Environmental Law in Australia* (Sydney: Butterworths, 1992), 2, 1.

[25] Geoffrey Bolton, *Spoils and Spoilers: Australians Make Their Environment* (Sydney: Allen & Unwin, 1981); William Lines, *Taming the Great South Land* (Berkeley: University of California, 1991), 278. Derek Whitelock, though, subtitled *Conquest to Conservation* (Netley: Wakefield, 1985) *History of Human Impact on the South Australian Environment.*

a century part of the European imagination, and then for two hundred years (until the early nineteenth century) had few and sporadic ties to the mother country. Isolation, *pace* the usual antipodean claims, was a more powerful force in North American development than in Australian. North Americans also lived for three hundred years with the knowledge that there was a land "beyond settlement" and that they were making their social world by destroying a rich and varied natural one. The antipodean settlers came to lands much less worked over by European imaginations, and they were isolated for a much shorter time. Half a century after settlement began, there were regular voyages between Australia and Europe. Fifty years more and drovers moving west from Queensland and east from Western Australia met in the Northern Territory. New Zealand had an even faster development, even more closely tied to Britain. A young adult who saw the Treaty of Waitangi signed in 1841, creating the colony, could have spent old age in a prosperous, settled Anglo New Zealand. In both countries native nature was less rich and more quickly and thoroughly changed by settler action. The lands were quickly populated by an array of wild and feral animals ranging from rabbits and cats to deer, and introduced plants are so abundant that "[t]he *Illustrated British Flora* is still the best available book for the identification of most temperate Australia weeds."[26]

How far this split will go no one can say. We all, still, face a common situation, seeking a place in lands now littered with the wrecks of earlier generations' hopes and dreams. There are in all our countries abandoned farmsteads, towns, irrigation works, factories, and mines. Some have sunk back into the soil, others are vast wrecks that hikers pick over for souvenirs. There are also the more enduring biological legacies. The skylark, singing as it soars into the Tasmanian sky, will be there long after the bronze and marble soldiers that commemorate the settler dead in Europe's wars have crumbled into dust. Hancock spoke of an end – when Australia "will be satisfied at long last, and . . . an Australian nation will in truth exist" – but both land and people have changed, and change seems endless. "One generation passeth away and another generation cometh, but the earth abideth forever" (Ecc. 1:4). So we hope, but it may be that only our capacity, in this generation, to listen to the land will allow later ones to explore in their lives what it means to be or become native to these lands.

---

[26] Tom McKnight, *Friendly Vermin*, University of California Publications in Geography, No. 21 (Berkeley: University of California Press, 1976). Quote from Jamie Kirkpatrick, *A Continent Transformed* (Melbourne: Oxford University Press, 1994), 84.

BIBLIOGRAPHY

Archival Collections

Australia

State Library of Victoria, Melbourne
  Australian Conservation Foundation
  Royal Australian Ornithological Union
  Tasmanian Wilderness Society (now the Wilderness Society)
State Library of New South Wales (Mitchell Library), Sydney
  Myles Dunphy Papers, Mitchell Library, Sydney
  Pamphlet file
National Library of Australia, Canberra
  Naturalists Society of New South Wales
  Francis Ratcliffe Papers
  Vincent Serventy Papers
  Griffith Taylor Papers
  Judith Wright Papers
Queensland State Archives
  Files relating to native birds, native animals, fauna protection, and fauna conservation

Canada

Public Archives Canada
  Records of the Canadian Wildlife Service, Record Group 109
  Records of the Dominion Park Service, Record Group 84
Saskatchewan Museum, Regina, Saskatchewan
  Whooping Crane files

New Zealand

Turnbull Library, Wellington
  Environmental and Conservation Groups of New Zealand, Inc.
  Royal Forest and Bird Protection Society Papers

New Zealand National Archives, Wellington
  Department of Internal Affairs, Wildlife files

United States

Smithsonian Institution Archives, Washington, D.C.
  Vernon Bailey Papers
Library of Congress, Washington, D.C.
  W. L. McAtee Papers
National Archives, Washington, D.C.
  Records of the United States Fish and Wildlife Service, Record Group 22
  Records of the United States Park Service, Record Group 79
Museum of Vertebrate Zoology, University of California, Berkeley
  Correspondence of the Museum; cited by permission of Dr. David Wake,
    Director
Conservation Center, Denver Public Library, Denver, Colorado
  Olaus Murie Papers
American Museum of Natural History, New York
  Files of the Department of Mammalogy

Interviews

Robin Barker, Division of Ecology and Wildlife, Commonwealth Scientific and
  Industrial Organization, 2 June 1992
John Calaby, Division of Ecology and Wildlife, Commonwealth Scientific and
  Industrial Organization, 2 June 1992
John Flux, retired from Ecology Division, Department of Scientific and Indus-
  trial Research, several interviews, 1990–7
Thomas Kimball, president of National Wildlife Federation, 19 December
  1979
Geoff Law, Wilderness Society, 17 July 1990
Starker Leopold, University of California, Museum of Vertebrate Zoology, 16
  June 1981
Patricia Mather, Queensland Museum, 5 August 1990
Alan Newsome, Division of Ecology and Wildlife, Commonwealth Scientific and
  Industrial Organization, 28 May and 2 June 1992
Mike Young, Division of Ecology and Wildlife, Commonwealth Scientific and
  Industrial Organization, 2 June 1992

Unpublished works

Bardwell, Sarah, "National Parks in Victoria, 1866–1956," Ph.D. dissertation,
  Monash University, 1974.
Cart, Theodore, "The Struggle for Wildlife Protection in the United States,
  1870–1900," Ph.D. dissertation, University of North Carolina, 1971.
Galbreath, Ross, "Colonisation, Science, and Conservation," Ph.D. dissertation,
  University of Waikato, 1989.

Gibb, John, "A Brief Review of Rabbit Research in Ecology Division since the 1950s," paper prepared for Ecology Division, New Zealand Department of Scientific and Industrial Research, March 1983. (This was in the Library of the Ecology Division, DSIR, Lower Hutt, New Zealand. It should be under the records of the new organization, Landcare New Zealand.)

Lewis, Harrison, "Lively: A History of the Canadian Wildlife Service," manuscript, Record Group 109, Records of the Canadian Wildlife Service, Public Archives Canada, Ottawa.

Lovely, Robert Allyn, "Mastering Nature's Harmony: Stephen Forbes and the Roots of American Ecology," Ph.D. dissertation, University of Wisconsin, 1995.

Millbrooke, Anne Marie, "State Geological Surveys of the Nineteenth Century," Ph.D. dissertation, University of Pennsylvania, 1981.

Moseley, J. G., "National Parks and Equivalent Reserves in Australia," draft prepared for the Australian Conservation Foundation, 1968, Box 1824, Australian Conservation Foundation Collection, State Library of Victoria, Melbourne.

Nowak, Ronald, "The Gray Wolf in North America," a preliminary report submitted to the New York Zoological Society and the U.S. Bureau for Sport Fisheries and Wildlife, 1 March 1974. Copy courtesy of Nowak.

Osborne, Michael A., "The Société Zoologique d'Acclimatisation and the New French Empire," Ph.D. dissertation, University of Wisconsin, 1987.

Stead, David, "Report on the Rabbit Menace in New South Wales (1925–1926)," 5 vols., Mitchell Library, Sydney.

Wallace, Ray, "A Body in Twain? The Royal Australasian Ornithological Union and the events of 1969," M.A. thesis, University of Melbourne, 1990.

Warren, A. F. J., "Management of Endangered Species: Policy Influences and Development in New Zealand," research project for master of science degree, University of Canterbury, 1987.

Articles, Book Chapters, Occasional Papers

"Abstract of the Report of the Membership Committee," *Journal of Wildlife Management,* 2 (April 1938), 70.

Adamson, D., and R. Bucanan, "Exotic Plants in Urban Bushland in the Sydney Region," cited in R. L. Amor and C. M. Piggin, "Factors Influencing the Establishment and Success of Exotic Plants in Australia," in Derek Anderson (editor), *Exotic Species in Australia: Their Establishment and Success* (Proceedings of the Ecological Society of Australia), *10* (1977), 15.

Albright, Horace M., "The National Park Service's Policy on Predatory Animals," *Journal of Mammalogy, 12* (May 1931), 185–6.

Anderson, Warwick, "Geography, Race and Nation: Remapping 'Tropical Australia,' 1890–1930," *Historical Records of Australian Science, 11,* 4 (1997), 457–68.

Bailey, Liberty Hyde, "What Is Nature Study?" in Anonymous, *Cornell Nature-Study Leaflets* (Albany, N.Y.; J. B. Lyon, 1904), 16.

Basalla, George, "The Spread of Western Science," *Science, 156* (5 May 1967), 611–22.

Bassett, C., and K. H. Miers, "Scientific Reserves in State Forests," *Journal of the Royal Society of New Zealand, 14,* 1 (1984), 29–35.

Berger, Carl, "The True North, Strong and Free," in Russell, *Nationalism in Canada,* 3–26.

Bernhardt, Peter, "Of Blossoms and Bugs: Natural History in May Gibb's Art," in Royal Botanic Gardens, *Gumnut Town,* 5–21.

Bocking, Stephen, "Conserving Nature and Building a Science: British Ecologists and the Origins of the Nature Conservancy," in Shortland, *Science and Nature,* 89–114.

Bordo, Jonathan, "Jack Pine – Wilderness Sublime or the Erasure of the Aboriginal Presence from the Landscape?" *Journal of Canadian Studies, 27* (Winter 1992–3), 98–128.

Bowen, James, "The Great Barrier Reef: Toward Conservation and Management," in Stephen Dovers (editor), *Australian Environmental History* (Melbourne: Oxford University Press, 1994), 235–56.

Branagan, David, and Elain Lim, "J. W. Gregory, Traveller in the Dead Heart," *Historical Records of Australian Science, 6,* 1 (1986), 71–84.

Brower, Matthew, "Framed by History," *Journal of Canadian Studies, 31* (Summer 1996), 178–82.

Brown, Bob, "Wilderness v. Hydro-Electricity in South West Tasmania," in Mosley and Messer, *Fighting for Wilderness,* 59–68.

Burgess, Robert L., *The Ecological Society of America* (Oak Ridge, Tenn.: Oak Ridge National Laboratory, n.d.), reprinted in Egerton, *History of American Ecology.*

Burn, Ian, "Beating about the Bush: The Landscapes of the Heidelberg School," in Anthony Bradley and Terry Smith (editors), *Australian Art and Architecture* (Melbourne: Oxford University Press), 83–98.

Burt, A. L., "If Turner Had Looked at Canada, Australia, and New Zealand When He Wrote about the West," in Wyman and Kroeber, *The Frontier in Perspective* 59–77.

Cahalane, Victor H., "The Evolution of Predator Control Policy in the National Parks," *Journal of Wildlife Management, 4* (July 1939), 229–37.

Cahn, Robert, and Patricia L. Cahn, "Reorganizing Conservation Efforts in New Zealand, *Environment, 31* (April 1989), 18–20, 40–5.

Caughley, Graeme, "Ecological Relationships," in Graeme Caughley, Neil Shepherd, and Jeff Short (editors), *Kangaroos: Their Ecology and Management in the Sheep Rangelands of Australia* (Cambridge University Press, 1987), 159–87.

Chambers, Robert (editor), *Transactions of the Eastern Coyote Workshop* (no publishing data, internal evidence suggests 1975).

Cittadino, Eugene, "Ecology and the Professionalization of Botany in America, 1890–1905," *Studies in the History of Biology, 4* (1980), 171–98.

Cockayne, Leonard, "Report on a Botanical Survey of Kapiti Island" (1907); "Report on a Botanical Survey of the Waipoua Kauri Forest" (1908); "Report on a Botanical Survey of the Tongariro National Park" (1908); "Report on a Botanical Survey of Stewart Island" (1909); "Report on the

Dune-Areas of New Zealand, Their Geology, Botany, and Reclamation" (1911). All were issued by the Department of Lands and Survey (Wellington: Government Printer).

Connell, D. W., "The Barrier Reef Conservation Issue," *Search,* 2 (June 1971), 188–92.

Cooch, F. Graham, "Canadian Connections," in U.S. Department of the Interior, *Flyways,* 304–9.

Cronon, William, "A Place for Stories: Nature, History, and Narrative," *Journal of American History, 78* (March 1992), 1347–76.

"*Telling Tales on Canvas,*" in Jules Prown et al. (editors), *Discovered Lands, Invented Pasts* (New Haven, Conn.: Yale University Press, 1992), 40–1.

"The Trouble with Wilderness," in Cronon (editor), *Uncommon Ground,* 61–90, reprinted in *Environmental History, 1* (January 1996), 7–28.

Davision, Graeme, "Sydney and the Bush: An Urban Context for the Australian Legend," in John Carroll (editor), *Intruders in the Bush* (Melbourne: Oxford University Press, 1982), 109–30.

"Destruction of Rabbits," *New Zealand Country Journal, 13* (May 1889), 209–13.

Dettleback, Michael, "Humboldtian Science," in N. Jardine, J. A. Secord, and E. C. Spary, *Cultures of Natural History* (Cambridge University Press, 1996), 287–304.

Dixon, Robert, "Nostalgia and Patriotism in Colonial Australia," in Hardy and Frost (editors), *Studies from Terra Australis to Australia,* 211–17.

Dugan, Kathleen G., "The Zoological Exploration of the Australian Region and Its Impact on Biological Theory," in Reingold and Rothenberg (editors), *Scientific Colonialism,* 79–100.

Dunlap, Thomas R., "Agriculture and the Concept of Climate," *Agricultural History, 63* (Spring 1989), 152–61.

"But What Did You Go Out Into the Wilderness to See?" *Environmental History, 1* (January 1996), 43–6.

"Ecology, Nature, and Canadian National Park Policy," *To See Ourselves/To Save Ourselves,* Proceedings of the Annual Conference of the Association for Canadian Studies, 1990 (Montreal: Association for Canadian Studies, 1991), 139–47.

"'The Old Kinship of Earth,'" *Journal of Canadian Studies,* 22 (Spring 1987), 104–20.

"Organization and Wildlife Preservation: The Case of the Whooping Crane in North America," *Social Studies of Science, 21* (1991), 197–221.

"'That Kaibab Myth,'" *Journal of Forest History, 32* (April 1988), 60–8.

"Wildlife, Science, and the National Parks, 1920–1940," *Pacific Historical Review, 59* (May 1990), 187–202.

Egerton, Frank, "Changing Concepts of the Balance of Nature," *Quarterly Review of Biology, 48* (June 1983), 320–50.

Elton, Charles, "On the Nature of Cover," *Journal of Wildlife Management, 3* (October 1939), 332–8.

Errington, Paul, "Bobwhite Winter Survival in an Area Heavily Populated with Grey Foxes," *Iowa State College Journal of Science, 8* (1933–4), 127–30.

"Modifications in Predation Theory Suggested by Ecological Studies of the Bobwhite Quail," in American Wildlife Institute, *Transactions of the Third North American Wildlife Conference* (Washington, D.C.: American Wildlife Institute, 1938), 736–40.

"Predation and Vertebrate Populations," *Quarterly Review of Biology, 21* (June 1946), 221–45.

"Some Contributions of a Fifteen-Year Local Study of the Northern Bobwhite to a Knowledge of Population Phenomena," *Ecological Monographs, 15* (January 1945), 3–34.

Evarts, Hal. C., "The Last Straggler," *Saturday Evening Post, 196* (14 July 1923), 48.

Faggion, A. E., "The Australian Groundwater Controversy, 1870–1910," *Historical Records of Australian Science, 10*, 4 (December 1995), 337–48.

Figgis, Penny, "Out of the Wilderness for the Wilderness Issue?" in Mosley and Messer (editors), *Fighting for Wilderness*, 227–44.

Fleming, Donald, "Science in Australia, Canada, and the United States: Some Comparative Remarks," *Actes du Dixième Congress International d'Histoire des Sciences* (Paris: Hermann, 1962), 179–96.

Flores, Dan, "Bison Ecology and Bison Diplomacy, *Journal of American History, 78* (September 1991), 465–85.

Frost, Alan, "Going Away, Coming Home," in Hardy and Frost (editors), *Studies from Terra Australis to Australia*, 219–31.

Flux, John E. C., "World Distribution," in Thompson and King, *The European Rabbit*, 8–21.

Forbes, Stephen F., "The Ecological Foundations of Applied Entomology," *Annals of the Entomological Society of America, 8* (1915), 1–9.

Fox, Francis, "Irrigation in Australia and What It May Accomplish," *Australia Today, 1911*, a special number of *Australasian Traveler* (1 November 1910), 85–94.

Froggart, Walter W., "Presidential Address: A Bureau of Biological Survey," *Australian Zoologist, 2* (January 1921), 2–8.

Gibb, J. A., and J. E. C. Flux, "Mammals," in Williams (editor), *The Natural History of New Zealand: An Ecological Survey*, 334–71.

Gibb, J. A., and J. Morgan Williams, "The Rabbit in New Zealand," in Thompson and King (editors), *The European Rabbit*, 158–204.

Gillbank, Linden, "The Acclimatisation Society of Victoria," *Victorian Historical Journal, 51* (1980), 255–70.

"The Life Sciences: From Collection to Conservation," in MacLeod (editor), *The Commonwealth of Science*, 99–129.

"The Origins of the Acclimatisation Society of Victoria: Practical Science in the Wake of the Gold Rush," *Historical Records of Australian Science, 6* (December 1986), 359–74.

Goddard, John, "A Real Whopper," *Saturday Night, 111* (May 1996), 46–54 et seq.

Goodman, Daniel, "The Theory of Diversity–Stability Relationships in Ecology," *Quarterly Review of Biology, 50* (September 1975), 237–64.

Gould, Stephen Jay, "Church, Humboldt, and Darwin: The Tension and Harmony of Art and Science," in Kelly (editor), *Frederic Edwin Church,* 94–107.

Griffiths, Tom, "'The Natural History of Melbourne': The Culture of Nature Writing in Victoria, 1880–1945," *Australian Historical Studies, 23* (October 1989), 339–65.

Grinnell, Hilda Wood, "Joseph Grinnell: 1887–1939," *Condor, 42* (January–February 1940), 3–34.

Grinnell, Joseph, "The Methods and Uses of a Research Museum," *Popular Science Monthly, 77* (August 1910), 163–169; reprinted in *Joseph Grinnell's Philosophy of Nature* (1943; reprinted, Freeport: Books for Libraries, 1968).

Hall, Colin Michael, "The 'Worthless Lands Hypothesis' and Australia's National Parks and Reserves," in Kevin J. Frawley and Noel M. Semple (editors), *Australia's Ever Changing Forests* (Canberra: Department of Geography and Oceanography and Australian Defense Force Academy, 1983), 441–56.

Harbo, Samuel J., Jr., and Frederick C. Dean, "Historical and Current Perspectives on Wolf Management in Alaska," in Ludwig N. Carbyn (editor), *Wolves in Canada and Alaska,* Canadian Wildlife Service Report Series No. 45 (Ottawa: Canadian Wildlife Service, 1983), 51–64.

Harris, Cole, "The Myth of the Land in Canadian Nationalism," in Russell, *Nationalism in Canada,* 27–43.

Heinimann, David, "Latitude Rising: Historical Continuity in Canadian Nordicity," *Journal of Canadian Studies, 28* (1993–4), 134–40.

Highsmith, R. Tod, "What's in a Name?" *Living Bird, 14* (Spring 1995), 30–5.

Hill, Dorothy, "The Great Barrier Reef Committee, 1922–1985, Part I, The First Thirty Years," and "Part II, The Last Three Decades," *Historical Records of Australian Science, 6,* Nos. 1 and 2. Copies courtesy Dr. Patricia Mather, Curator, Queensland Museum, South Brisbane. First part was unpaged, second is pages 195–211.

Holden, Robert, "Gumnut Town: Fact, Fantasy & Folklore," in Royal Botanic Gardens, *Gumnut Town,* 29–47.

Hopper, John L., "Opportunities and Handicaps of Antipodean Scientists: A. J. Nicholson and V. A Bailey on the Balance of Animal Populations," *Historical Records of Australian Science, 7* (1989), 179–88.

Howard, Walter E., *The Rabbit Problem in New Zealand,* DSIR Information Series No. 16 (Wellington: Department of Scientific and Industrial Research, 1958).

Hunt, H. A., "Climate of Australia," in G. H. Knibbs (editor), *Federal Handbook* (Melbourne: Government Printer, 1914), 122–62.

Inkster, Ian, and Jan Todd, "Support for the Scientific Enterprise, 1850–1900," in Home (editor), *Australian Science in the Making,* 102–32.

Irwin, David, "An English Home in the Antipodes," in Hardy and Frost (editors), *Studies from Terra Australis to Australia,* 195–210.

Jack, Sybil, "Cultural Transmission: Science and Society to 1850," in Home (editor), *Australian Science in the Making,* 45–66.

Kadlec, John A. "Wildlife Training and Research," in Howard P. Brokaw (editor), *Wildlife in America* (Washington, D.C.: U.S. Government Printing Office, 1978), 485–97.

Kingsland, Sharon E., "Defining Ecology as a Science," in Leslie A. Real and James H. Brown (editors), *Foundations of Ecology* (Chicago: University of Chicago Press, 1991), 1–13.

Kohlstedt, Sally Gregory, "International Exchange and National Style: A View of Natural History Museums in the United States, 1850–1900," in Reingold and Rothenberg (editors), *Scientific Colonialism*, 167–90.

"Nature Study in North America and Australasia, 1890–1945," in R. W. Home (editor), *The Scientific Savant in Nineteenth Century Australia*, Historical Studies of Australian Science, 11, No. 3 (Canberra: Australian Academy of Science, 1997), 439–55.

"The Nineteenth-Century Amateur Tradition," in G. Holton and W. A. Blanpied (editors), *Science and Its Public* (Dordrecht: Reidel, 1976), 173–90.

Laney, Lewis, "New War Born 1080 Coyote Poison May Kill All Predators," *New Mexico Stockman* (May 1948), 75.

Lantis, Margaret, "The Reindeer Industry in Alaska," *Arctic, 3* (April 1950), 27–44.

Laut, P., "Changing Patterns of Land Use in Australia," in Australian Bureau of Statistics, *Year Book Australia, 1988*, 547–56.

Law, Geoff, "The 1980s: A Great Decade for Wilderness," *Wilderness, 11* (March 1990), 5–6.

Leopold, Aldo, "The State of the Profession," *Journal of Wildlife Management, 4* (July 1940), 343–6.

Lyle K. Sowls, and David L. Spencer, "A Survey of Over-Populated Deer Ranges in the United States," *Journal of Wildlife Management, 11* (April 1947), 162–77.

"Life Goes to the Arapahoe Hunt," *Life, 26,* (30 May 1949), 106–8.

Lincoln, Frederic C., "Bird Banding in America," *Annual Report of the Smithsonian Institution, 1927* (Washington, D.C.: U.S. Government Printing Office, 1928), 331–54.

"A Decade of Bird Banding in America: A Review," *Annual Report of the Smithsonian Institution, 1932* (Washington, D.C.: U.S. Government Printing Office, 1933), 327–52.

Macdonald, Donald, "The Sportsman in Australia with Rod and Gun," *Australia Today,* 1911, a special number of *The Australian Traveller, 1* (November 1910), 143–50.

MacKenzie, John M., "Hunting and the Natural World in Juvenile Literature," in Jeffrey Richards (editor), *Imperialism and Juvenile Literature* (New York: Manchester University Press, 1989), 144–72.

Macleod, Roy, "Organizing Science under the Southern Cross," in Macleod (editor), *Commonwealth of Science*, 19–39.

McMichael, D. F., "New South Wales," in "Further Case Studies in Selecting and Allocating Land for Nature Conservation," in A. B. Costin and R. H. Groves (editors), *Nature Conservation in the Pacific* (Morges: International Union for the Conservation of Nature, 1973), 53–6.

McNally, J., "Koala Management in Victoria," Wildlife Circular No. 4, Fisheries and Game Department, Victoria, 1957.

Meinig, Donald, "Introduction," in Donald Meinig (editor), *The Interpretation of Ordinary Landscapes* (New York: Oxford University Press, 1979), 1.

Mobbs, C. J. (editor), *Nature Conservation Reserves in Australia*, Occasional Paper No. 19 (Canberra: Australian National Parks and Wildlife Service, 1989).

Muir, John, "The Wild Parks and Forest Reservations of the West," *Atlantic Monthly*, January 1898, 15, reproduced in William Cronon (editor), *Muir: Nature Writings* (New York: Library of America, 1997), 721.

Murie, Olaus J., "Food Habits of the Coyote in Jackson Hole, Wyo.," USDA Circular 362 (Washington, D.C.: U.S. Government Printing Office, 1935).

Myers, K., I. Parer, D. Wood, and B. D. Cooke, "The Rabbit in Australia," in Thomson and King, *The European Rabbit*, 108–57.

Newby, K. R., "The Fitzgerald River National Park, Western Australia: Some Conservation Issues," in Mosley and Messer (editors), *Fighting for Wilderness*, 83–95.

Newland, B. C., "From Game Laws to Fauna Protection Acts in South Australia: The Evolution of an Attitude," *South Australian Ornithologist*, 23 (March 1961), 52–63.

Newland, Elizabeth Dalton, "Dr. George Bennett and Sir Richard Owen: A Case Study in the Colonization of Early Australian Science," in Home and Kohlstedt (editors), *International Science and National Scientific Identity*, 55–74.

Norman, F. I., and C. A. D. Young, "Short-Sighted and Doubly Short-Sighted Are They," *Journal of Australian Studies*, 7 (1980), 2–24.

Norrdahl, Kai, "Population Cycles in Northern Small Mammals," *Biological Review*, 70 (1995), 621–37.

Osborne, Michael A., "A Collaborative Dimension of European Empire," in Home and Kohlstedt (editors), *International Science and National Scientific Identity*, 97–119.

Palmer, T. S., "Chronology and Index to the More Important Events in American Game Protection, 1776–1911," Biological Survey Bulletin 41 (Washington, D.C.: U.S. Government Printing Office, 1912).

"The Danger of Introducing Noxious Animals and Birds," in U.S. Department of Agriculture, *Yearbook of the United States Department of Agriculture, 1898* (Washington, D.C.: U.S. Government Printing Office, 1899), 87–110.

"Extermination of Noxious Animals by Bounties," in U.S. Department of Agriculture, *Yearbook of the United States Department of Agriculture, 1896* (Washington, D.C.: U.S. Government Printing Office, 1897), 55–68.

Parkes, John, "Rabbits as Pests in New Zealand: A Summary of the Issues and Critical Information," Landcare Research Contract Report LC9495/141 (Lincoln, New Zealand, 1995).

Phillips, John, "Wild Birds Introduced or Transplanted in North America," U.S. Department of Agriculture, Technical Bulletin 61 (Washington, D.C.: U.S. Government Printing Office, 1928).

Paszkowski, Lech, "Dr. Jan Danycz and the Rabbits of Australia," *Australian Zoologist*, 15 (August 1969), 109–20.

"Physical Geography and Climate of Australia," Australian Bureau of Statistics, *Year Book Australia, 1988*, 202–56.

Potts, D. C., "The Crown of Thorns Starfish: Man-Induced Pest or Natural Phenomenon?" in R. L. Kitching and R. E. Jones (editors), *The Ecology of Pests: Some Australian Case Histories* (Melbourne: Commonwealth Scientific and Industrial Research Organization, 1981), 55–86.

Powell, Joseph, "Protracted Reconciliation: Society and the Environment," in MacLeod (editor), *The Commonwealth of Science*, 249–71.

Proctor, James D., "Whose Nature? The Contested Moral Terrain of Ancient Forests," in Cronon (editor), *Uncommon Ground*, 269–97.

"Rabbit Destruction," *Journal of the Department of Agriculture of Western Australia, 14* (20 October 1906), 281–4.

"Rabbit Extermination," printed in *Webster's Tasmanian Agricultural and Machinery Gazette*, reprinted in *New Zealand Country Journal, 13* (March 1889), 119–20.

Ransom, William, "Wasteland to Wilderness," in O. J. Mulvaney (editor), *The Humanities and the Australian Environment* (Canberra: Australian Academy of the Humanities, 1991), 10–11.

Riney, Thane, "New Zealand Wildlife Problems and Status of Wildlife Research," *New Zealand Science Review, 10* (March 1952), 26–32.

Robin, Libby, "Of Desert and Watershed: The Rise of Ecological Consciousness in Victoria, Australia," in Shortland (editor), *Science and Nature*, 115–49.

Rodgers, Margaret, "Exploring the Quintessential," *Journal of Canadian Studies, 31* (Summer 1996), 174–8.

Russell, Edmund P., "'Speaking of Annihilation'; Mobilizing for War Against Human and Insect Enemies, 1914–1945," *Journal of American History, 82* (March 1996), 1505–29.

Schedvin, Boris, "Environment, Economy, and Australian Biology, 1890–1939," in Reingold and Rothenberg (editors), *Scientific Colonialism: A Cross-Cultural Comparison*, 101–26.

Schiebinger, Linda, "Why Mammals Are Called Mammals: Gender Politics in Eighteenth-Century Natural History," *American Historical Review, 98* (April 1993), 382–411.

Shultis, John, "Improving Wilderness," *Forest and Conservation History, 39* (July 1995), 121–9.

Stafford, Robert A., "The Long Arm of London: Sir Roderick Murchison and Imperial Science in Australia," in Home (editor), *Australian Science in the Making*, 69–101.

Stanbury, P. J., "The Discovery of the Australian Fauna and the Establishment of Collections," in Bureau of Flora and Fauna, *Fauna of Australia*, 202–26.

Stefansson, Vilhjalmur, "Central Australia: Report by Dr. Vilhjalmur Stefansson" (Melbourne: Victorian Government Printer, 1924; for Commonwealth Government).

Storer, Tracy I., "Mammalogy and the American Society of Mammalogists," *Journal of Mammalogy, 50*, 4 (1969), 785–93.

"Summary of the Report of the Second Annual Meeting," *Journal of Wildlife Management, 2* (April 1938), 67.

Taylor, Griffith, *The Australian Environment, Especially as Controlled by Rainfall*, Advisory Council of Science and Industry, Memoir No. 1 (Melbourne: Government Printer, 1918).

The Climatic Control of Australian Production, Bulletin 11, Commonwealth Bureau of Meteorology (Melbourne: Victorian Government Printer, 1915).

The Control of Settlement by Temperature and Humidity, Bulletin 14, Commonwealth Meteorological Service (Melbourne: Victorian State Printer, 1916).

"Nature versus the Australian," *Science and Industry*, 2 (August 1920), 459–72.

"The Physical and General Geography of Australia," in G. H. Knibbs (editor), *Federal Handbook* (Melbourne: Government Printer, 1914), 86–121.

Teale, Edwin Way, "The Great Companions of Nature Literature," *Bird-Lore*, *46* (November–December, 1944), 363–6.

Thomson, J. M., J. L. Long, and D. R. Horton, "Human Exploitation of and Introductions to the Australian Fauna," in Bureau of Flora and Fauna, *Fauna of Australia*, 227–49.

Thoreau, Henry David, "Walking," in Thoreau, *Walden* (1854; reprinted, New York: Modern Library, 1937).

U.S. Park Service, "Report of the U.S. National Park Service," in U.S. Congress, Senate, Special Committee on the Conservation of Wildlife Resources, *The Status of Wildlife in the United States* (Senate Report 1203, 76th Congress, 3d Session) (Washington, D.C.: U.S. Government Printing Office, 1940), 350–2.

Van Dersal, William, "The Viewpoint of Employers in the Field of Wildlife Conservation," *Transactions of the Seventh North American Wildlife Conference, 1942* (Washington, D.C.: American Wildlife Institute, 1942), 506–17.

Walsh, P., "The Effect of Deer," in *Transactions of the New Zealand Institute*, *25* 435–9.

Ward, Elizabeth, "A Child's Reading in Australia," *Washington Post, Book World*, 4 November 1990, 17, 22.

Weaver, John C., "Beyond the Fatal Shore: Pastoral Squatting and the Occupation of Australia, 1826–1852," *American Historical Review*, *101* (October 1996), 981–1007.

Webb, Melody, "Arctic Saga: Vilhjalmur Stefansson's Attempt to Colonize Wrangel Island," *Pacific Historical Review, 61* (May 1992), 215–39.

White, Richard, "Are You an Environmentalist or Do You Work for a Living?" in Cronon (editor), *Uncommon Ground*, 171–85.

Williams, H. W., "Maori Bird Names," *Journal of the Polynesian Society, 15* (December 1906), 193–208.

Wilson, George R., "Cultural Values, Conservation and Management Legislation," in Bureau of Flora and Fauna, *Fauna of Australia*, 250–60.

Wodzicki, Kazimierz, "Status of Some Exotic Vertebrates in the Ecology of New Zealand," in H. G. Baker and G. L. Stebbins (editors), *The Genetics of Colonizing Species* (New York: Academic Press, 1965), 432–5.

Wright, Mabel Osgood, "The Law and the Bird," *Bird-Lore, 1* (December 1899), 203–4.

Young, Stanley Paul, "The War on the Wolf," Part II, *American Forests, 48* (December 1942), 574.

Books

Abbey, Edward, *Desert Solitaire* (New York: McGraw-Hill, 1968).

Albright, Horace M., with Robert Cahn, *The Birth of the National Park Service* (Salt Lake City: Howe Brothers, 1985).

Allee, W. C., Orlando Park, Alfred E. Emerson, Thomas Park, and Karl P. Schmidt, *Principles of Animal Ecology* (Philadelphia: Saunders, 1949).

Allen, David, *The Naturalist in Britain* (London: Allen Lane, 1976).

Allen, Durward L., *Wolves of Minong: Their Vital Role in a Wild Community* (Boston: Houghton Mifflin, 1979).

Allen, T. F. H., and Thomas B. Starr, *Hierarchy: Perspectives for Ecological Complexity* (Chicago: University of Chicago Press, 1982).

Allen, Thomas B., *Guardian of the Wild: The Story of the National Wildlife Federation, 1936–1986* (Bloomington: Indiana University Press, 1987).

Anderson, Derek, *Exotic Species in Australia: Their Establishment and Success*, Proceedings of the Ecological Society of Australia, Vol. 10 (Adelaide: Ecological Society of Australia, 1977).

Anderson, Edgar, *Plants, Man, and Life* (Berkeley: University of California, Press, 1971).

Anderson, H. Allen, *The Chief* (College Station: Texas A&M University Press, 1986).

Atran, Scott, *Cognitive Foundations of Natural History* (Cambridge University Press, 1993).

Atwood, Margaret, *Survival* (Toronto: Anansi, 1972).

Strange Things: The Malevolent North in Canadian Literature (Oxford: Oxford University Press, 1995).

Australian Bureau of Flora and Fauna, *Fauna of Australia* (Canberra: Australian Government Publishing Service, 1987), Vol. 1A.

Australian Bureau of Meteorology, *Climate of Australia* (Canberra: Australian Government Printing Service, 1989).

Australian Bureau of Statistics, *Australia's Environment* (Canberra: Government Printer, 1993).

Year Book Australia, 1988 (Canberra: Australian Bureau of Statistics, 1988).

Australian Conservation Foundation, *Conservation Directory* (Melbourne: Australian Conservation Foundation, 1973).

Green Pages, 1991–1992 (Melbourne: Australian Conservation Foundation, 1992).

Australian Department of the Arts, Sport, the Environment, Tourism and Territories, *Report on the Review of the Australian National Parks and Wildlife Service* (Canberra: Australian Government Printing Service, 1989).

Australian Environmental Council, *Guide to Environmental Legislation and Administrative Arrangements* in Australia, *Second Edition*, Report No. 18 (Canberra: Australian Government Publishing Service, 1986).

Australian Parliament, House of Representatives, Select Committee on Wildlife Conservation, *Wildlife Conservation* (Report from the House of Representatives Select Committee) (Canberra: Australian Government Publishing Service, 1972).

Australian Parliament, Senate, Select Committee on Animal Welfare, *Kangaroos* (Canberra: Australian Government Publishing Service, 1988).

Barber, Lynn, *The Heyday of Natural History, 1820–1870* (London: Jonathon Cape, 1980).

Barnes, John (editor), *The Writer in Australia* (Melbourne: Oxford University Press, 1969).

Barrett, James (editor), *Save Australia: A Plea for the Right Use of our Flora and Fauna* (Melbourne: Macmillan, 1925).

Barrow, Mark V., Jr., *A Passion for Birds* (Princeton, N.J.: Princeton University Press, 1998).

Barsness, Larry, *The Bison in Art* (Fort Worth, Tex.: Amon Carter Museum, 1977).

Bates, G. M., *Environmental Law in Australia* (Sydney: Butterworths, 1992).

Bean, Michael, *The Evolution of Federal Wildlife Law* (New York: Praeger, 1983).

Beard, Dan, *The American Boys' Handbook of Camplore and Woodcraft* (Philadelphia: Lippincott, 1920).

Bella, Leslie, *Parks for Profit* (Montreal: Harvest House, 1987).

Bentley, Arthur, *An Introduction to the Deer of Australia with Special Reference to Victoria* (Melbourne: Koetong Trust Service Fund, Forests Commission, Victoria, 1978).

Berger, Carl, *Science, God and Nature in Victorian Canada* (Toronto: University of Toronto Press, 1983).

Berlin, Brent, *Ethnobiological Classification* (Princeton, N.J.: Princeton University Press, 1992).

Berton, Pierre, *The Arctic Grail* (Canada: McClellard and Stewart, 1988; reprinted New York: Penguin, 1988).

Blainey, Geoffrey, *The Tyranny of Distance* (rev. ed., South Melbourne: Macmillan, 1982).

Boardman, Robert, *International Organization and the Conservation of Nature* (Bloomington: Indiana University Press, 1981).

Boime, Albert, *The Magisterial Gaze* (Washington, D.C.: Smithsonian Institution, 1991).

Bonnifield, Paul, *The Dust Bowl* (Albuquerque: University of New Mexico Press, 1979).

Bordais, D. M., *Stefansson: Ambassador of the North* (Montreal: Harvest House, 1963).

Bowen, Margarita, *Empiricism and Geographical Thought* (Cambridge University Press, 1981).

Brady, Edwin J., *Australia Unlimited* (Melbourne: George Robertson, 1918).

Breckwoldt, Roland, *A Very Elegant Animal: The Dingo* (Melbourne: Angus & Robertson, 1988).

Brewer, T. M. (editor), *Wilson's American Ornithology* (edition with notes by Jardine and other material by Brewer) (1840; reprinted New York: Arno, 1970).

Brockway, Lucille, *Science and Colonial Expansion: The Role of the British Royal Botanic Gardens* (New York: Academic Press, 1979).

Brown, David E., and Neil B. Carmony (editors), *Aldo Leopold's Southwest* (1990; reprinted Albuquerque: University of New Mexico Press, 1995).

Browne, Thomas, *Religio Medici* (1642; Oxford: Oxford University Press, 1964).

Bolton, Geoffrey C., *Britain's Legacy Overseas* (London: Oxford University Press, 1973).

Spoils and Spoilers (Sydney: Allen & Unwin, 1981).

Buell, Lawrence, *The Environmental Imagination* (Cambridge, Mass.: Harvard University Press, 1995).

Buhrs, Ton, and Robert V. Bartlett, *Environmental Policy in New Zealand* (Auckland: Oxford University Press, 1993).

Burrell, Harry, *The Platypus* (Sydney: Angus & Robertson, 1927).

Butler, David, and Don Merton, *The Black Robin: Saving the World's Most Endangered Bird* (Auckland: Oxford University Press, 1992).

Callenbach, Ernest, *Ecotopia* (1975; reprinted New York: Bantam, 1977).

Callicott, J. Baird, *In Defense of the Land Ethic* (Albany: State University of New York Press, 1989).

Callicott, J. (editor), *Companion to "A Sand County Almanac"* (Madison: University of Wisconsin Press, 1987).

Cameron, Jim, *The Canadian Beaver Book* (Burnstown, Ontario: General Store Publishing, 1991).

[Canadian] Federal Environmental Assessment Review Office, *Northern Diseased Bison* (Ottawa: Minister of Supply and Services Canada, 1990).

Cannon, Susan Faye, *Science in Culture* (New York: Dawson and Science History Publications, 1978).

Carson, Rachel, *Silent Spring* (Boston: Houghton Mifflin, 1962).

Carter, Paul, *The Road to Botany Bay* (Chicago: University of Chicago Press, 1987).

Cartmill, Matt, *From a View to a Death in the Morning* (Cambridge, Mass.: Harvard University Press, 1993).

Caughley, Graeme, *The Deer Wars* (Auckland: Heinemann, 1983).

Cayley, Neville, *What Bird is That?* (Sydney: Angus and Robertson, 1931).

Chapman, Frank M., *Color Key to North American Birds* (New York: Appleton, 1903; 2d ed., 1912, same text).

What Bird Is That? (New York: Appleton, 1920).

Chase, Alston, *In a Dark Wood* (Boston: Houghton Mifflin, 1995).

Playing God in Yellowstone (Boston: Atlantic Monthly Press, 1986).

Chisholm, Alec H., *Mateship with Birds* (Melbourne: Whitcombe & Tombs, 1922).

Chisholm, Alec N. (editor), *Land of Wonder* (Sydney: Angus & Robertson, 1964).

Chitty, Dennis, *Do Lemmings Commit Suicide?* (New York: Oxford University Press, 1996).

Choate, Ernest A., *American Bird Names* (Boston: Gambit, 1973).

Clark, Andrew Hill, *The Invasion of New Zealand* (New Brunswick, N. J.: Rutgers University Press, 1949).

Clark, Manning, *A Short History of Australia* (New York: Penguin, 1987).

Clarke, F. G., *The Land of Contrarieties* (Clayton: Melbourne University Press, 1977).

Clements, Frederic E., *Plant Succession and Indicators: A Definitive Edition of Plant Succession and Plant Indicators* (New York: H. W. Wilson, 1928).

Clements, Frederic E., and Ralph W. Chaney, *Environment and Life in the Great Plains* (Washington, D.C.: Carnegie Institution, 1937).

Cockayne, Leonard, *Monograph on the New Zealand Beech Forests, II. The Forests from the Practical and Economic Standpoints* (Wellington: Government Printer, 1928).

The Vegetation of New Zealand (Leipzig: Engelmann, 1921).

Cohen, Michael P., *The History of the Sierra Club* (San Francisco: Sierra Club, 1988).

Commission on Conservation, Canada, *Annual Report* (Toronto: various printers, annual).

Conference of Authorities on Australian Fauna and Flora (Hobart: Government Printer, 1949).

Conway, Jill Ker, *The Road From Coorain* (New York: Knopf, 1990).

Coues, Elliott, *Key to North American Birds* (1872; 3d ed., Boston: Dana Estes, 1903; reprinted New York: Arno, 1974).

Craighead, Frank C., *Track of the Grizzly* (San Francisco: Sierra Club, 1979).

Crevecoeur, J. Hector St. John, *Letters from an American Farmer* (1782; reprinted New York: Doubleday, n.d.).

Croker, Robert A., *Pioneer Ecologist: The Life and Work of Victor Ernest Shelford, 1877–1968* (Washington, D.C.: Smithsonian Institution Press, 1991).

Cronon, William (editor), *Uncommon Ground* (New York: Norton, 1995).

Crosby, Alfred, *The Columbian Exchange* (Cambridge University Press, 1990).

Ecological Imperialism (Cambridge University Press, 1986).

Crowcroft, Peter, *Elton's Ecologists: A History of the Bureau of Animal Population* (Chicago: University of Chicago Press, 1991).

Cutright, Paul Russell, *Theodore Roosevelt: The Making of a Conservationist* (Urbana: University of Illinois Press, 1985).

Daniels, Stephen, *Fields of Vision* (Princeton, N.J.: Princeton University Press, 1993).

Dargavel, John, *Fashioning Australia's Forests* (Melbourne: Oxford University Press, 1995).

Dasmann, Raymond, *The Destruction of California* (New York: Macmillan, 1965).

Davies, Robertson, *Bred in the Bone* (New York: Viking, 1985).

The Merry Heart (New York: Viking, 1997).

Davison, Frank Dalby, *Dusty* (Sydney: Angus & Robertson, 1946).

Manshy (Sydney: Angus & Robertson, 1931).

Dawson, Sarah (editor), *The Penguin Australian Encyclopedia* (Melbourne: Penguin, 1990).

Dean, Warren, *With Broadax and Firebrand* (Berkeley: University of California Press, 1995).

Dice, Lee R., *Natural Communities* (Ann Arbor: University of Michigan Press, 1952).

Dillard, Annie, *Pilgrim at Tinker Creek* (New York: Harper's, 1974).

Diubaldo, Richard J., *Stefansson and the Canadian Arctic* (Montreal: McGill-Queen's University Press, 1978).

Donne, T.E., *The Game Animals of New Zealand* (London: John Murray, 1924).

Doughty, Robin, *Feather Fashions and Bird Preservation* (Berkeley: University of California Press, 1975).

Dow, Hume, *Frank Dalby Davison* (Melbourne: Oxford University Press, 1971).

Drummond, James, *Nature in New Zealand* (Christchurch: Whitcombe & Tombs, n.d.).

Dunlap, Thomas R., *Saving America's Wildlife* (Princeton N.J.: Princeton University Press, 1988).

Dubos, René, *The Mirage of Health* (New York: Doubleday, 1959).

Dupree, A. Hunter, *Asa Gray* (Cambridge, Mass: Harvard University Press, 1959).
    Science and the Federal Government (Cambridge, Mass.: Harvard University Press, 1957).

Edquist, Alfred G., *Nature Studies in Australasia* (Melbourne: Lothian, 1916).

Egerton, Frank N. (editor), *History of American Ecology* (New York: Arno, 1977).

Eiseley, Loren, *All the Strange Hours* (New York: Scribner's, 1975).

Ellis, Vivienne Rea, *Louisa Anne Meredith* (Sandy Bay: Blubber Head Press, 1979).

Elton, Charles, *Animal Ecology* (London: Sidgwick & Jackson, 1927; reprinted London: Methuen, 1966).
    Voles, Mice and Lemmings: Problems in Population Dynamics (Oxford: Clarendon Press, 1942).

Errington, Paul, *Of Predation and Life* (Ames: Iowa State University Press, 1962).

Fairchild, David, *The World Was My Garden* (New York: Scribner's, 1938).

Falla, R. A., R. B. Sibson, and E. G. Turbott, *Collins Guide to the Birds of New Zealand* (Auckland: HarperCollins, 1981).

Farager, John Mack, *Daniel Boone* (New York: Henry Holt, 1992).

Farber, Paul Lawrence, *The Emergence of Ornithology as a Scientific Discipline, 1760–1850* (Boston: Reidel, 1982).

Fenner, Frank, and Francis N. Ratcliffe, *Myxomatosis* (Cambridge University Press, 1965).

Finlayson, H. H., *The Red Centre* (Sydney: Angus & Robertson, 1935).

Finney, Colin, *Paradise Revealed: Natural History in Nineteenth Century Australia* (Melbourne: Museum of Victoria, 1993).

Fisher, D. E., *Natural Resources Law in Australia* (Sydney: Law Book Company, 1987).

Flader, Susan, *Thinking Like a Mountain* (Columbia: University of Missouri Press, 1975).

Flader, Susan, and J. Baird Callicott (editors), *The River of the Mother of God* (Madison: University of Wisconsin Press, 1991).

Flannery, Tim, *The Future Eaters* (London: Secker & Warburg, 1996).

Fleming, C. A., *Science, Settlers and Scholars* (Wellington: Royal Society of New Zealand, 1987).

Foord, Malcolm, *The New Zealand Descriptive Animal Dictionary* (Dunedin: Malcolm Foord, 1990).

Forbes, Stephen A., *Ecological Investigations of Stephen Alfred Forbes* (New York: Arno, 1977).

Foreman, David, *Confessions of an Eco-Warrior* (New York: Harmony Books, 1991).

Foster, Janet, *Working for Wildlife* (Toronto: University of Toronto Press, 1978).

Friedman, Robert Marc, *Appropriating the Weather* (Ithaca, N.Y.: Cornell University Press, 1989).

Frith, H. J., *Wildlife Conservation* (rev. ed., Sydney: Angus & Robertson, 1979).

Frith, H. J., and J. H. Calaby, *Kangaroos* (Melbourne: F. W. Cheshire, 1969).

Galbreath, Ross, *Walter Buller: The Reluctant Conservationist* (Wellington: GP Books, 1989).

Working for Wildlife (Wellington: Bridget Williams, 1993).

Gascoigne, John, *Joseph Banks and the English Enlightenment* (Cambridge University Press, 1994).

George, John, *The Program to Eradicate the Imported Fire Ant* (New York: Conservation Foundation, 1958).

Gibb, May, *Little Obelia* (Sydney: Angus & Robertson, 1921).

Little Ragged Blossom (Sydney: Angus & Robertson, 1920).

Snugglepot and Cuddlepie (Sydney: Angus & Robertson, 1918).

Gibbons, Felton, and Deborah Strom, *Neighbors to the Birds* (New York: Norton, 1988).

Gibson, Ross, *The Diminishing Paradise* (Sydney: Angus & Robertson, 1984).

Gillies, William, and Robert Hall, *Nature Studies in Australia* (Melbourne: Whitcombe & Tombs, 1903; 2d ed. is undated).

Glacken, Clarence, *Traces on the Rhodian Shore* (Berkeley: University of California Press, 1967).

Gleick, James, *Chaos: Making a New Science* (New York: Penguin, 1987).

Glover, James M., *A Wilderness Original* (Seattle: The Mountaineers, 1986).

Godman, John, *American Natural History* (Philadelphia; Carey & Lea, 1826; reprinted New York: Arno, 1974).

Golley, Frank, *A History of the Ecosystem Concept* (New Haven, Conn.: Yale University Press, 1993).

Graham, Frank, Jr., *The Audubon Ark* (New York: Knopf, 1990).

Gould, Stephen Jay, *Ontogeny and Phylogeny* (Cambridge, Mass.: Harvard University Press, 1977).

Gregory, J. W., *Australia* (Cambridge University Press, 1916).

The Dead Heart of Australia (London: John Murray, 1906).

Griffiths, Tom, *Hunters and Collectors* (Cambridge University Press, 1996).

Secrets of the Forest (Sydney: Allen & Unwin, 1992).

Grinnell, Elizabeth (editor), and Joseph Grinnell, *Gold Hunting in Alaska* (Chicago: David Cook, 1901).

Grinnell, Joseph, Harold C. Bryant, and Tracy I. Storer, *The Game Birds of California* (Berkeley: University of California Press, 1918).

Grove, Richard, *Green Imperialism* (Cambridge University Press, 1995).

Guiler, Eric R., *Thylacine: The Tragedy of the Tasmanian Tiger* (Melbourne: Oxford University Press, 1985).

Guthrie-Smith, H., *Tutira: A New Zealand Sheep Station* (London: Blackwood, 1921).

Hagen, Joel B., *An Entangled Bank* (New Brunswick, N.J.: Rutgers University Press, 1992).

Hall, D. J., *Clifford Sifton* (Vancouver: University of British Columbia Press, 1985).

Hamelin, Louis-Edmond, *About Canada: The North and Its Conceptual Referents* (Ottawa: Minister of Supply and Services, 1986).

Hardy, John, and Alan Frost (editors), *Studies from Terra Australis to Australia*, Occasional Paper No. 6 (Canberra: Australian Academy of the Humanities, 1989).

Harper, J. Russell, *Painting in Canada* (2d ed., Toronto: University of Toronto Press, 1977).

Hayden, Sherman Strong, *The International Protection of Wild Life* (New York: Columbia University Press, 1942).

Hays, Samuel P., *Conservation and the Gospel of Efficiency* (Cambridge, Mass.: Harvard University Press, 1959).

Beauty, Health, and Permanence (Cambridge University Press, 1987).

Herbert, William Henry (Frank Forester), *Complete Manual for Young Sportsman* (1856; reprinted New York: Arno Press, 1974).

Hewitt, C. Gordon, *The Conservation of the Wild Life of Canada* (New York: Scribner's, 1921).

Hodgins, Bruce W., et al., *Federalism in Canada and Australia* (Peterborough: Frost Centre, 1989).

Holden, Philip, *The Deerstalkers* (Auckland: Hodder & Stoughton, 1987).

Home, R. W. (editor), *Australian Science in the Making* (Cambridge University Press, 1988).

Home, R. W., and Sally Gregory Kohlstedt (editors), *International Science and National Scientific Identity: Australia between Britain and America* (Dordrecht: Kluwer, 1991).

Hopkins, Harry, *The Long Affray: The Poaching Wars, 1760–1914* (London: Secker & Warburg, 1985).

Hornaday, William, *Wildlife Conservation in Theory and Practice* (1914; reprinted New York: Arno, 1974).

Houston, Douglas B., *The Northern Yellowstone Elk: Ecology and Management* (New York: Macmillan, 1982).

Howard, Leland O., *A History of Applied Entomology*, Smithsonian Miscellaneous Collections, Vol. 84 (Washington, D.C.: Smithsonian Institution Press, 1930).

Hunt, William B., *Stef: A Biography of Vilhjalmur Stefansson, Canadian Arctic Explorer* (Vancouver: University of British Columbia Press, 1986).

Hurst, J. Willard, *Law and Economic Growth* (Cambridge, Mass.: Harvard University Press, 1964).

Hutton, Frederick Wollaston, *Catalogue of the Birds of New Zealand* (Wellington: Government Printer, 1871).

Hutton, Frederick Wollaston, and James Drummond, *The Animals of New Zealand* (Christchurch: Whitcombe & Tombs, 1905).

Inglis, Gordon, *Sport and Pastime in Australia* (London: Methuen, 1912), 73.

Jackson, Wes, *Becoming Native to this Place* (Lexington: University Press of Kentucky, 1994).

Jenks, Edward, *A History of the Australasian Colonies* (Cambridge University Press, 1912).

Johnson, Dick, *The Alps at the Crossroads* (Melbourne: Victorian National Parks Association, 1974).

Lake Pedder (Lake Pedder Action Committee of Victoria and Tasmania and the Australian Union of Students, 1972).

Jones, Frederic Wood, *The Mammals of South Australia* (Adelaide: Government Printer, 1923), Vol. 1.

Kalm, Peter, *Travels in North America* (1753; reprint of English translation, New York: Dover, 1966).

Kelly, Franklin, *Frederic Edwin Church and the National Landscape* (Washington, D.C.: Smithsonian Institution Press, 1988).

Keneally, Thomas, *Woman of the Inner Sea* (New York: Doubleday, 1993).

Kevin, J. C. G. (editor), *Some Australians take Stock* (London: Longmans, 1939).

Killan, Gerald, *Protected Places* (Toronto: Queen's Printer, 1993).

King, Carolyn, *Immigrant Killers* (Auckland: Oxford University Press, 1984).

King, Carolyn (editor), *Handbook of New Zealand Mammals* (Auckland: Oxford University Press, 1990).

Kirkpatrick, Jamie, *A Continent Transformed* (Melbourne: Oxford University Press, 1994).

Lack, David, *The Natural Regulation of Animal Numbers* (London: Oxford University Press, 1954).

Lamond, Henry, *White Ears the Outlaw* (Sydney: Angus & Robertson, 1949).

Land Conservation Council, *Wilderness: Special Investigation, Descriptive Report* (Melbourne: Land Conservation Council, 1990).

Langston, Nancy, *Forest Dreams, Forest Nightmares* (Seattle: University of Washington Press, 1995).

Leach, J. A., *An Australian Bird Book* (Melbourne: Whitcombe & Tombs, 1911).

Nature Study: A Descriptive list of the Birds Native to Victoria (Melbourne: State Printer, 1908).

Lee, Martha F., *Earth First! Environmental Apocalypse* (Syracuse: Syracuse University Press, 1995).

Lennie, Campbell, *Landseer: The Victorian Paragon* (London: Hamish Hamilton, 1976).

Lever, Christopher, *Introduced Birds of the World* (London: Longman, 1987).

Introduced Mammals of the World (London: Longman, 1985).

The Naturalized Animals of the British Isles (London: Hutchinson, 1977).

They Dined on Eland (London: Quiller, 1992).

Levere, Trevor H., *Science and the Canadian Arctic: A Century of Exploration, 1818–1918* (Cambridge University Press, 1993).

Leopold, Aldo, *Game Management* (New York: Scribners, 1933).

A Sand County Almanac (1949; reprinted New York: Ballantine, 1967).

Leopold, A. Starker, and F. Fraser Darling, *Wildlife in Alaska* (New York: New York Zoological Society and the Conservation Foundation, 1953).

Limerick, Patricia, *Desert Passages* (Albuquerque: University of New Mexico Press, 1985).

Lindsay, Debra, *Science in the Subarctic: Trappers, Traders, and the Smithsonian Institution* (Washington, D.C.: Smithsonian Institution Press, 1993).

Lines, William, *An All-Consuming Passion: Orignis, Modernity, and the Australian Life of Georgiana Molloy* (Berkeley: University of California Press, 1996).

Taming the Great South Land (Berkeley: University of California, 1991).

Long, Charles R., *Stories of Australian Exploration* (Melbourne: Whitcombe & Tombs, 1903).

Long, Charles R. (editor), *Review of the State Schools Exhibition* (Melbourne: Government Printer, 1908).

Lothian, W. F., *A History of Canada's National Parks* (Ottawa: Environment Canada and predecessor agencies, 1977–87).

Lowe, Philip, and Jane Goyder, *Environmental Groups in Politics* (London: Allen & Unwin, 1983).

Lutts, Ralph, *The Nature Fakers* (Golden, Colo.: Fulcrum, 1990).

McCandless, Robert, *Yukon Wildlife: A Social History* (Edmonton: University of Alberta Press, 1985).

Macdonald, Donald, *The Bush Boy's Book* (Melbourne: Endacott, 1911).

Gum Boughs and Wattle Bloom (Melbourne: Cassell, 1888).

McGregor, Gaile, *The Wacousta Syndrome* (Toronto: University of Toronto Press, 1985).

EcCentric Visions (Waterloo: Wilfrid Laurier University Press, 1994).

MacKenzie, John M., *The Empire of Nature* (New York: Manchester University Press, 1988).

MacLeod, Roy (editor), *The Commonwealth of Science: ANZAAS and the Scientific Enterprise in Australia, 1888–1898* (Melbourne: Oxford University Press, 1988).

McIntosh, Robert P., *Background to Ecology* (Cambridge University Press, 1985).

McKnight, Tom, *Friendly Vermin*, University of California Publications in Geography, No. 21 (Berkeley: University of California Press, 1976).

Mahoney, T. E., *The Rabbit in Australia* (Ararat: privately printed, n.d.).

Mandelbaum, Maurice H., *History, Man and Reason: A Study in Nineteenth-Century Thought* (Baltimore: Johns Hopkins University Press, 1971).

Marks, Stuart, *Southern Hunting in Black and White* (Princeton, N.J.: Princeton University Press, 1991).

Marsh, George Perkins, *Man and Nature* (1864; reprinted Cambridge, Mass.: Harvard University Press, 1965).

Marshall, A. J. (editor), *The Great Extermination* (Melbourne: Heinemann, 1966).

Martin, Calvin, *Keepers of the Game* (Berkeley: University of California Press, 1978).

Mather, Patricia, *A Time for a Museum* (South Brisbane: Queensland Museum, 1986).

Matthams, James, *The Rabbit Pest in Australia* (Melbourne: Specialty Press, 1921).

Mayr, Ernst, *The Growth of Biological Thought* (Cambridge, Mass.: Harvard University Press, 1982).

Principles of Systematic Zoology (2d ed., New York: McGraw-Hill, 1991).

This Is Biology (Cambridge, Mass.: Harvard University Press, 1996).

Mech, L. David, *The Wolf: The Ecology and Behavior of an Endangered Species* (Garden City, N.Y.: Natural History Press, 1970).

The Wolves of Isle Royale, Fauna Series No. 7, National Park Service, Department of the Interior (Washington, D.C.: U.S. Government Printing Office, 1966).

Meine, Kurt, *Aldo Leopold* (Madison: University of Wisconsin Press, 1988).

Meinig, Donald, *On the Margins of Good Earth* (Chicago: Association of American Geographers, 1962).

Menon, P. Ali, and Harvey C. Perkins (editors), *Environmental Planning in New Zealand* (Palmerston North: Dunmore, 1993).

Mighetto, Lisa, *Wild Animals and American Environmental Ethics* (Tucson: University of Arizona Press, 1991).

Mighetto, Lisa (editor), *Muir Among the Animals* (San Francisco: Sierra Club Books, 1986).

Mills, Stephanie, *Whatever Happened to Ecology?* (San Francisco: Sierra Club, 1989).

Mitchell, Lee Clark, *Witnesses to a Vanishing America: The Nineteenth-Century Response* (Princeton, N.J.: Princeton University Press, 1981).

Mitman, Gregory, *The State of Nature* (Chicago: University of Chicago, 1992).

Moncrieff, Perrine, *New Zealand Birds and How to Identify Them* (Auckland: Whitcombe & Tombs, 1925).

Morrell, Jack, and Arnold Thackray, *Gentlemen of Science* (Oxford: Clarendon Press, 1981).

Moore, N. W., *The Bird of Time* (Cambridge University Press, 1987).

Morris, Edward E., *Austral English: A Dictionary of Australasian Words, Phrases, and Usages* (London: Macmillan, 1898).

Mosley, J. G., and J. Messer (editors), *Fighting for Wilderness* (Fontana: Australian Conservation Foundation, 1984).

Mott, Frank Luther, *A History of American Magazines* (5 vols.; Cambridge, Mass.: Harvard University Press, 1930–68).

Mowat, Farley, *Never Cry Wolf* (Boston: Little, Brown, 1963).

Moyal, Ann, *"A Bright and Savage Land": Scientists in Colonial Australia* (Sydney: Collins, 1986).

Mungall, Elizabeth Cary, and William J. Sheffield, *Exotics on the Range* (College Station: Texas A & M University Press, 1994).

Murie, Adolph, *Ecology of the Coyote in the Yellowstone*, Fauna Series No. 4, National Park Service, Department of the Interior (Washington, D.C.: U.S Government Printing Office, 1940).

 The Wolves of Mt. McKinley, Fauna Series No. 5, National Park Service, Department of the Interior (Washington, D.C.: U.S. Government Printing Office, 1944).

Murray, Keith, *Reindeer and Gold* (Bellingham, Wash.: Center for Pacific Northwest Studies, 1988).

Musgrave, Ruth S., and Mary Anne Stein, *State Wildlife Laws Handbook* (Rockville, Md.: Government Institutes, 1993).

Nasgaard, Roald, *The Mystic North* (Toronto: University of Toronto Press, 1984).

Niall, Brenda, *Australia Through the Looking-Glass: Children's Fiction, 1830–1980* (Melbourne: Melbourne University Press, 1984).

Nash, Roderick, *Wilderness and the American Mind* (3d ed., New Haven, Conn.: Yale University Press, 1982).

New Zealand Official Yearbook (Wellington: Bureau of Statistics, 1992).

Nicholson, E. M., *The Art of Bird-watching* (New York: Scribner's, 1932).

Novak, Barbara, *Nature and Culture* (1980; rev. ed., New York: Oxford University Press, 1995).

Nowak, Ron, *Walker's Mammals of the World,* Fifth Edition (Baltimore: Johns Hopkins University Press, 1991).

Nuttall, Thomas, *A Manual of Ornithology* (Cambridge, Mass.: Hilliard & Brown, 1932; reprinted New York: Arno 1974).

Official Report of the Federal Conference on Education Convened by the League of the Empire (London: League of the Empire, 1908).

Oldroyd, David R., *The Highlands Controversy* (Chicago: University of Chicago Press, 1990).

O'Neil, Curry, *Classic Australian Paintings* (Melbourne: Curry O'Neil Ross Pty, 1983).

Opie, John, *Ogallala: Water for a Dry Land* (Lincoln: University of Nebraska Press, 1993).

Ormond, Richard, *Sir Edwin Landseer* (New York: Rizzoli, 1981).

Osborn, Fairfield, *Our Plundered Planet* (Boston: Little, Brown, 1948).

Osborne, Michael A., *Nature, the Exotic and the Science of French Colonialism* (Bloomington: Indiana University Press, 1994).

Owram, Doug, *Promise of Eden* (Toronto: University of Toronto Press, 1980).

Pedley, Ethel M., *Dot and the Kangaroo* (London: Thomas Burleigh, 1899).

Peterson, Roger Tory, *Field Guide to the Birds* (1st ed., Boston: Houghton Mifflin, 1934).

Pick, Jock H., *Australia's Dying Heart: Soil Erosion in the Inland* (Melbourne: Melbourne University Press, 1942).

Pinchot, Gifford, *The Fight for Conservation* (1910; reprinted Seattle: University of Washington Press, 1967).

Pisani, Donald, *To Reclaim a Divided West* (Albuquerque: University of New Mexico Press, 1992).

Pizzey, Graham, *Crosbie Morrison* (Melbourne: Victoria Press, 1992).

A Field Guide to the Birds of Australia (Sydney: Collins, 1980).

Powell, John Wesley, *Report on the Lands of the Arid Region,* U.S. Congress, House of Representatives, 45th Congress, 2d Session, H.R. Exec. Doc. 73 (Washington, D.C.: U.S. Government Printing Office, 1878).

Powell, Jospeh, *Environmental Management in Australia, 1788–1914* (Melbourne: Oxford University Press, 1976).

Griffith Taylor and "Australia Unlimited" (Brisbane: University of Queensland Press, 1988).

An Historical Geography of Modern Australia (Cambridge University Press, 1988).

Plains of Promise, Rivers of Destiny (Brisbane: Boolarong, 1991).

Watering the Garden State (Sydney: Allen & Unwin, 1989).

Preston, Brian J., *Environmental Litigation* (Sydney: Law Book Co., 1989).

Price, A., Grenfell, *White Settlers in the Tropics* (New York: American Geographical Society, 1939.

Pyne, Stephen, *Burning Bush* (New York: Henry Holt, 1991).

Quartermaine, Peter, *Thomas Keneally* (London: Edward Arnold, 1991).

"Rabbit Conference, The," Report of the Proceedings of a Conference Respecting the Rabbit Pest in NSW, presented to the Parliament (Sydney: Government Printer, 1895).

Ratcliffe, Francis, *Flying Fox and Drifting Sand* (1938; Australian ed., Sydney: Angus & Robertson, 1947).

The Rabbit Problem (Melbourne: Commonwealth Scientific and Industrial Research Organization, 1951).

Regis, Pamela, *Describing Early America* (DeKalb: Northern Illinois University Press, 1992).

Reiger, John, *American Sportsmen and the Origins of Conservation* (1975; reprinted Norman: University of Oklahoma Press, 1986).

Reingold, Nathan, and Marc Rothenberg (editors), *Scientific Colonialism: A Cross-Cultural Comparison* (Washington, D.C.: Smithsonian Institution Press, 1987).

Richardson, Elmo, *The Politics of Conservation, 1897–1913* (Berkeley: University of California Press, 1962).

Ritvo, Harriet, *The Animal Estate* (Cambridge, Mass.: Harvard University Press, 1987).

Roberts, Charles G. D., *The Kindred of the Wild* (Boston: Page, 1902).

Watchers of the Trails (Boston: Page, 1904).

Rolls, Eric, *They All Ran Wild* (Sydney: Angus & Robertson, 1979).

Royal Botanic Gardens, Sydney, *Gumnut Town: Botanic Fact and Bushland Fantasy* (Sydney: Royal Botanic Gardens, 1992).

Rudd, Robert, *Pesticides: Their Use and Toxicity in Relation to Wildlife*, Bulletin 7, California Department of Fish and Game (Sacramento: California Department of Fish and Game, 1956).

Pesticides and the Living Landscape (Madison: University of Wisconsin Press, 1964).

Runte, Alfred, *National Parks: The American Experience* (2d ed., Lincoln: University of Nebraska Press, 1987).

Russell, Peter (editor), *Nationalism in Canada* (Toronto: McGraw-Hill, 1966).

Sanderson, Marie, *Griffith Taylor: Antarctic Scientist and Pioneer Geographer* (Ottawa: Carleton University Press, 1988).

Sax, Joseph, *Defending the Environment* (New York: Knopf, 1971).

Schedvin, Boris, *Shaping Science and Industry* (Sydney: Allen & Unwin, 1987).

Scheffer, Victor, *The Shaping of Environmentalism in America* (Seattle: University of Washington Press, 1991).

Schwartz, William (editor), *Voices for the Wilderness* (New York: Ballantine, 1969).

Sears, Paul, *Deserts on the March* (Norman: University of Oklahoma Press, 1935).

Seton, Ernest Thompson, *The Book of Woodcraft and Indian Lore* (New York: Doubleday, 1912).

Wild Animals I have Known (New York: Scribners's, 1898).

Shankland, Robert, *Steve Mather of the National Parks* (3d ed., New York: Knopf, 1970).

Sheail, John, *Seventy-five Years in Ecology: The British Ecological Society* (Oxford: Blackwell, 1987).

Shortland, Michael (editor), *Science and Nature* (Oxford: British Society for the History of Science, 1993).

Serventy, Vincent, *Around the Bush with Vincent Serventy* (Sydney: Australian Broadcasting Commission, 1970).

A Continent in Danger (Sydney: Ure Smith, 1966).

Shannon, Fred, *Farmers' Last Frontier* (New York: Holt, Rinehart, & Winston, 1945).

Sheets-Pyenson, Susan, *Cathedrals of Science* (Kingston: McGill-Queens University Press, 1988).

Shelford, Victor E., *Animal Communities in Temperate America* (Chicago: Geographic Society of Chicago, 1913).

Shelford, Victor E., and Frederic Clements, *Bioecology* (New York: Wiley, 1939).

Silver, Timothy, *A New Face on the Countryside* (Cambridge University Press, 1990).

Sinclair, Keith, *A History of New Zealand* (Auckland: Penguin, 1980).

Slater, Peter, *The Birdwatcher's Notebook* (Sydney: Weldon, 1988).

Smith, Bernard William, *The Antipodean Manifesto* (Melbourne: Oxford University Press, 1976).

Australian Painting, 1788–1970 (Melbourne: Oxford University Press, 1971).

European Vision and the South Pacific, 1768–1850 (London: Oxford University Press, 1960).

Smith, Henry Nash, *Virgin Land* (Cambridge, Mass.: Harvard University Press, 1950).

Smith, Michael, *Pacific Visions* (New Haven, Conn.: Yale University Press, 1987).

Smuts, Jan Christian, *Holism and Evolution* (London: Macmillan, 1926).

Snyder, Gary, *The Practice of the Wild* (San Francisco: North Point Press, 1990).

Spencer, Baldwin, *Wanderings in Wild Australia* (London: Macmillan, 1928), Vol. 1.

Splatt, William, and Susan Bruce, *Australian Landscape Painting* (Melbourne: Viking O'Neil, 1989).

Splatt, William, and Barbara Burton, *Masterpieces of Australian Painting* (Melbourne: Viking O'Neil, 1987).

Stead, David, *The Rabbit in Australia* (Sydney: Winn, 1935).

The Rabbit Menace in New South Wales (Sydney: Government Printer, 1928).

Stefansson, Vilhjalmur, *The Friendly Arctic* (New York: Macmillan, 1921).

The Northward Course of Empire (New York: Harcourt, Brace, 1922).

Stegner, Wallace, *Beyond the Hundredth Meridian* (Boston: Houghton Mifflin, 1953).

Stein, Sara, *Noah's Garden* (Boston: Houghton Mifflin, 1993).

Sterling, Keir, *Last of the Naturalists* (New York: Arno, 1977).

Stilgoe, John, *Common Landscapes of America, 1580–1845* (New Haven, Conn.: Yale University Press, 1982).

Stoddard, Herbert L., *The Bobwhite Quail* (New York: Scribner's, 1931).

Memoirs of a Naturalist (Norman: University of Oklahoma Press, 1969).

Swain, Donald, *Federal Conservation Policy, 1921–1933* (Berkeley: University of California Press, 1963).

Wilderness Defender (Chicago: University of Chicago Press, 1970).

Takacs, David, *The Idea of Biodiversity* (Baltimore: Johns Hopkins University Press, 1996).

Tansley, A. G., *Practical Plant Ecology* (London: Allen & Unwin, 1923).

Taylor, Griffith, *The Australian Environment, Especially as Controlled by Rainfall*, Advisory Council of Science and Industry, Memoir No. 1 (Melbourne: Government Printer, 1918).

The Control of Settlement by Temperature and Humidity, Bulletin 14, Commonwealth Meteorological Service (Melbourne: Victorian State Printer, 1916).

Journeyman Taylor (London: Robert Hale, 1958).

Thom, David, *Heritage: The Parks of the People* (Auckland: Lansdowne, 1987).

Thomas, Keith, *Man and the Natural World* (New York: Pantheon, 1983).

Thompson, Harry V., and Carolyn M. King (editors), *The European Rabbit* (Oxford: Oxford University Press, 1994).

Thompson, Patrick (editor), *Myles Dunphy: Selected Writings* (Sydney: Ballagirin, 1986).

Thomson, George M., *The Naturalisation of Animals and Plants in New Zealand* (Cambridge University Press, 1922).

Thoreau, Henry David, *Walden* (1854; reprinted New York: Modern Library, 1937).

Tobey, Ronald, *Saving the Prairies* (Berkeley: University of California, 1981).

Tooby, Michael (editor), *The True North: Canadian Landscape Painting, 1896–1939* (London: Barbican Art Gallery, 1991).

Trefethen, James B., *An American Crusade for Wildlife* (New York: Winchester, 1975).

Turner, James, *Reckoning with the Beast* (Baltimore: Johns Hopkins University Press, 1980).

U.S. Congress, Senate, Special Committee on the Conservation of Wildlife Resources, *The Status of Wildlife in the United States*, Report No. 1203, Senate, 76th Congress, 3d Session) (Washington, D.C.: U.S. Government Printing Office, 1940).

U.S. Department of Agriculture, *Climate and Man*, Yearbook of Agriculture, 1941 (Washington, D.C.: U.S. Government Printing Office, 1941).

U.S. Department of the Interior, *Flyways* (Washington, D.C.: U.S. Government Printing Office, 1984).

U.S. Park Service, *Fading Trails: The Story of Endangered American Wildlife* (New York: Macmillan, 1943).

Vogt, William, *Road to Survival* (New York: William Sloane, 1948).

Wadham, S. M., and G. L. Wood, *Land Utilization in Australia* (2d ed., Melbourne: Melbourne University Press, 1950).

Waiser, William, *The Field Naturalist* (Toronto: University of Toronto Press, 1989).

Wallace, Alfred Russell, *My Life* (London: Chapman & Hall, 1905), Vol. 1.

Ward, Russel, *The Australian Legend* (2d ed., Melbourne: Oxford University Press, 1978).

Warren, Louis S., *The Hunter's Game* (New Haven, Conn.: Yale University Press, 1997).

Watson, Ian, *Fighting over the Forests* (Sydney: Allen & Unwin, 1990).

White, Gilbert, *A Natural History of Selbourne* (1787; reprinted, New York: Penguin, 1977).

White, Richard, *Inventing Australia* (Sydney: Allen & Unwin, 1981).

White, Stewart Edward, *The Adventures of Bobby Orde* (New York: Doubleday, 1901).

Whitelock, Derek, *Conquest to Conservation* (Netley: Wakefield, 1985).

Williams, D. A. R., *Environmental Law* (Wellington: Butterworth of New Zealand, 1980).

Williams, Gordon R. (editor), *The Natural History of New Zealand: An Ecological Survey* (Wellington: Reed, 1973).

Wilson, E. O., *Biophilia* (Cambridge, Mass.: Harvard University Press, 1994).

Naturalist (Washington, D.C.: Island Press, 1994).

Wilson, Roger, *From Manapouri to Aramoana: The Battle for New Zealand's Environment* (Wellington: Earthworks, 1982).

Wodzicki, Kazimierz, *Introduced Mammals of New Zealand* (Wellington: Department of Scientific and Industrial Research, 1950).

Wood, Peter, *Black Majority* (New York: Random House, 1974).

Worster, Donald, *Dust Bowl* (New York: Oxford University Press, 1979).

Nature's Economy (1977; reprinted Cambridge University Press, 1985).

Wright, George M., Joseph S. Dixon, and Ben H. Thompson, *Fauna of the National Parks of the United States*, Fauna Series No. 1, National Park Service, Department of the Interior (Washington, D.C.: U.S. Government Printing Office, 1933).

Wright, Judith, *The Coral Battleground* (Melbourne: Nelson, 1977).

Wyman, Walker D., and Clifton B. Kroeber (editors), *The Frontier in Perspective* (Madison: University of Wisconsin Press, 1965).

Yaffee, Steven Lewis, *Prohibitive Policy* (Cambridge, Mass.: MIT Press, 1982).

Yeo, Richard, *Defining Science* (Cambridge University Press, 1993).

Young, Stanley Paul, and Hartley H. T. Jackson, *The Clever Coyote* (1951; reprinted. Lincoln: University of Nebraska Press, 1978).

Younger, R. M., *Kangaroo Images Through the Ages* (Melbourne: Hutchinson, 1988).

Zeller, Suzanne, *Inventing Canada* (Toronto: University of Toronto Press, 1987).

INDEX